T0176655

Antineoplastic Drugs

Antineoplastic Drugs: Organic Synthesis

DANIEL LEDNICER

WILEY

This edition first published 2015
© 2015 John Wiley & Sons, Ltd

Registered Office
John Wiley & Sons, Ltd, The Atrium, Southern Gate, Chichester, West Sussex, PO19 8SQ, United Kingdom

For details of our global editorial offices, for customer services and for information about how to apply for permission to reuse the copyright material in this book please see our website at www.wiley.com.

The right of the author to be identified as the author of this work has been asserted in accordance with the Copyright, Designs and Patents Act 1988.

All rights reserved. No part of this publication may be reproduced, stored in a retrieval system, or transmitted, in any form or by any means, electronic, mechanical, photocopying, recording or otherwise, except as permitted by the UK Copyright, Designs and Patents Act 1988, without the prior permission of the publisher.

Wiley also publishes its books in a variety of electronic formats. Some content that appears in print may not be available in electronic books.

Designations used by companies to distinguish their products are often claimed as trademarks. All brand names and product names used in this book are trade names, service marks, trademarks or registered trademarks of their respective owners. The publisher is not associated with any product or vendor mentioned in this book.

Limit of Liability/Disclaimer of Warranty: While the publisher and author have used their best efforts in preparing this book, they make no representations or warranties with respect to the accuracy or completeness of the contents of this book and specifically disclaim any implied warranties of merchantability or fitness for a particular purpose. It is sold on the understanding that the publisher is not engaged in rendering professional services and neither the publisher nor the author shall be liable for damages arising herefrom. If professional advice or other expert assistance is required, the services of a competent professional should be sought.

The advice and strategies contained herein may not be suitable for every situation. In view of ongoing research, equipment modifications, changes in governmental regulations, and the constant flow of information relating to the use of experimental reagents, equipment, and devices, the reader is urged to review and evaluate the information provided in the package insert or instructions for each chemical, piece of equipment, reagent, or device for, among other things, any changes in the instructions or indication of usage and for added warnings and precautions. The fact that an organization or Website is referred to in this work as a citation and/or a potential source of further information does not mean that the author or the publisher endorses the information the organization or Website may provide or recommendations it may make. Further, readers should be aware that Internet Websites listed in this work may have changed or disappeared between when this work was written and when it is read. No warranty may be created or extended by any promotional statements for this work. Neither the publisher nor the author shall be liable for any damages arising herefrom.

Library of Congress Cataloging-in-Publication Data applied for.

ISBN: 9781118892541

A catalogue record for this book is available from the British Library.

Set in 10/12pt Times by SPi Publisher Services, Pondicherry, India
Printed and bound in Singapore by Markono Print Media Pte Ltd

1   2015

This book is dedicated to all of us Lednicers—Anne, Beryle, Daniella, David, Lisa Grace, Oliver, Ruth, and Sylvie

Contents

Preface

A recent perusal of the USAN Dictionary for new generic, or more precisely nonproprietary, names for drugs awarded over the past several decades, quite unexpectedly, turned up a sizeable group of recently named antineoplastic agents. The chemical structure of many of these new drug candidates comprised a collection of carbo- and heterocyclic moieties strung together in the form of a chain. The mechanisms by which those agents attack cancer cells were also quite novel.

The search for compounds for treating malignant cancer dates back to the early twentieth century. This effort has been at best a daunting task for chemists engaged in the search. The fact that cancer cells show few, if any, biochemical differences from normal cells complicated the task. When the search began in earnest in the late 1940s, chemists concentrated on the then-accepted means for finding new drugs: synthesizing assay candidates one at a time in pure form or alternatively supplying biologists with pure compounds isolated from plants, molds, or other natural sources. Those products were then tested one at a time *in vitro* or *in vivo* against neoplastic cells. This approach was rewarded with only moderate success. Many plant-derived antineoplastic drugs trace their origin to that period as do the agents which act by alkylating DNA. Management in both the private and public sectors eventually came to the conclusion that this method for finding new and better tolerated antineoplastic agents was giving scant return for the effort expended. Several changes, one in the method for synthesizing test material and the other to the screening assays, have inarguably resulted in the large list of new names in the USAN Dictionary.

Legislation and implementing regulations for the FDA approval process have also provided additional impetus for developing new antineoplastic drugs. The Orphan Drug Act recognized that pharmaceutical firms were loath to expend time and effort on drugs that would be used by a very small number of patients. In addition to some monetary awards, the Act provides special rules for approval of drugs for treating diseases suffered by 200,000 or fewer patients. The FDA Fast Track Development Program offers expedited requirements and review for drugs for treating serious, life-threatening medical conditions for which no other drug exists.

The appearance of the name in the USAN Dictionary served as the screen for selecting compounds in this compendium. Listing in the Dictionary requires only that the agent in question carries a generic name. The existence of such a designation is taken to indicate that the sponsor considers that the activity of the compound shows sufficient promise to be groomed for testing in the clinic.

Discussions of newly named potential therapeutic drugs have customarily sorted the compounds in chemical structure-based chapters. The structures of many of the antineo-plastic agents in this monograph—a string of carbo- and heterocyclic moieties—would however make the conventional arrangement difficult. An alternative method for sorting compounds was chosen. Chapters in this volume list compounds that share the same mechanism by which they attack neoplastic cells. The numbers of compounds listed in a given chapter, using that criterion, vary markedly: 41 pages for protein kinase and 5 pages for agents that inhibit histone deacetylase. The recently discovered kinase inhibitors comprise a major portion of the contents of this monograph. To provide context, the several opening chapters deal mainly with older neoplastic drugs and only a few of the newer antineoplastic drug candidates.

Many of the drugs for treating cancer, popularly known as "chemo," comprise natural products. These widely used antineoplastic agents have been omitted since those drugs and their derivatives involve few, if any, chemical transformations. Drugs given short shrift include mainstay chemo compounds from plants and fermentation products such as the vinca alkaloids, doxorubicin, maytansine, and most recently paclitaxel.

This volume focuses on the chemistry to prepare antineoplastic agents rather than a detailed account of the biology of those drugs. Since this is mainly a chemistry monograph, the bibliography is confined to sources for the description of the chemistry. The Internet provided the gist for the brief thumbnail description of the biological activity of the potential antineoplastic agent; they are thus not referenced. Not having a license to practice medicine, I take no blame or credit for the accuracy of the short notes about clinical trials that precede discussions of most of the compounds in this account. In the same vein, the chemistry is focused on the drug itself rather than its salts.

One more caveat is called for. The synthetic sequences that follow represent those presented in publication that have appeared in journals, largely the *Journal of Medicinal Chemistry, Bioorganic & Medicinal Chemistry*, and *Bioorganic & Medicinal Chemistry Letters*, as well as US Patents. It is more than likely that there will be more efficient schemes than those devised for drugs approved by the FDA by chemists in the sponsors' laboratories.

Daniel Lednicer

Introduction

Cancer has a long history as a scourge for mankind. Some prehistoric fossilized human bones, in fact, show growths that have been interpreted as malignant tumors. The term cancer actually encompasses a group of closely related diseases that have in common unregulated cell division. Many vital processes such as growth require the synthesis of new proteins. This process calls on instructions from DNA found in genes. In rough outline, cell division is normally directed by protein factors that are in turn controlled by two opposing genes. Proto-oncogenes control proteins that encourage cell proliferation, while those controlled by tumor suppressor genes tend to oppose the process. Any one of a number of stimuli, for example, chronic exposure to carcinogenic chemicals, can cause a proto-oncogene to mutate and become an oncogene. That oncogene then causes the proteins involved in cell division to become overactive. The cells whose growth has up to now been controlled escape the restraints on cell division and lose controls on proliferation. The now-cancerous cells often also lose many of functions they had played prior to becoming neoplastic. The absence of restraints in addition causes those cells to divide much more quickly than the normal progenitor; they then go on to form a malignant tumor. Untreated cancer virtually always causes premature death.

For centuries, the only means for treating the malignant tumors consisted of surgical extirpation of the lesion. Texts dating from Greco-Roman times describe excision of cancerous lesions; many of these sources refer to the recurrence of cancer within a short time after the surgery. Cancers of the circulatory system such as leukemias and lymphomas were considered a death warrant up to quite recent times because there was no visible tumor that could be removed. Today's greatly advanced surgical technique and adjuncts such as the sterile operating field and anesthesia made surgical removal of malignant tumors practical; surgery for treating solid tumors is now still the first-line treatment after a carcinoma has been identified. Unless caught at a very early stage, many cancerous lesions spread to other parts of the body by splitting off malignant daughter cells in a process called metastasis. Metastases spread throughout the body via the lymphatic and sometimes the circulatory system. The fact that surgeons now take special measures to insure that all cancer cells are excised helps avoid the spread of the cancer to other locations. The principal targets of antineoplastic drugs now comprise first the circulatory system cancer tumors not susceptible to surgical excision such as leukemia; metastases from solid tumors comprise an equally important target for these drugs. Antineoplastic agents are in addition also used following surgery to kill any cancer cells that had been left behind. These drugs are also not infrequently used to shrink tumors prior to surgery.

The beginning of antineoplastic therapy can be ironically traced back to the First World War when the Germans followed up their use of chlorine as a poison gas by what came to be called sulfur mustard (**I-1**). The name is said to come from the yellow-brown appearance of the substance while still liquid and the mustard-like odor. Exposure to this gas, now

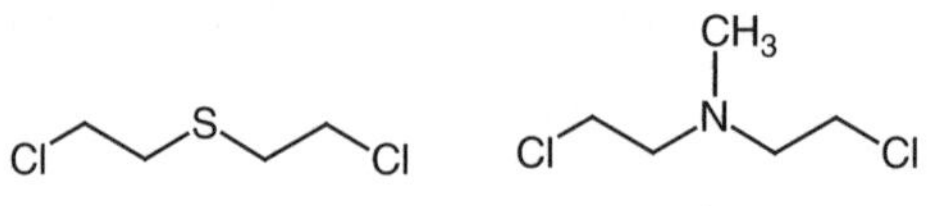

I 1
Sulfur mustard

I 2
Nitrogen mustard

Scheme 1 *Methchloramine.*

classed as a cytotoxic agent, caused large painful skin blisters; afflicted troops often lost eyesight. (A very moving larger-than-life-size John Singer Sargent painting depicts a line of gassed and blinded Great War soldiers.) The inhalation of the gas led to blister-like lesions in the lung. Postwar studies on individuals who were exposed to mustard gas showed a lowering of hematopoiesis—that is, the formation of blood cells. This was confirmed during the early 1940s by the examination of individuals who had been exposed to an inadvertent release of mustard gas.

Sulfur mustard is a liquid with a low boiling point that is difficult and dangerous to handle. The nitrogen analogue (**1.2**) is a solid as its hydrochloride salt is much easier to handle and thus safer. This prompted pharmacologists Goodman and Gilman to launch a study to determine whether this compound, subsequently dubbed methchloramine, had the same effect on cancer as its sulfur predecessor. They consequently studied the effect of this compound on lymphomas, malignancies of blood cells that had been implanted in mice. They found that methchloramine markedly reduced the mass of cancerous tissue in that *in vivo* disease model. They and a group of physicians went on to administer the drug to a lymphoma patient. The drug now granted the generic name mustine dramatically reduced the mass of cancerous tissues. The 1946 paper announcing that result is now considered to mark the beginning of antineoplastic drug therapy [2]. Methloramine (Mustargen®) is still commonly used as a chemotherapy drug. (The class of anticancer compounds that act by alkylating DNA will be found in Chapter 1.) That section deals largely with older compounds since there is currently little research devoted to antineoplastic agents that act by alkylating DNA.

The central circumstance that makes the search for new antineoplastic agents so difficult lies in the fact that the properties of cancer cells are almost identical to those of their cancer-free counterparts [1]. In addition to alkylating agents, several other classes of antineoplastic drugs rely on the fact that cancerous tissue turns over at a considerably higher rate than normal tissue. As a result, cytotoxic chemicals will to some degree have a greater effect on cancerous tissues than on normal cells. The common side effects of the administration of many antineoplastic agents, such as loss of hair, dry mouth, and dry tear ducts, demonstrate that the selectivity of those drugs is not perfect; the drugs also attack normal cells that are turning over quickly.

An alternate approach for treating cancer involves the use of antimetabolites. Folic acid and some of its metabolites are an essential factor for many bodily processes. This class of compounds, known as folates, is essential for building and repairing DNA. A group of antineoplastic drugs, most of which have chemical structures that mimic folates, act as metabolic inhibitors of folate synthesis. Chapter 2 treats antineoplastic drugs that act by inhibiting that process. Each of the two purines and three pyrimidines that comprise the coding bases in DNA in genes and the RNA that controls the construction of proteins is synthesized in the body by a set of specialized enzymes. Life as we know it is totally dependent on the six bases that form DNA and RNA. A collection of anticancer agents that inhibit the enzymes for building those substances is found in the same chapter.

Many of organs that comprise the sexual complex of women and men are studded with receptors for the agents that control their functions: the estrogens in women and androgens in men. Many, but not all, cancers of those organs retain those receptors and have become estrogen or androgen dependent. Chapter 3 describes hormone antagonists that have shown activity against hormone-dependent tumors. A sizeable number of those antineoplastic agents were elaborated in the 1970s up to the early 1990s as shown by the corresponding dates of the references.

When not involved in replication, DNA, a physically extremely long molecule, is supercoiled. The process of generating a new protein requires access to a relatively short sequence for copying to RNA that may be buried within the coil. The enzyme topoisomerase I expedites the process of bringing the required segment to the fore by cutting a strand in double-stranded DNA. The enzyme then temporarily marks the location of the cut and then reconnects the ends when the sequence has served its function. Closely related topoisomerase II cuts both strands at the same time. Topoisomerase inhibitors are discussed in Chapter 4.

The process of replication, called mitosis, involves the separation of the doubled cell nuclei. Chapter 5 describes drugs that interfere with this process. A set of very small fibers termed microfibers in the cell nucleus derived from the protein tubulin connect the doubled nuclei where they aid the separation of those entities. These structural elements are absorbed once mitosis is complete. One set of microtubules stabilizes the microfibers so that they are no longer absorbed, in effect halting mitosis. A second group of agents inhibit the formation of the microtubules.

A series of unrelated anticancer agents act at the level of the DNA within the cell nucleus. That DNA is tightly wrapped around a series of proteins that form a spindle-like structure known as histones. Reading the DNA code in response to a signal that calls for the production of a new protein is controlled by the series of acetyl groups attached to the histones. The enzyme histone deacetylase regulates the addition and deletion of those acetyl groups. Chapter 6 describes a number of inhibitors of the deacetylase enzyme that interfere with instructions for reading the genome.

Metalloproteinases are a family of related metal-containing enzymes that act on the extracellular matrix that holds cells together and in place. The process of dispersion of cancer to locations remote from the original tumor requires the disruption of the matrix. Chapter 7 describes a small group of compounds that inhibit those enzymes.

Kinases comprise a group of enzymes that connect a phosphate group to a specific amino acid on regulatory proteins. The resulting phosphorylated substrate then controls various cellular processes. The process of adding phosphate groups also serves as a means for signaling the start or ending of a process. The largest section by far in this compendium comprises two sizeable chapters on drugs that inhibit the action of kinases.

Chapter 8 describes a very large group of compounds that inhibit the binding to tyrosine kinases. It is noteworthy that close to half of those drugs have been approved by the FDA for treating patients with a narrowly defined cancers. The still sizeable number of compounds that inhibit other kinases and related proteins is to be found in Chapter 9.

No book of this nature is complete without a chapter that deals with compounds that cannot be included in the previous classes. Chapter 10, titled Miscellaneous Agents, describes a handful of such potential drugs.

The preponderant mode for prescribing drugs for treating most diseases called for prescribing a single drug that had been approved for that use by regulatory agencies. The administration of antineoplastic drugs, on the other hand, not infrequently leads to an initial shrinkage of the tumor. This is however too frequently followed by recurrence of the disease as the tumor develops resistance to the drug. This has led oncologists to administer a cocktail of drugs, each of which killed cells by different mechanisms. Before, too long cancer chemotherapy came to rely on sets of defined groups of drugs, cocktails, designated by acronyms. First-line treatment of Hodgkin's disease, for example, relied for a long time on the regime MOPP: mustine, Oncovin, procarbazine, and prednisone. It is of interest that a similar regime, administering a collection of antivirals that attack the virus by different mechanisms, is now used for treating HIV.

The US Food Administration generally grants fairly broad approvals for new drugs. This is applied to antineoplastic drugs as well. A new antineoplastic might, for example, be licensed for treating non-small cell lung cancer. Pressure from Congress and cancer support groups changed that practice in order to speed the approval of new antineoplastic agents. Approval currently more directly reflects the results of a clinical trial, or trials, where the drug in question showed a statistically significant more favorable outcome than that observed with other available treatments. The new drug will be typically approved for treating patients whose treatment with paclitaxel had failed. This volume is not a prescribing guide and thus steers away from those very detailed specifications; it merely states that a compound has been approved, occasionally indicating the organ.

References

[1]   Anon. http:/scienceeducation.nih.gov/supplements/nih1/cancer/guide/understanding2.htm (accessed on September 3, 2014).
[2]   Goodman LS, Wintrobe MM, Dameshek W, Goodman MJ, Gilman A, McLennan MT, *JAMA.* **251**(17), 2255–2261 (1984, May 4).

1

Alkylating Agents

An impressive number of cytotoxic compounds whose antineoplastic activity is due to their reactions with DNA have been studied in the clinic. Many of these comprise drugs that currently form part of the combinations used to treat neoplastic disease. This account however includes only a limited number of alkylating agents since this area has been well covered elsewhere.

1.1 bis-Chloroethyl Amines

As noted in the Introduction, antineoplastic agents that include in their structure highly reactive chemical moieties comprise the earliest class of drugs for treating malignant tumors. This applies particularly to those cancers that afflict the system for producing and maintaining blood-forming tissues such as leukemia and lymphoma. The first of these agents, mustine (**1.1**), also known as mechlorethamine, was, as noted in the Introduction, actually developed empirically. An understanding of the mechanism by which alkylating agents kill cancer cells awaited the discovery of the structure of DNA in the 1950s as well as elaboration of the chemistry for studying that substance. The relatively large group of alkylating anticancer drugs was actually synthesized before their mode of action was fully understood. Many of those anticancer agents were designed as analogues of prior compounds that sported the chemically reactive chloroethyl group or some other highly reactive function.

The alkylating agents as a class attack many tissues in the body that contains basic nitrogen. Those agents target all cells that are susceptible to alkylation, be they cancerous cells or unrelated normal cells. The latter circumstance leads to many classical side effects manifested by alkylating antineoplastic drugs, such as loss of hair, dry mouth, and dry eyes,

Antineoplastic Drugs: Organic Synthesis, First Edition. Daniel Lednicer.
© 2015 John Wiley & Sons, Ltd. Published 2015 by John Wiley & Sons, Ltd.

experienced by patients exposed to this class of antineoplastic agents. The effects on neoplastic cells are however more relevant to this discussion. Reaction with DNA is not a random process; it has been shown that alkylating agents react preferentially with the more electron-rich, more basic nitrogen atoms in DNA. The stacked bases between the two strands of that macromolecule in the helical arrangement constitute a particularly favorable configuration for attack on each of the two separate strands of DNA. Drugs that incorporate two alkylating moieties form a covalent bridge between the two strands of DNA. This effect is demonstrated by the significantly lower concentration of bifunctional agents required to kill cancer cells *in vitro* than that of molecules that include only a single reactive group. That cross-link inactivates alkylated DNA since almost all functions of DNA, such as replication, require access to a single strand. RNA, the counterpart directly involved in synthesizing new protein, can only read a single DNA strand. The affected cell then simply ceases to function and dies. Although very large number of alkylating drugs has been studied since the early 1940s, the present account is restricted to five subsets that illustrate research in this field.

The chloroethyl group found in this subset of alkylating drugs does not react with DNA as administered. Instead, the basic nitrogen in mustine (**1.1**) displaces the side chain chlorine to form an aziridinium salt (**1.2**). The reaction of this activated species with nitrogen in DNA leads to ring-opened DNA adduct. The repetition of that sequence with the second chloroethyl function followed by the reaction of the new aziridinium function with alkylated DNA leads to cross-linked DNA.

The first recorded preparation of this rather venerable antineoplastic agent involves the reaction of methyl-bis(hydroxylethane) (**2.1**) with thionyl chloride. The starting diol is speculatively available from the reaction of methylamine and ethylene oxide. The resulting product, **mustine (2.2)**, needs to be handled as a positively charged salt to prevent *ex vivo* aziridinium formation [1].

Cyclophosphamide is one of the best known and widely used antineoplastic agents. The drug comprises "C" in a large number of multidrug cocktails for treating cancer. One of the

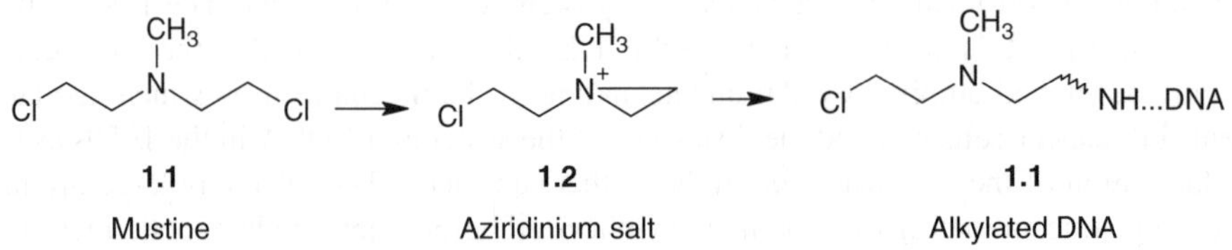

Scheme 1.1   *Aziridinium salt formation.*

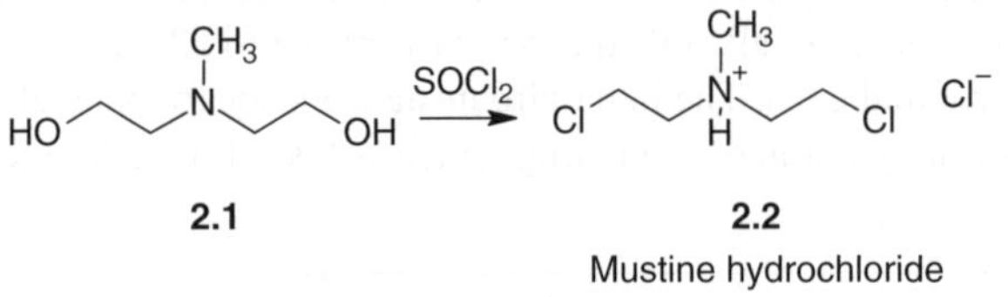

Scheme 1.2   *Synthesis of mustine.*

several schemes for preparing this compound starts with the condensation of aminoalcohol (**3.1**) with phosphorus oxychloride to afford the oxazaphosphorine derivative (**3.2**) through stepwise displacement of halogens in phosphorus oxychloride by the base and alkoxide group in (**3.1**). The still reactive chlorine in that product is then displaced with 2-chloroethylamine (**3.3**). The same reagent is then used to add a second chloroethyl function. This brief sequence affords **cyclophosphamide** (**3.5**) [2].

This drug is actually not the active alkylating species. Instead, enzymes open the ring by first hydroxylating the carbon bearing oxygen. The resulting hemiacetal then hydrolyzes to afford the phosphoramide mustard species (**3.6**). This has been approved for clinical use by many regulatory bodies. It is available as a generic drug since the patent covering this entity expired many years ago.

It is widely known that a large proportion of human female breast and possibly other genital tissues are equipped with receptors for estrogens. Binding of estrogens such as estrone, estradiol (**4.3**), and other related estrogenic compounds with those receptors stimulates growth of estrogen-positive tissues; those hormones will most likely cause the malignant tumors to flourish. Before the discovery of the estrogen antagonist drugs, the treatment of

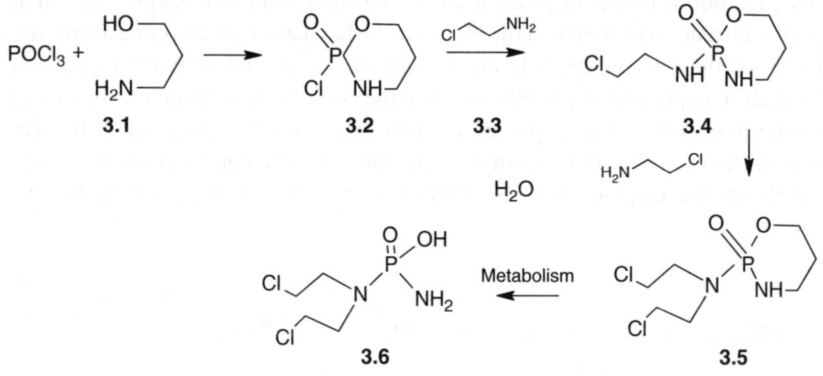

Scheme 1.3   *Cyclophosphamide.*

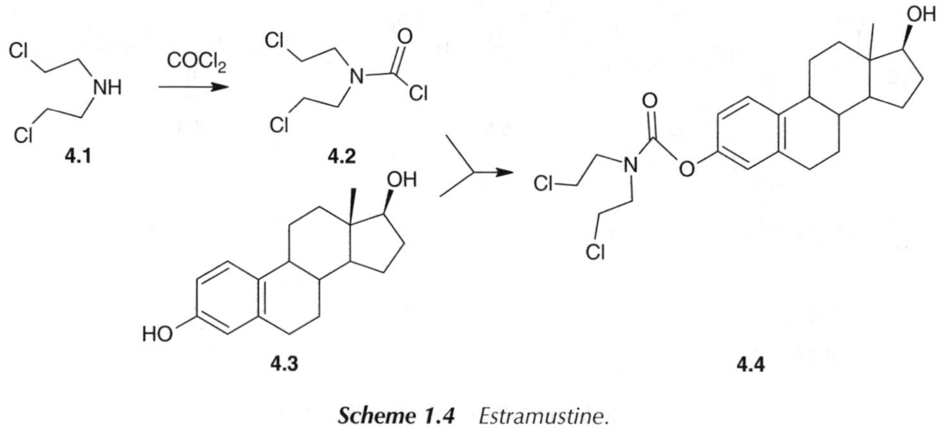

Scheme 1.4   *Estramustine.*

breast and related cancers often consisted of surgery followed by the administration of androgenic drugs in a vain hope that they would decrease estrogen-induced proliferation. The administration of alkylating anticancer drug seemed at the time to be the only alternative for treating breast cancer. One strategy for avoiding the severe side effects from the administration of those drugs comprises limiting exposure of the drug to the malignant tissue. Estrogen receptors in breast and related tissues are at first sight prime targets for directed antineoplastic agents. One approach for steering the alkylating mustine moiety consisted attaching the moiety to an estrogen. It should be noted in passing that the current treatment of estrogen receptor-positive cancers consists of surgery followed by the administration of one of the handful of estrogen antagonists such as tamoxifen or raloxifene (*see* Chapter 3).

The straightforward preparation of estramustine (**4.4**) starts with the acylation of bis(2-chloroethyl)amine with phosgene to afford the corresponding carbamoyl chloride (**4.2**). The acylation of estradiol (**4.3**) with that reactive intermediate affords **estramustine** (**4.4**) [3]. There is some evidence that this drug also disrupts tubulin, a precursor of tubules, an essential structure for cell division.

The drug is now available as a generic from a selection of vendors.

A more recent compound based on the same rationale comprises a mustine-equipped dipeptide glutathione mimic intended to direct the compound to a receptor for glutathione. The specific instance was based on the observation that malignant cells often have relatively high levels of the enzyme glutathione transferase, compared to normal cells, and that enzyme leads to expulsion of glutathione from the body. Attaching the mustine moiety to a glutathione-like moiety was expected to steer that agent to malignant cells. The drug has shown activity in the clinic against several cancers. The construction of this cytotoxic agent starts by the displacement of chlorine in phosphorus oxychloride by means of

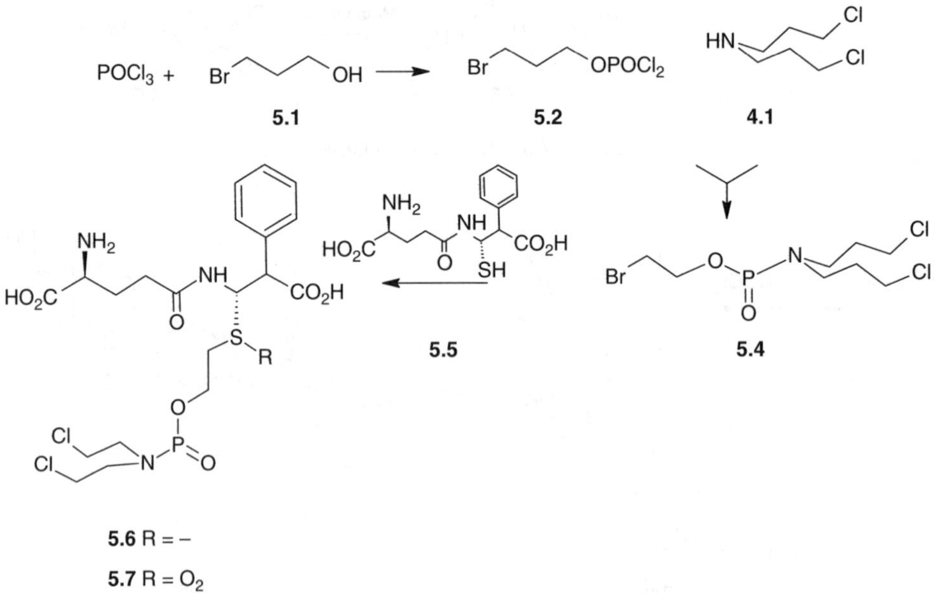

Scheme 1.5 *Canfosfamide.*

bromoalcohol (**5.1**). The product is next treated with bis(chloroethyl)amine (**5.3**); the amine in that reagent displaces the remaining halogen to afford phosphoramide (**5.4**). That intermediate is next reacted with the glutathione analogue in which phenylglycine replaces glycine found in the prototype. The mercaptan in reagent (**5.5**) then displaces bromine to give the condensation product; oxidation of sulfur with hydrogen peroxide completes the synthesis of **canfosfamide** (**5.6**) [4, 5].

1.2 Several Other Chloroethyl Agents

A pair of closely related compounds that act by a similar mechanism can be prepared by a relatively short sequence of reaction. The condensation of 2-chloroethyl-1-amine (**6.1**) with isocyanide (**6.2**) leads to the corresponding urea (**6.3**). The treatment of that product with nitrous acid leads to *N*-nitrosourea (**6.4**), **carmustine**, also known by the trivial acronym BCNU.

The same sequence starting with cyclohexylamine (**6.5**) gives **lomustine** (**6.8**) or CCNU. These nitrosoureas decompose in aqueous media by a sequence that involves loss of nitrogen from the N-nitroso moiety. The decomposition of both (**6.4**) and (**6.7**) apparently proceeds via a transient chloroethyl carbocation [6]. These drugs, which are approved for use in the clinic, also cross-link both DNA and RNA even though each yields a species with single reactive center. Both drugs have been approved by the FDA for the treatment of neoplastic disease. BCNU was used for many years as monotherapy of brain cancer.

A rather more complex chain of heteroatoms supports the chloroethyl side chain in the alkylating agent cloretazine. The reaction of 2-hydrazinoethanol (**7.1**) with methanesulfonyl chloride sulfonates the two hydrazine atoms as well as the hydroxyl group to afford the tris-sulfonated intermediate (**7.2**). Heating that intermediate with lithium chloride displaces the O-sulfonate by chlorine, thus establishing the requisite chloroethyl side chain. The condensation of intermediate (**7.3**) with methyl isocyanate converts the last free nitrogen to a urea, yielding the alkylating agent **cloretazine**, also known as **laromustine** (**7.4**) [7, 8].

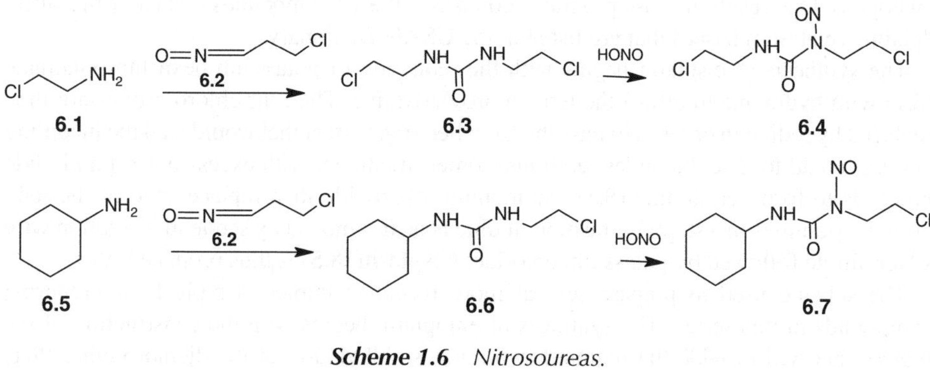

Scheme 1.6 *Nitrosoureas.*

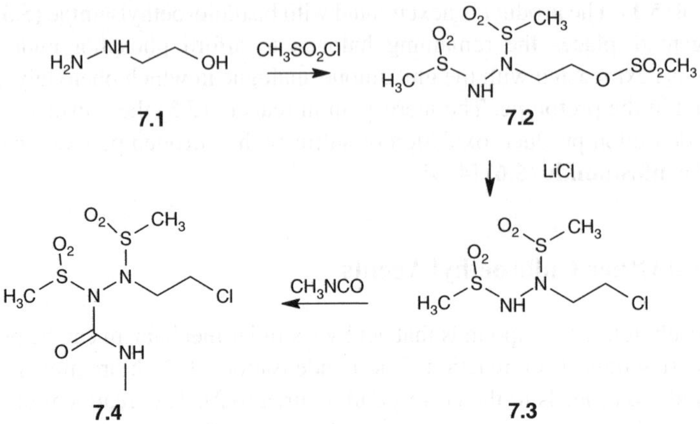

Scheme 1.7 Cloretazine.

1.3 Platinum-Based Antineoplastic Agents

The oft-told story of the discovery of cisplatinum provides an outstanding example of serendipity. Intending to ascertain the effect of electric currents on bacteria, the investigator, Barnet Rosenbloom, inserted a platinum electrode into a bacteria-seeded culture bath. He found that bacterial growth had indeed been inhibited. That effect was however not due to the electric current; they found that the inhibition could be attributed to a compound formed by electrolytic dissolution of the electrode. It was later found that the newly formed compound also inhibited the proliferation of cancer cells. The isolated platinum compound, cisplatin (**8.2**), was later found to be cytotoxic to human cancer cells. The drug is now widely used in combination therapy. The letter P in the acronym of a typical combination of antineoplastic agent refers to this drug. Cisplatin and its analogues, like other alkylating agents, act by inactivating DNA; in this specific case, each of the chlorine atoms in cisplatin is displaced by a base on the neighboring strands of DNA. The specificity for cancerous cells of platinum drugs depends on the faster turnover of cancer cells compared to that of normal cells. The harsh side effects of cisplatin may be a reflection of that limited specificity for neoplasms. This has led to major programs for preparing better tolerated cisplatin analogues. One result of those programs comprises the 18 compounds that carry the suffix "platin" (platinum based) that are listed in the USAN Dictionary.

The synthesis of cisplatin begins with the reduction of potassium hexachloroplatinate (**8.1**) with hydrazine to afford the tetrachloro derivative. The four chloro groups are then replaced by iodine in order to bypass the so-called trans effect that would lead the incoming amines to add to give the undesired trans isomer; treatment with excess potassium iodide proceeds to form tetraiodide (**8.3**). Ammonium hydroxide then replaces two of the iodo groups by amines in a stepwise fashion. Iodine is next removed by sequential reaction with silver nitrate followed by potassium chloride. **Cisplatin (8.5)** is thus produced [9].

The scheme used to prepare several more recent examples is typical for preparing compounds in this series. The synthesis of enloplatin begins with the construction of the moiety that will provide the required two amines. Alkylation of the dichloro ether (**9.1**)

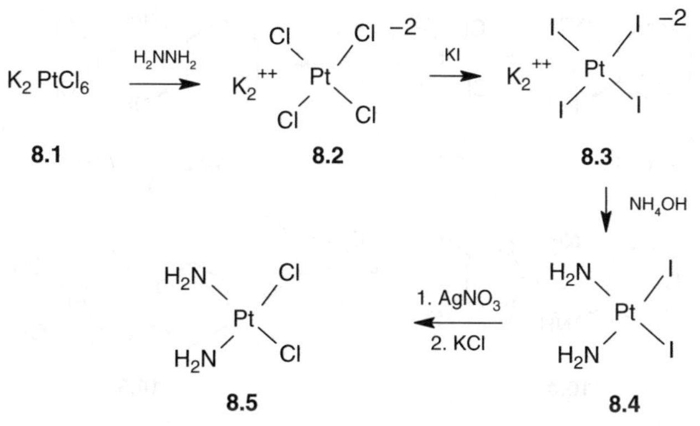

Scheme 1.8   *Cisplatin synthesis.*

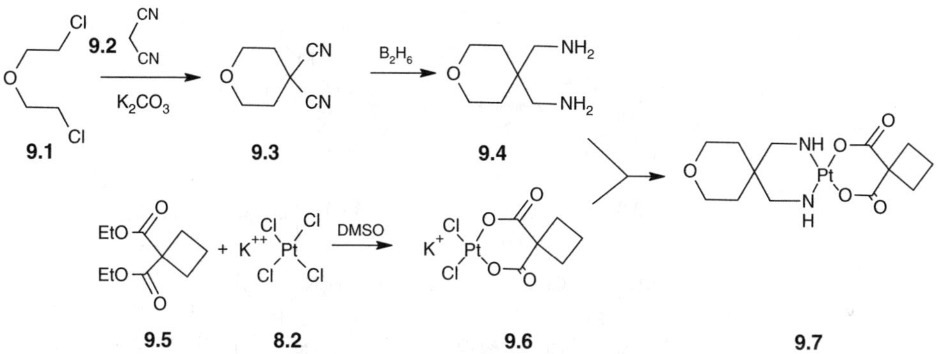

Scheme 1.9   *Enloplatin.*

affords annulated perhydropyran (**9.3**). The treatment of that intermediate (**9.3**) with diborane leads to one of the ligands (**9.4**). In a converging step, malonate (**9.5**) is condensed with the tetrachloro platinum intermediate (**8.2**) in dimethyl sulfoxide. The ester groups are saponified in the course of the reaction. Those carboxylic acids then displace chlorines from the tetrachloro starting material (**8.2**) to form the platinum compound (**9.6**). The treatment of that intermediate (**9.6**) with diamine (**9.4**) causes the amines to displace the remaining chlorine groups in the platinum intermediate (**9.6**) to form **enloplatin** (**9.7**) [9, 10].

In much the same vein, the condensation of cyclohexane-*trans*-diamine (**10.1**) with the ubiquitous tetrachloroplatinum derivative (**8.2**) affords the intermediate (**10.2**) in which basic nitrogen has displaced two of the halogen atoms. Reaction with aqueous silver nitrate precipitates the remaining chloride atoms (**10.3**). Treatment with oxalic acid then yields **oxaliplatin** (**10.4**) [11], available under the trade name Eloxatin®.

Cytotoxic activity is retained in a platinum-based compound that features two different basic amine ligands. The synthesis of one such unsymmetrical agent starts by the displacement of one of the iodine atoms in the intermediate (**8.3**) by 2-picoline. Steric hindrance about the newly formed bond by the ortho methyl group on newly introduced pyridine

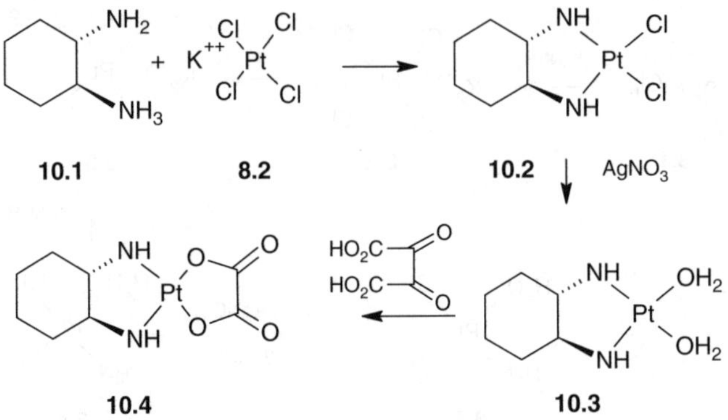

Scheme 1.10   *Oxaliplatin.*

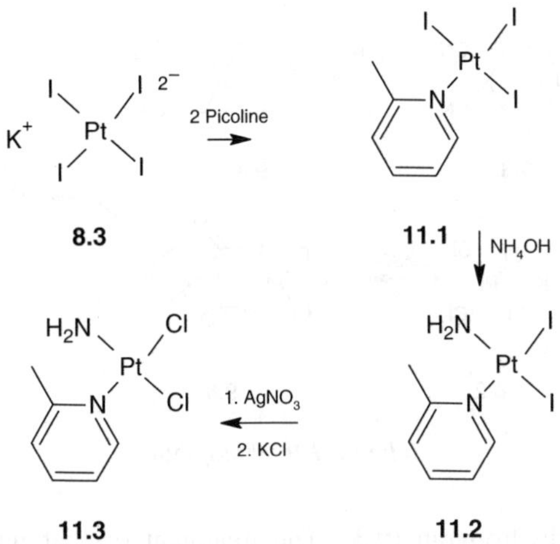

Scheme 1.11   *Picoplatin.*

derivative apparently prevents attack by a second picoline. The other basic nitrogen is then introduced by treating the intermediate (**11.1**) with aqueous ammonia (**11.2**). The sequential reaction of that product (**11.2**) with silver nitrate and then potassium chloride restores the chlorine ligands and thus affords **picoplatin** (**11.3**) [12].

1.4   Miscellaneous Alkylating Agents

The chemically reactive compound imexon (**12.4**) binds to the intracellular thiols that normally scavenge oxidants in cells. The overall effect of this drug thus comprises increased levels of oxidizing species. This compound has shown activity against a range of human cancers in cell culture as well as cancers implanted in rodents. Where approved by regulatory agencies, the drug carries the trade name Amplimexon®.

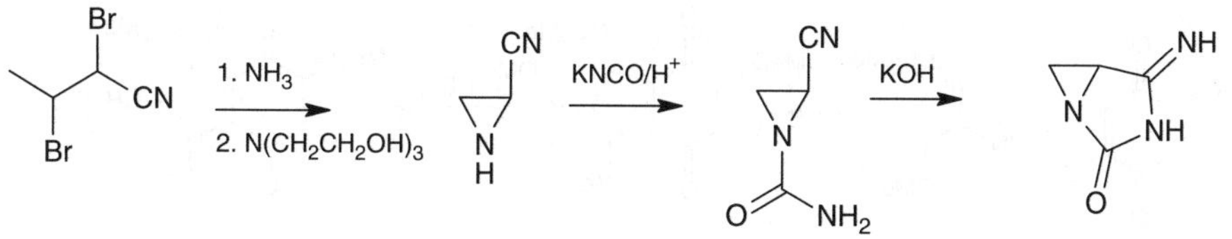

Scheme 1.12 *Imexon.*

The concise preparation of imexon starts with the preparation of cyanoaziridine (**12.2**). The reaction of 2,3-dibromopropionitrile with ammonia leads to aziridine (**12.3**). The treatment of that intermediate with potassium isocyanate and acid—actually isocyanic acid—converts the primary amine to a carboxamide. Under the basic conditions provided by potassium hydroxide, amide nitrogen adds to the nitrile to form a fused five-membered ring. **Imexon (12.4)** is thus obtained [13].

Drug delivery can be a major issue in drug development. Many drugs are, for example, not absorbed when administered orally, while others may be metabolized so quickly that they may never reach effective concentrations at the critical site. In addition to research to deal with those problems, there has been some work devoted to the opposite scenario: developing drugs that need to be metabolized near the site of action. Such drugs are often referred to as prodrugs. The antineoplastic agent apaziquone represents an example of such a prodrug. Apaziquone (**13.7**) is a synthetic analogue of the antineoplastic fermentation product mitomycin. The drug is reduced to active metabolites in hypoxic locations such as the interiors of tumors. The compound has been extensively tested for treating bladder cancer.

The condensation of the substituted quinone (**13.1**) with the enamine from ethyl acetoacetate (**13.2**) leads to pyrroloquinone (**13.3**). The reaction can be visualized by the displacement of bromine in the quinone by the enamine nitrogen. This then transforms the quinone itself to an enamine; that then adds to the newly formed side chain and closes the ring (**13.3**). The treatment of that intermediate with dichlorodicyanoquinone (DDQ) oxidizes the methoxymethyl group on the pyrrole to the corresponding aldehyde. The reaction of this last intermediate with ylide from Emmons reagent and triethylamine leads to the derivative with a lengthened side chain (**13.6**). The presence of lithium chloride in the reaction medium enforces the tendency to form a trans olefin. The treatment of this last intermediate with diisobutylaluminum hydride reduces both esters to alcohols. The displacement of the quinone methoxyl by aziridine may well involve an addition–expulsion reaction. Whatever the mechanism, the reaction affords **apaziquone (13.7)** [14].

The nitrogen-rich heterocyclic compound temozolomide is approved by the FDA for treating several brain cancers. The drug is available under several trade names: Temodar®, Temodal®, and Temcad®. The concise synthesis starts with the treatment of imidazole (**14.1**) with nitrous acid. The resulting diazonium salt (**14.2**) is then allowed to react with methyl isocyanate of ill fame. The initial product comprises the urea from addition of the reagent to one of the ring nitrogen atoms (**14.3**). The exocyclic urea nitrogen then attacks the charged azo group to close the ring, forming **temozolomide (14.4)** [15]. The parent drug has little if any activity as such. This compound actually undergoes a number of metabolic transformations that terminates in azomethane (**14.5**). The carbocation from loss of nitrogen then alkylates RNA guanine.

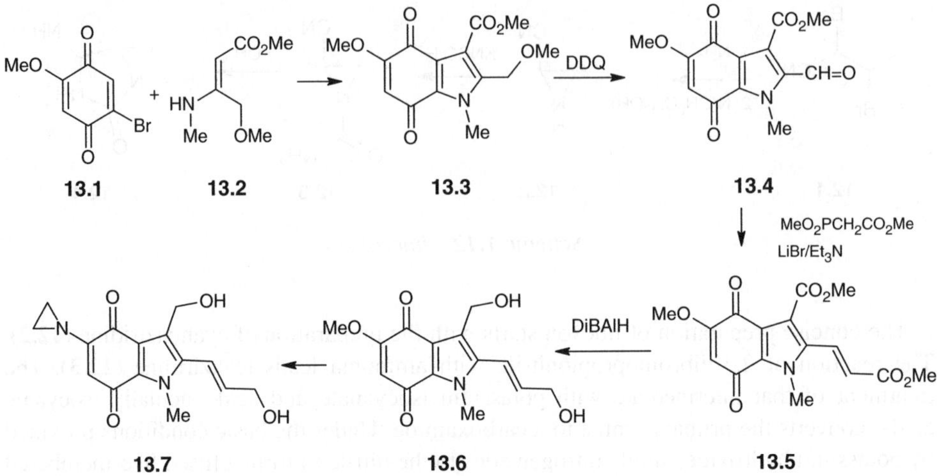

Scheme 1.13   Apaziquone.

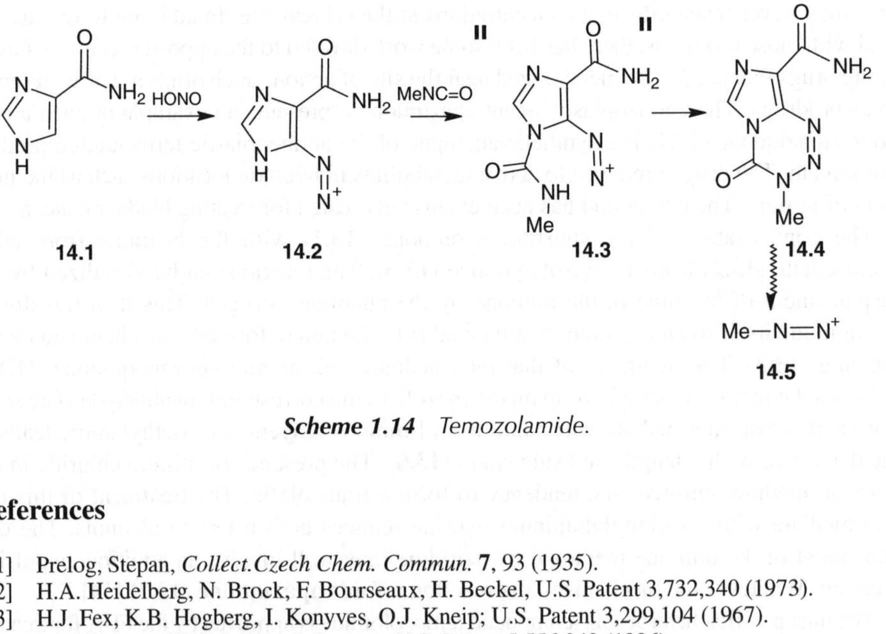

Scheme 1.14   Temozolamide.

References

[1]   Prelog, Stepan, *Collect.Czech Chem. Commun.* **7**, 93 (1935).
[2]   H.A. Heidelberg, N. Brock, F. Bourseaux, H. Beckel, U.S. Patent 3,732,340 (1973).
[3]   H.J. Fex, K.B. Hogberg, I. Konyves, O.J. Kneip; U.S. Patent 3,299,104 (1967).
[4]   L.M. Kauvar, M.H. Lyttle, A. Satyam, U.S. Patent 5,556,942 (1996).
[5]   A. Satyam, M.D. Hocker, K.A. Kane-Maguire, A.S. Morgan, H.O Villar, M.H. Lyttle, *J. Med. Chem.* **39**, 1736 (1996).
[6]   J.A. Montgomery, R. James, G.S. McCaleb, M.C, Kirk, T.P. Johnston, *J. Med. Chem.* **18**, 568 (1975).
[7]   A.P. Schroff, U.S. Patent 4,027,019 (1977).
[8]   A.C. Sartorelly, S. Krishnamurthy, P.G. Penketh, U.S. Patent 5,637,619 (1997).
[9]   G.B. Kaufman, D.O. Cowan, *Inorg. Syn.* **7**, 239 (1963).
[10] P. Bitha, S.G. Carvajal, R.V. Citavella, R.V. Child, E.F.D. Santos, T.s. Dunne, F.E. Durr, J.J. Hlavka, S.A. Lang, H.L. Lindsay, *J. Med. Chem.* **32**, 2015 (1989).

[11] M. Galanski, A. Yasemi, S. Slabi, M. Jakupec, V.B. Arion, M. Rausch, A. Nazarov, B.K. Keppler, *Eur. J. Med. Chem.* **39**, 707 (2004).
[12] B.A. Murrer, U.S. Patent 5,665,771 (1997).
[13] B.S. Iyengar, R.T. Dorr, D.S. Alberts, E.M. Hersh, S.E. Salmon, W.A. Remers, *J. Med. Chem.* **42**, 510 (1999).
[14] E. Comer, W.S. Murphy, ARKIVOC, **286** (2006).
[15] M.G.F. Stevens, J.A. Hickman, R. Stone, N.W. Gibson, G.U. Baig, E. Lunt, C.G. Newton, *J. Med. Chem.* **27**, 196 (1984).

2

Antimetabolites

2.1 Introduction

The process by which cells proliferate requires a collection of chemical compounds that are necessary for the synthesis of new vital cell constituents such as nucleic acids. An alternate approach for selectively wiping out cancerous cells without harming normal tissues involves administering the so-called false substrates for the compounds normally used for building new DNA and/or RNA. False substrate is a general term embracing compounds whose chemical structure is close to that of the compound that normally used for some endogenous biochemical process. If the chemical structure of the mimic is close enough to that of the authentic substrate, it will be accepted by the enzymes involved in the process. The false substrate cannot, however, substitute operationally and thus brings the process to a halt. The cell, thus denied the authentic product, simply expires. Note however that antimetabolite cancer chemotherapy, in common with that using alkylating agents, also relies for preferential effect on cancer cells on the significantly more rapid turnover of malignant cells over that of normal cells. Many of the side effects from administering these drugs are due to operation of the same process on normal cells.

Folic acid, for example, or more properly its reduction product, dihydrofolic acid (**1.1**), is an essential compound for building and repairing DNA. Folic acid mimics comprise an important set of false substrate chemotherapy agents, pyrimidines (**1.2**) and purines (**1.3**), coupled to pentose sugars that also play an essential role in the construction of DNA and RNA. Compounds based on these two heterocyclic systems also comprise the bases found in the interior of the DNA double helix.

Antineoplastic Drugs: Organic Synthesis, First Edition. Daniel Lednicer.
© 2015 John Wiley & Sons, Ltd. Published 2015 by John Wiley & Sons, Ltd.

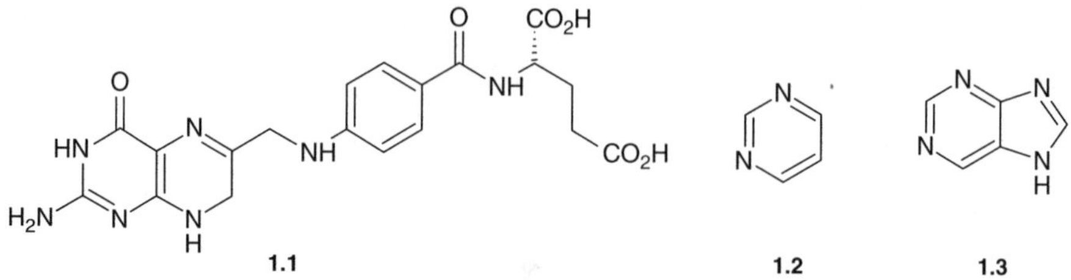

Scheme 2.1   *Targets of Antimetabolites.*

2.2   Folate Antagonists

2.2.1   Compounds with Glutamate Side Chain

Research on folate antagonists began with the near-serendipitous discovery of the activity of the pteridine methotrexate (**2.2**) in 1942 [1]. Investigations on the antineoplastic activity as well as synthesis of modified pteridines have stayed quite active to the present day [2]. It is of interest that methotrexate differs from folic acid (**1.3**) only by the presence of a methyl group on the side chain nitrogen atom.

One concise synthesis of this compound begins with the condensation of commercially available tetra-aminopyrimidine (**2.1**) with dibromoaldehyde (**2.2**). The formation of pteridine (**2.3**) can be visualized as starting by the displacement of the halogen α to the carbonyl group followed by the condensation of the aldehyde with adjacent amine. (The steps in reversed order will give the same product.) In a convergent step, glutamate (**2.4**) is then prepared by the acylation of glutamic acid with *N*-methyl-4-aminobenzoate. The displacement of bromine from pteridine (**2.3**) by the amine in benzoate (**2.4**) leads to *N*-methyldihydrofolate (**2.5**). Simple air oxidation then affords **methotrexate** (**2.6**) [3]. This folate-like compound inhibits the enzyme tetrahydrofolate dehydrogenase which is essential for the formation of thymidylate. The latter substance is essential for the formation of new DNA. The patent having expired, this drug is available as a generic from several vendors under several trade names.

A folate analogue in which the N-methyl moiety in the side chain is replaced by a neutral ethyl group and the amine present in folic acid is replaced by carbon interestingly still retains antimetabolic activity or more specifically inhibition of dihydrofolate reductase, the enzyme that adds a pair on hydrogen atoms to the pyrazine moiety.

The preparation of edatrexate starts with the fully unsaturated pteridine (**3.1**). Thus, the reaction of bromomethylpteridine (**3.1**) with triphenylphosphine affords the corresponding phosphonium salt (**3.2**). This salt is next converted to the ylide by means of butyl lithium. The condensation of this last intermediate with the substituted propiopheone (**3.3**) gives condensation product **3.4**. Catalytic hydrogenation removes the newly introduced side chain double bond; the fused pyrazine ring is also reduced in an undesired side reaction. Treatment with hydrogen peroxide restores that unsaturation (**3.5**). The benzoate ester is next saponified. Dicyclohexylcarbodiimide-mediated condensation with diethyl glutamate attaches the aminoester. A second saponification step completes the synthesis of **edatrexate** (**3.6**) [4]. This antitumor drug, like methotrexate, inhibits the enzyme dihydrofolate reductase and thus the addition of two hydrogen atoms to the pyrazine ring.

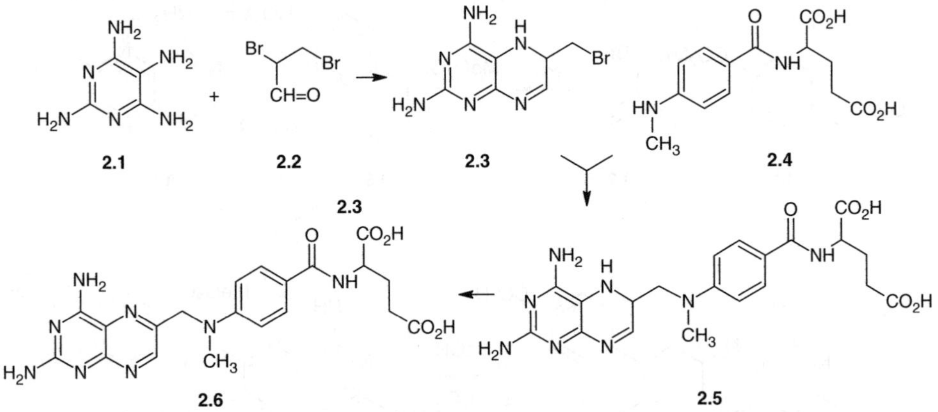

Scheme 2.2 *Methotrexate.*

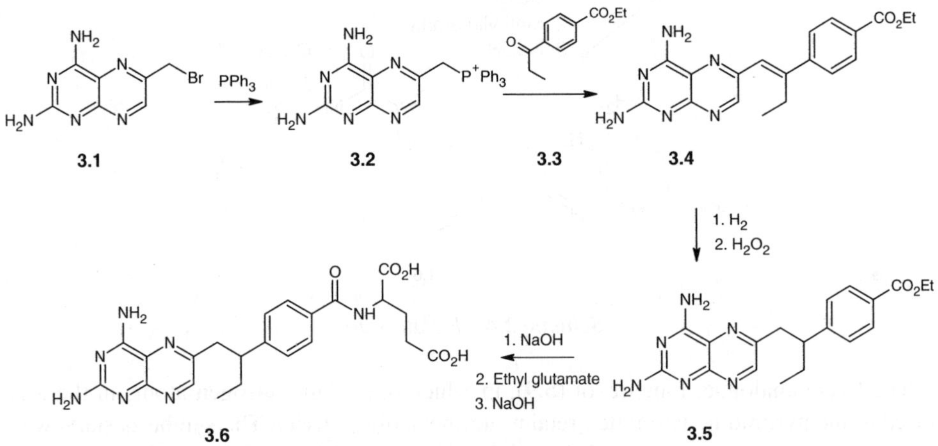

Scheme 2.3 *Edatrexate.*

Yet another analogue that is based on a fully unsaturated pteridine features a propargyl group at the position that is occupied by nitrogen in folic acid. The scheme for preparing this compound starts with the alkylation of the carbanion from treating diester (**4.1**) with potassium hydride with propargyl bromide (**4.2**). The product is again treated with base; the resulting new carbanion is next alkylated with bromomethyl pteridine (**3.1**) to give an intermediate (**4.4**) that now includes the full folate skeleton. The superfluous carbonyl ester on the quaternary position is first saponified with base; mild heat causes the carboxylic acid to depart as carbon dioxide to afford intermediate **4.5**. The remaining carboxylic acid is next converted to the corresponding acid chloride. This product, which features an activated carboxyl group, is then used to acylate dimethyl glutamate. The saponification of the ester groups yields the antimetabolite **pralatrexate** (**4.6**) [5]. This drug is approved for treating T-cell lymphomas and is available under the trade name Folotyn®.

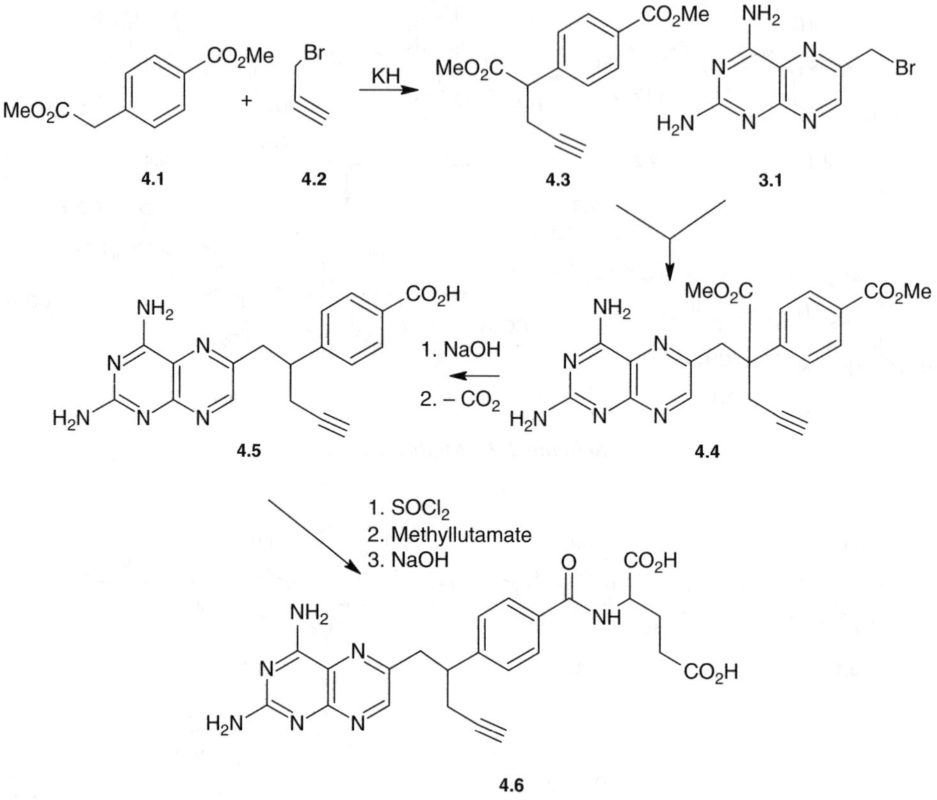

Scheme 2.4 *Pralatrexate.*

The folate analogue, lometrexol (**5.9**), in which one of the nitrogen atoms in the ring fused to the pyrimidine is omitted retains antimetabolic activity. The synthesis starts with the condensation of pyrimidine (**5.1**) with bromomalonaldehyde (**5.2**) to afford pyrimido-pyridine (**5.3**), an intermediate that carries a bromine atom at the site of the future side chain. This condensation can be envisaged as involving imine formation between aldehyde (**5.2**) with the amine on the pyrimidine; aldol-like addition of the carbon adjacent to the carbonyl closes the ring (a sequence reversing the steps will give the same product). The amine in the resulting product **5.3** is next protected as its pivaloyl amide (**5.4**) by acylation with pivaloyl chloride. Palladium-catalyzed coupling of the intermediate (**5.4**) with bromotrimethylsilyl acetylene (**5.5**) adds the requisite two-carbon atom side chain (**5.6**). Treatment with tetrabutylammonium fluoride removes the trimethylsilyl group at the end of acetylene (**5.6**). A second palladium-catalyzed coupling reaction, this time between the newly revealed carbon in (**5.7**) with the iodine in benzoyl glutamamide (**5.8**) completes construction of the skeleton. Catalytic hydrogenation then reduces the acetylenic function; the fused pyridine is also hydrogenated in this step. Saponification then removes both the pivaloylamide protecting group and the glutamine esters to afford **lometrexol** (**5.9**) [6]. In common with the major portion of folate antimetabolites, lometrexol inhibits dihydrofolate reductase, the enzyme that adds a pair of hydrogen atoms to the aromatic ring.

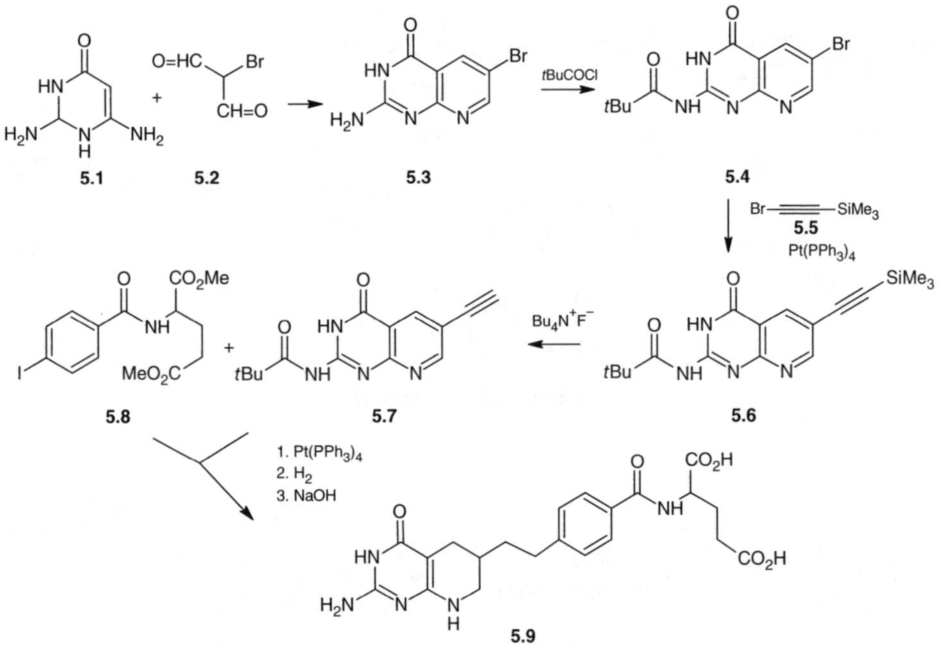

Scheme 2.5 *Lometrexol.*

There exist a fair number of instances where biological activity is retained after replacement of a benzene ring by thiophene. (This correspondence is called isosteric; it hinges in this case on the near-identical size of benzene and thiophene rings and the same number of π electrons.) This relation also applies to the folate false substrates. One arm of the convergent synthesis of raltitrexed (**6.7**) starts with the construction of the fused pyrimidine ring. Thus, the reaction of anthranilic acid (**6.1**) with ethyl iminoacetate affords quinazoline (**6.2**). This reaction can be visualized by the addition of aniline nitrogen to the end of the imino function and expulsion of ethoxide, followed by the formation of the cyclic amide.

The treatment of quinazoalone (**6.2**) with *N*-bromosuccinimide (NBS) attacks the methyl group on the benzene ring to form the reactive moiety (**6.3**). The construction of the other parts of the antimetabolite begins by the acylation of diethyl glutamate with nitrothiophene acid chloride (**6.4**). Catalytic hydrogenation of the intermediate (**6.5**) converts the nitro group to the corresponding amine (**6.6**). That product is then N-methylated by any one of several procedures (**6.7**). The reaction of this intermediate with quinazolone (**6.3**) leads to the replacement of bromine by the basic amine in **6.7** and connection of the two moieties. The saponification of the ester groups in the coupled intermediate converts the glutamic esters to the corresponding carboxylic acids and thus to **raltitrexed** (**6.7**) [7]. The drug is available under the trade name Tomudex®. This compound, like others in this class, inhibits the enzyme dihydrofolate reductase.

The preparation of the folate antimetabolite pelitrexol starts with the palladium-catalyzed coupling of the ethynyl function in quinazolone (**5.7**), used in the synthesis of lometrexol, with the bromo thiophene (**7.1**). Catalytic hydrogenation reduces the side chain ethynyl function

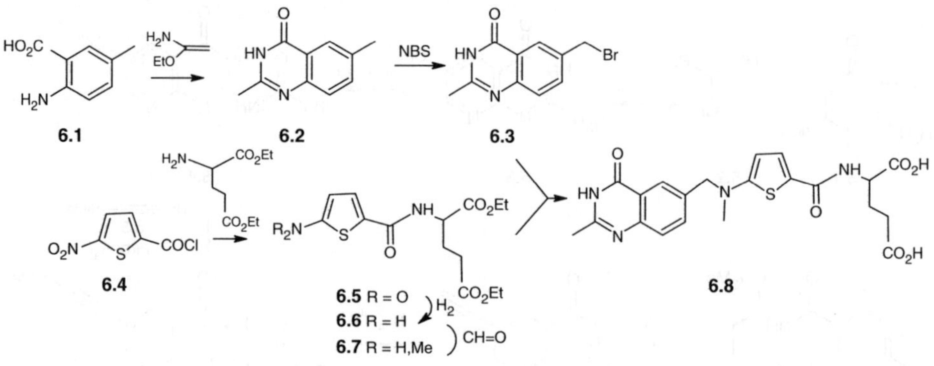

Scheme 2.6   *Ralitrexed.*

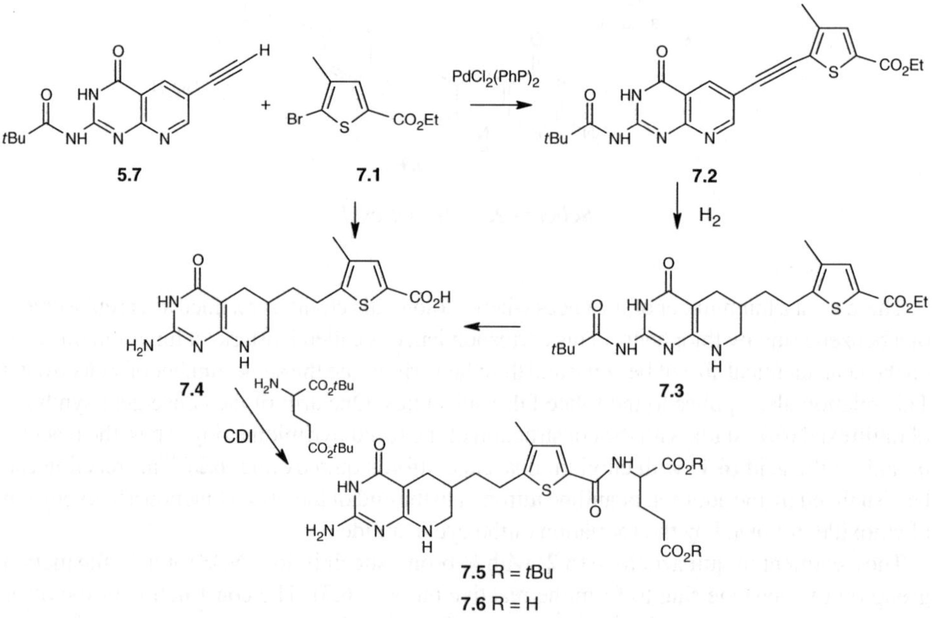

Scheme 2.7   *Pelitrexol.*

as well as the fused pyridine ring (**7.3**). Hydrolysis followed by saponification removes the pivaloyl amide protecting group as well as the thiophenyl ester. Diimidazole-mediated amide formation between thiophene carboxylic acid (**7.4**) and the amine group in the *tert*butyl ester of glutamic acid gives amide (**7.5**). Simple hydrolysis then leads to **pelitrexol** (**7.6**) [8–10]. This agent inhibits activity of glycinamide ribonucleotide formyltransferase (GARFT), the enzyme that catalyzes the synthesis of purines.

Antimetabolic activity is interestingly maintained in a compound where the right-hand ring is shrunk to a fused pyrrole. The scheme for preparation of this drug starts with the preparation of the pyrimidone by the well-precedented condensation of the

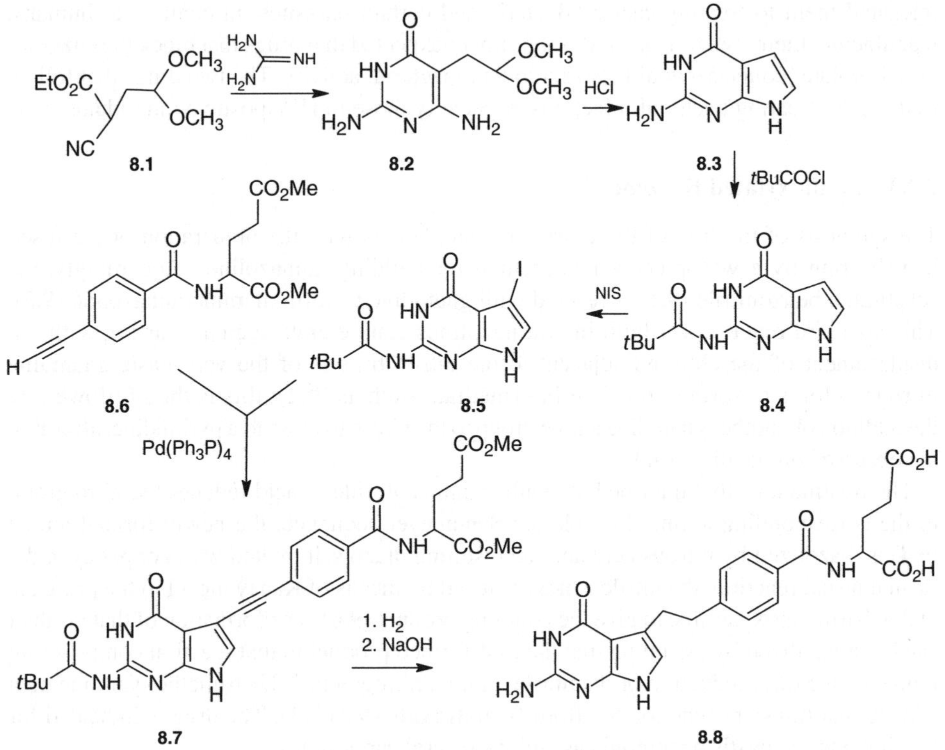

Scheme 2.8 *Pemetrexed.*

cyano-carbethoxy starting material (**8.1**) with guanidine. Acid hydrolysis of the product (**8.2**) opens the acetal to the corresponding aldehyde; that hypothetical transient intermediate cyclizes to a fused pyrrole under reaction conditions to afford a pteridine-like nucleus (**8.3**).

The amine is next protected as its amide (**8.4**) by acylation with pivaloyl chloride. The reaction of this last product with *N*-iodosuccinimide introduces the halogen at the position where the other major component will be attached. The condensation of the iodo derivative (**8.5**) with acetylene (**8.6**) in the presence of platinum tetrakis triphenylphosphine leads to the linked product (**8.7**) which now contains the desired skeleton. The triple-bonded moiety in the side chain is next removed by catalytic hydrogenation. Saponification removes the pivaloyl protecting group and concurrently frees the glutamoyl acids. The folate antimetabolite compound **pemetrexed** (**8.8**) is thus obtained [6]. This pteridine with an abbreviated ring, like of the other pteridines in this section, acts as an antagonist to dihydrofolate reductase. The drug was approved by the FDA in 2013 and is sold under the trade name Alimta®.

2.2.2 Compounds Lacking the Glutamate Moiety

The broad tolerance for structural variations of folate receptors is further illustrated by compounds that lack the glutamate moiety present in the previous compounds. The list of side effects elicited by those drugs, especially at the required higher doses, has largely relegated them to treating cancer. Bacteria and certain parasites, in contrast to humans, manufacture their own folic acid. It is thus not unexpected that drugs that block the enzymes dihydrofolate reductase would also exhibit antibacterial activity. The two drugs that follow find use for treating bacterial infections, especially those in HIV-positive individuals.

2.2.3   Methoxylated Benzenes

The synthesis of the first of these, trimetrexate, begins with the preparation of the fused bicyclic ring by a well-precedented method for building quinazolines. Accordingly, the substituted benzonitrile (**9.1**) is reacted with guanidine to form diaminoquinazoline (**9.3**). This standard reaction for building quinazolines can be envisaged as starting with the displacement of the chlorine adjacent to the nitrile by one of the very basic guanidine nitrogen atoms to form a transient intermediate such as (**9.2**); this is then followed by the addition of another guanidine amine group to the nitrile to close to a pyrimidine, affording the observed quinazoline (**9.3**).

The treatment of that intermediate with stannic chloride in acid reduces the nitro group to the corresponding amine. In a classic Sandmeyer sequence, the newly formed amine (**9.4**) is diazotized by nitrous acid and the resulting diazo salt treated with copper cyanide. In an unusual reaction, the nitrile is next reduced by means of Raney nickel in the presence of 3,4,5-trimethoxyaniline to give the coupling product (**9.6**). The formation of that product can be rationalized by assuming that the first-formed product in that reaction comprises an imine from partial reduction of the nitrile; aniline nitrogen in 2,3,4-trimethoxyaniline then adds to that transient function to afford **trimetrexate** (**9.6**) [11]. The drug is indicated for treating pneumocystis pneumonia as well as several carcinomas.

A yet further simplified quinazoline whose structure includes a less methoxylated benzene group than trimetrexate also inhibits dihydrofolate reductase. A classic sodium ethoxide-mediated condensation of ethyl acetoacetate (**10.1**) with dimethoxybenzalde-hyde (**10.2**) proceeds to add the anion from the position between the carbonyl groups to

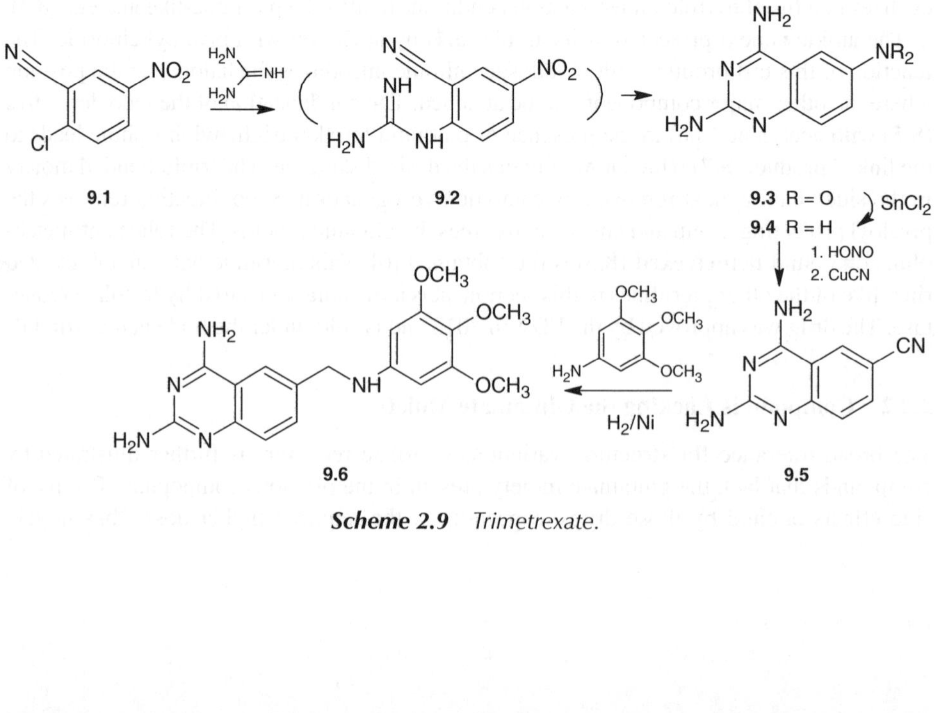

Scheme 2.9   *Trimetrexate.*

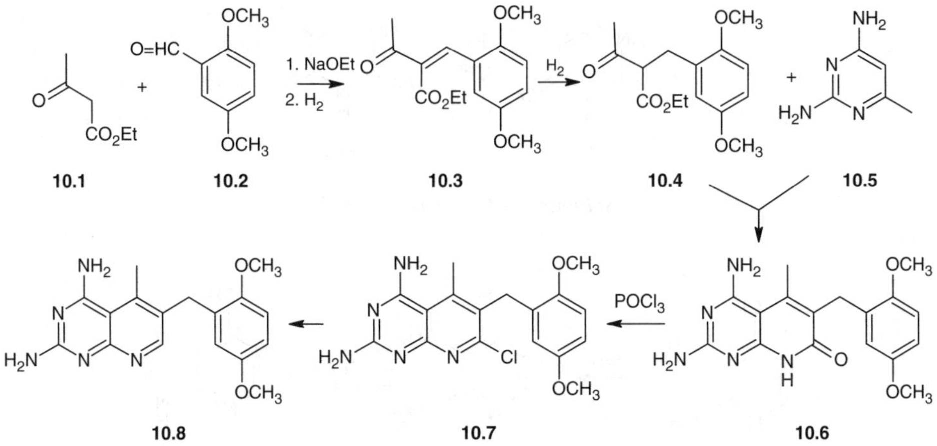

Scheme 2.10   *Piritrexim.*

the aldehyde to form a transient carbinol; that species dehydrates under reaction conditions to afford the observed product **10.3**. Catalytic hydrogenation then removes the superfluous unsaturation (**10.4**). The reaction of this last intermediate with diaminopyrimidine leads to the fused bicyclic intermediate (**10.6**). This process can be visualized by assuming that one of the amine groups in heterocycle (**10.5**) displaces the ethoxyl moiety in (**10.4**); the remaining amine then forms an imine with the ketone to form (**10.6**) (the order of those steps may well be reversed). Phosphorus oxychloride then converts the amide to its enol chloride (**10.7**). Catalytic hydrogenation over palladium removes the ring halogen to afford **piritrexim** (**10.8**) [12]. This FDA-approved drug is used principally for treating a list of carcinoma.

2.3   Pyrimidines and Purines

2.3.1   Aglycones

Research on the use of the heterocyclic bases represented by the letters of the genetic code for treating cancer actually harked back to the days when the DNA double helix had just been discovered. The chemical structure of these agents usually comprises either a monocyclic pyrimidine or a two-ring purine (*see* Schemes 11.1 and 12.3) attached to sugar. The first two drugs to be used as antineoplastic agents consist of the bases minus the sugar moiety; these are often referred to as aglycones. It is of passing interest that 5-fluorouracil and 6-mercaptopurine were each developed in the mid-1950s in different laboratories. Both compounds are still in use as antineoplastic agents.

The first of those two drugs, 5-fluorouracil (**11.2**) is prepared by the straightforward, if hazardous, direct reaction of uracil (**11.2**) with nitrogen-diluted elemental fluorine gas [13]. The drug, **fluorouracil**, also known as 5-FU, may be one of the best known false substrates. On administration, the drug is incorporated in growing DNA and RNA halting the process.

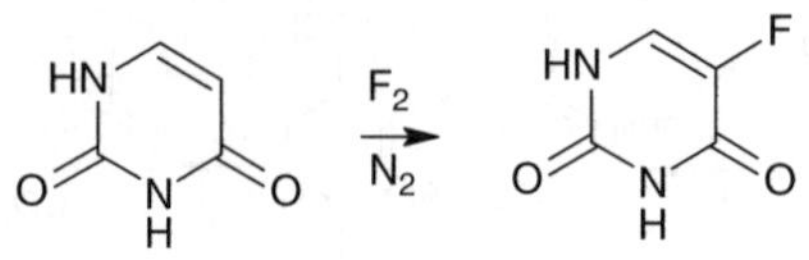

Scheme 2.11   *5-Fluorouracil.*

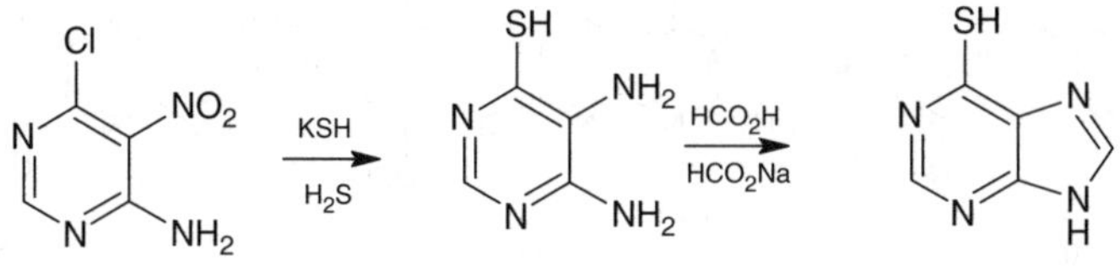

Scheme 2.12   *6-Mercaptopurine.*

Because of the side effects that come with parenteral use, the drug is used mainly for treating topical tumors. The patent on 5-FU expired many years ago. This now-generic drug is available from more than a few suppliers under a variety of trade names.

One of the schemes for the synthesis of 6-mercaptopurine starts with nucleophylic aromatic displacement of chlorine in the highly substituted pyrimidine (**12.1**) with potassium sulfide in the presence of hydrogen sulfide gas. The nitro group is apparently reduced to an amine under those conditions. The reaction of the product (**12.2**) with formic acid in the presence of sodium formate closes the fused ring to afford the purine **6-mercaptopurine** (**12.3**).

2.3.2   Saccharide-Linked Compounds

The pyrimidines and purines that form part of DNA and RNA are linked to pentose sugars in those molecules. (Sugars of adjacent bases are linked by phosphates, providing the backbone for DNA and RNA.) The covalent connecting link from the sugar to the base comprises an α-amino ether, a functional group akin to an acetal. The heterocyclic moiety in the following examples comprises cytosine, the unnatural triazine, and, in the case of purines, adenine. Structural changes occur mainly on the sugar moiety.

2.3.2.1   Pyrimidones

Simply inverting the configuration of the hydroxyl group at position 2 in the sugar leads to cytarabine, a compound that acts as a false substrate for the endogenous synthesis of DNA or RNA. This compound has been found to inhibit DNA polymerase; it has found use in the clinic for treating leukemias.

One concise preparation starts by heating citidine, one of the pyrimidines found in DNA, with polyphosphoric acid (PPA). This dehydrating reagent leads to the formation of an ether linkage between the enol form of the pyrimidone and the hydroxyl at position 2 in the

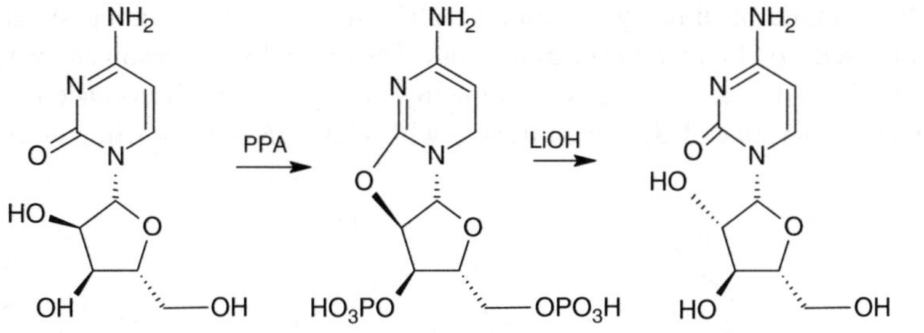

Scheme 2.13   *Cytarabine.*

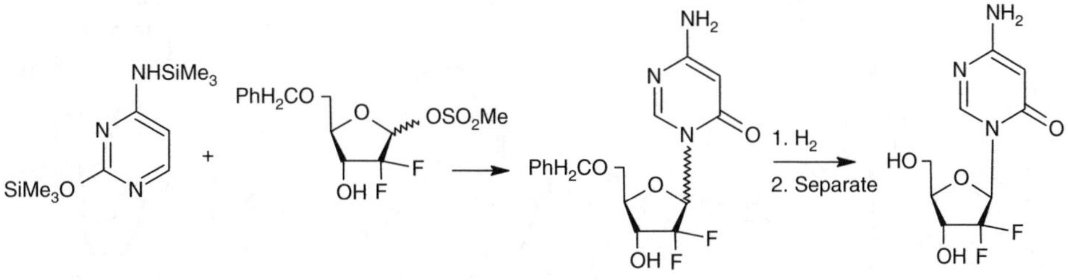

Scheme 2.14   *Gemcitabine.*

sugar; the reagent additionally converts the other two hydroxyl groups to their phosphates (**13.2**). Aqueous base then opens the internal ether by the attack of hydroxyl on the backside of the sugar. This in essence reverses to configuration of the hydroxyl at position 2. The phosphates hydrolyze in the process to afford **cytarabine** (**13.3**) [14]. This drug has been approved by the FDA for the treatment of leukemias.

The reaction of cytosine with trimethylsilyl chloride locks the heterocycle in its enol tautomer (**14.1**). The reactivity of the ring nitrogen in this silyl derivative is increased over that of cytosine; the basicity of now-silated primary amine is significantly decreased. This derivative is next allowed to react with the methanesulfonate function of an anomeric mixture (**14.2**) of the difluoro sugar. Straightforward displacement of the methanesulfonate group affords the glycosylated intermediate (**14.3**). Catalytic hydrogenation over a platinum catalyst removes the benzyl protecting group on the pentose to give the coupled product as a mixture of the two anomers. The desired compound in which the linkage has the same β-configuration as the sugar found in nature affords the antineoplastic agent **gemcitabine** (**14.4**) [15]. Gemcitabine is currently approved for treating a selection of malignant solid tumors.

The preparation of tezacitabine (**1.7**) involves a rather more complex scheme. The synthesis begins with the protection by the reaction of citidine (**15.2**) with the silyl reagent triisopropylsilyl dichloride (TIPSCl$_2$) (**15.1**). This forms the cyclic bis-silyl derivative (**15.3**) that covers the two hydroxyl groups not involved in the next set of transforms. The amine on pyrimidine (**15.3**) was next converted to its amidine (**15.4**) by reaction with dimethylformamide in dimethyl acetamide. Oxidation of the remaining sugar hydroxyl group had posed a problem in the original work. The difficulty was resolved by a procedure that involved oxidation by means of dimethyl sulfoxide in trifluoroacetic acid [16]. The carbonyl derivative (**1.5**) is then treated with ylide (**15.6**) from trimethoxyphosphon

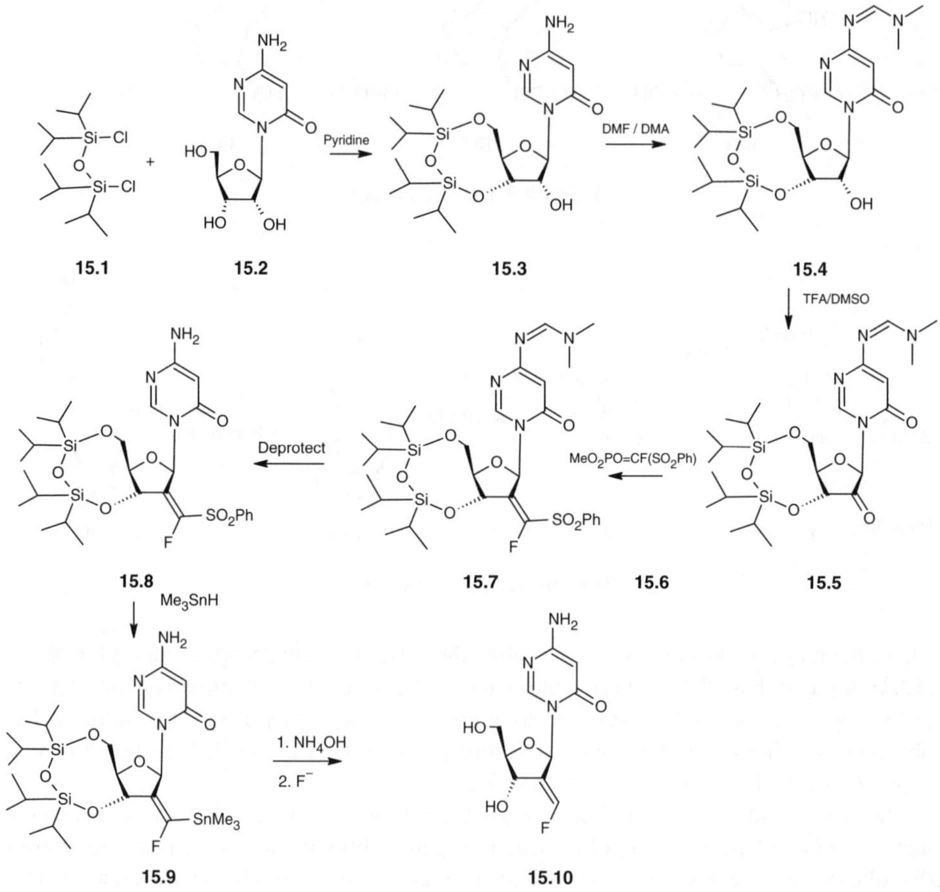

Scheme 2.15   *Tezacitabine.*

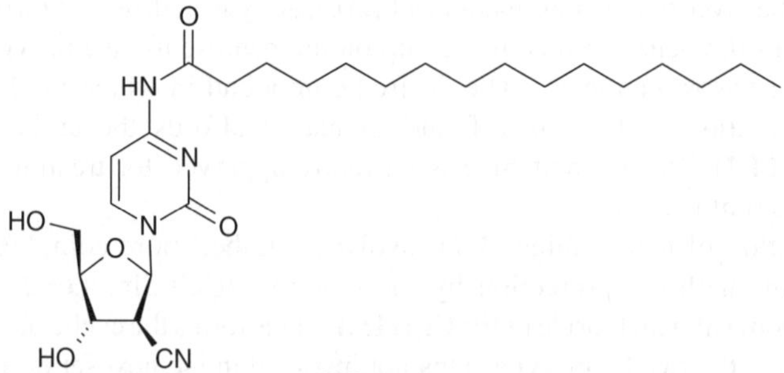

Scheme 2.16   *Sapacitabine.*

yl(fluoro(phenylsulfonyl))methane. Hydrolysis in dilute acid serves to remove the amine protecting group (**15.8**). Note that the cyclic silyl protecting group is still in place; the remaining chore involves removal of the phenylsulfonyl function introduced by the Horner–Emmons reagent. The reaction of (**15.8**) with trimethyltin hydride displaces that function and yields the organotin derivative (**15.9**). The ammonolysis of the latter serves to replace tin by hydrogen. The cyclic silyl protecting group is then removed by treatment with a fluoride salt such as tetrabutylammonium fluoride. The RNA reductase inhibitor **tezacitabine** (**15.10**) is thus obtained [17]. This drug binds irreversibly to the enzyme ribonucleotide reductase which is expected to hinder formation of DNA.

Pyrimidone (**16.1**) features two atypical structural features; these include a nitrile function on the sugar and a 16-carbon amide on the pyrimidine amine, the latter arguably added to increase lipophilicity. The mode of action of this agent is also unusual. The drug **sapacitabine** (**16.1**) has been found to cause single-strand DNA breaks. Most of the preceding compounds inhibit DNA and RNA by acting as false substrates. The drug has been granted "orphan drug" status by the FDA in 2009. No source is available at this writing for presenting a detailed description of the synthetic scheme for preparing sapacitabine.

2.3.2.2 Triazines

Antineoplastic activity is retained in compounds that include an additional nitrogen atom in the heterocyclic moiety. The triazine-based antineoplastic drug decitabine (Dacogen®) was approved by the FDA in 2006. The drug inhibits DNA methyltransferase, an action that inactivates the DNA.

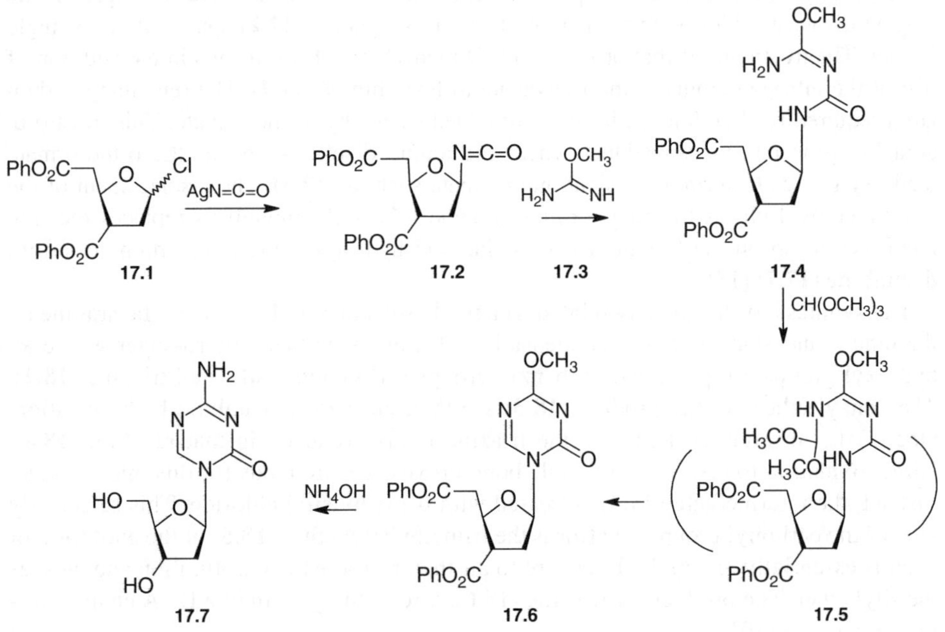

Scheme 2.17 *Decitabine.*

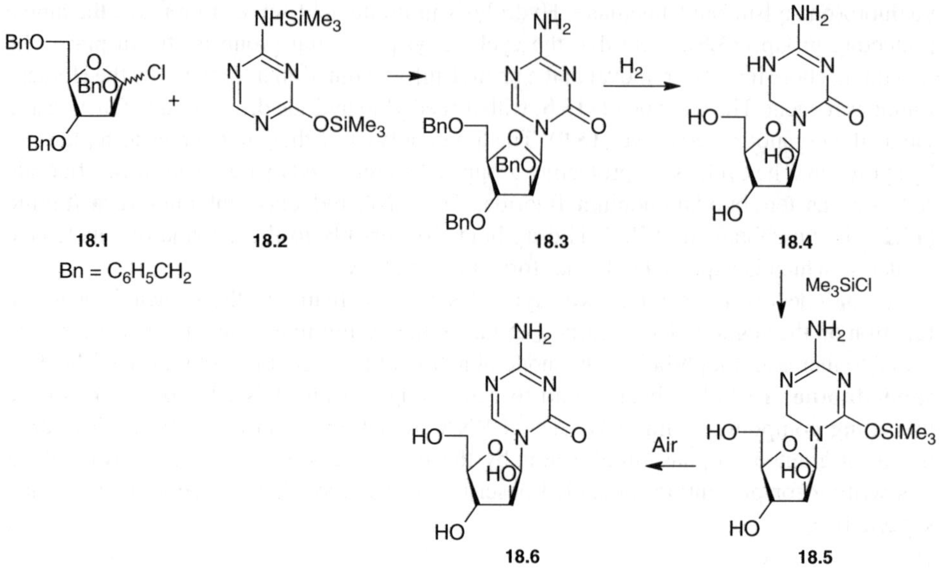

Scheme 2.18   *Fazarabine.*

The published scheme for preparing that drug on the other hand dates back to 1979. The synthesis uses a different strategy from those discussed previously in that the heterocyclic ring is built with the pentose in place. The synthesis starts with the reaction of the pair of anomers at position 2 of chloro pentose (**17.1**) with silver isocyanate. Nitrogen in the reagent displaces chlorine in the pentose to form isocyanate (**17.2**) apparently as a single isomer. The treatment of that product with O-methylurea (**17.3**) results in the addition of one of the nitrogen atoms to the isocyanate to form biuret (**17.4**). The remaining carbon atom required for building a triazine is provided by methyl orthoformate. This transform arguably proceeds via an adduct where the amine displaces one of the orthoformate methoxy groups to form a transient intermediate such as (**17.5**). The displacement of the other methoxyl closes the ring forming triazine (**17.6**). Ammonolysis replaces the ring methoxyl group on the triazine by an amine and hydrolyzes the ester groups to afford **decitabine** (**17.7**) [18].

The synthesis of the closely related triazine-based antineoplastic agent fazarabine on the other hand starts with the condensation of a preassembled chloro-sugar where all hydroxyl groups are protected by benzyl groups (**18.1**) and a silylated triazine (**18.2**). The benzyl ethers in the product (**18.3**) are then cleaved by catalytic hydrogenation. One of the unsaturated bonds in the triazine is also reduced in that reaction (**18.4**). Direct oxidation to restore that double bond proves too strenuous for this intermediate. Instead, the intermediate (**18.4**) is treated with trimethylsilyl chloride. This selectively attacks the carbonyl group and affords the trimethylsilyl ether (**18.5**) of the enol form of what is essentially an amide. Exposure to air then restores the unsaturation and cleaves the silyl ether. The product **fazarabine** (**18.6**), arrests the growth of a DNA chain, causing cells to die [19].

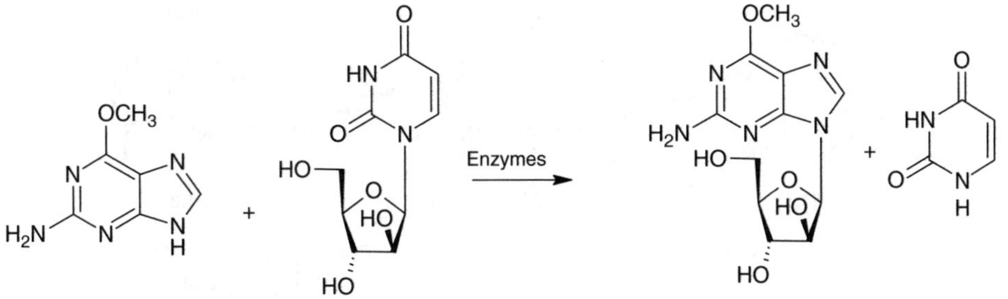

19.1 **19.2** **19.3** **19.4**

Scheme 2.19 *Nelarabine.*

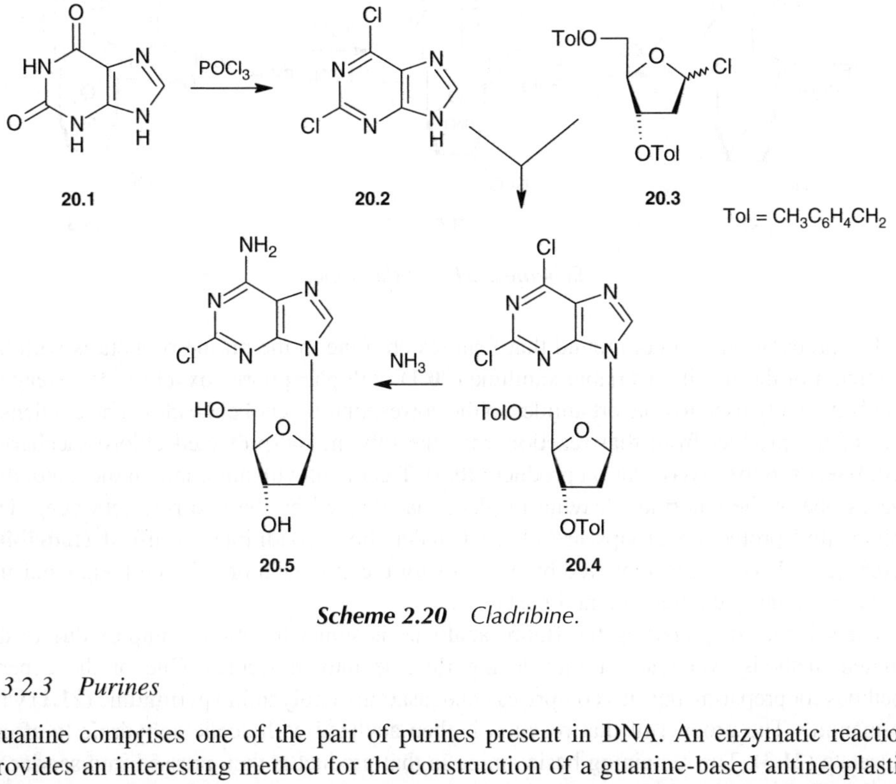

20.1 **20.2** **20.3** **20.4** **20.5**

Tol = $CH_3C_6H_4CH_2$

Scheme 2.20 *Cladribine.*

2.3.2.3 Purines

Guanine comprises one of the pair of purines present in DNA. An enzymatic reaction provides an interesting method for the construction of a guanine-based antineoplastic compounds. The treatment of a mixture of the guanine derivative (**19.1**) and uracil arabinoside (**19.2**) with the enzymes uridine phosphorylase and purine nucleoside phosphorylase affects the transfer of the sugar from pyrimidone (**19.2**) to the guanine [20]. The resulting compound **nelarabine (19.3)**, is demethylated in the body after it has been absorbed. The demethylated product is a classical false substrate; it is incorporated in the DNA where it inhibits DNA synthesis and stops the synthesis of DNA required for proliferation. The FDA approved the drug in 2005 for treating patients suffering T-cell leukemia. The drug is sold under the trade name Arranon®.

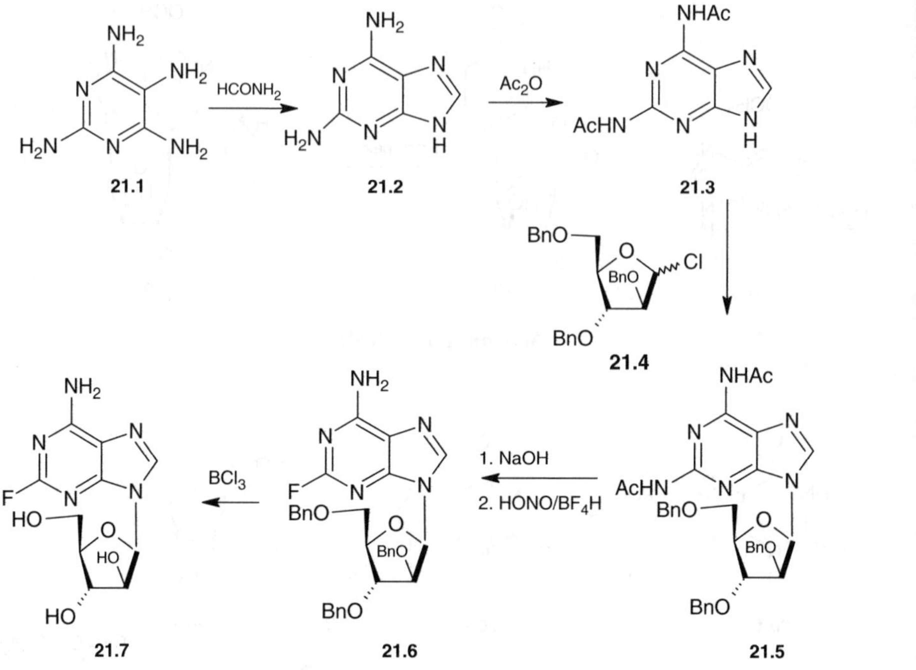

Scheme 2.21   *Fludarabine.*

The preparation of a compound that features chlorine in the purine ring starts with the treatment of the readily available xanthine (**20.1**) with phosphorus oxychloride, a reagent that is typically used to convert amides to the corresponding enol chlorides. The condensation of the product from that reaction with the tolylmethyl-protected chloro-saccharide (**20.3**) leads to the glycosidation product (**20.4**). Treatment with ammonia in methanol displaces one of the chlorines, leaving in place that flanked by the two ring nitrogens. The tolylmethyl protecting groups are cleaved under those conditions to afford **cladribine** (**20.5**) [21]. This drug is approved by the FDA for the treatment of hairy cell leukemia and is available under the trade name Leustatin®.

The scheme for preparing the fluoro analogue is somewhat more complex due to the limited methods available for introducing fluorine into molecules. One of the general methods for preparing purines comprises condensation of polyamino pyrimidine (**21.1**) with formamide. The product of that reaction is then acylated with acetic anhydride to afford diacetate (**21.3**). The remaining basic amine in that product is then pressed into service to displace the anomeric chlorine in the benzyl-protected pentose (**21.4**). The acetyl groups are then saponified back to primary amines. Treatment with nitrous acid selectively converts the amine flanked by two-ring nitrogens to the corresponding diazo salt. The presence of fluoroboric acid in the diazotization medium in quick order converts the salt to a fluoro substituent. Boron trichloride cleaves to benzyl ethers to afford **fludarabine** (**21.7**) [22]. This compound too is approved by the FDA for treating leukemias and is available commercially as Oforta®. This agent, like many of the preceding examples acts as a false substrate. Its presence in DNA deactivates that molecule and causes the death of affected cells.

References

[1]   For an interesting account of the history of cancer chemotherapy, *see* V.T. DeVita, E. Chu, *J. Cancer Res.* **68**, 8643 (2008).

[2]   Many of the final folate antimetabolites in this section include chiral centers in addition to that in the glutamate moiety. This account does not deal with the problem of diastereomers.

[3]   D.R. Seeger, D.B. Cosulich, J.M. Smith, M.E. Hultquist, *J. Am. Chem. Soc.* **71**, 1753 (1949).

[4]   J.R. Piper, C. Johnson, F.M. Sirotnak, *J. Med. Chem.* **35**, 3002 (1992).

[5]   J.J. DeGraw, W.T. Colewell, J.R. Piper, F.M. Sirotnak, *J. Med. Chem.* **36**, 2228 (1993).

[6]   E.C. Taylor, D. Kuhnt, C. Shih, S.M. Rinzel, G.B. Grindey, J. Baredo, M. Jannatipour, R.G.M. Moran, *J. Med. Chem.* **35**, 4450 (1992).

[7]   P.R. Marsham, L.R. Hughes, A.L. Jackman, A.J. Hayter, J. OLdfield, J.M. Wadleworth, J.A. Bishop, B.M. O'Connor, A.H. Calvert, *J. Med. Chem.* **34**, 1594 (1991).

[8]   E. Duvalsantos, E.J. Flahive, B.J. Alden, M.B. Mitchell, W.R.L. Notz, A. O'Neil-Slawetski, Q. Tian, U.S. Patent Application 20040,266,796 (2004).

[9]   For the synthesis of a single diastereomer *see* L. Rahman, U.S. Patent Application 2005,085, 492 (2005).

[10] For a chiral chemoenzymatic process *see* S. Hu, S. Kelly, S. Lee, J. Tao, E. Flahive, *Org. Lett.* **8**, 1653 (2006).

[11] E.F. Elslager, J.L. Johnson, L. Werbel, *J. Med. Chem.* **26**, 1753 (1983).

[12] A. Rosowski, R.A. Forsch, U.S. Patent Application 2006/0,142,315 (2006).

[13] S.A. Giller, A. Lazdinsh, A. Karlovitch, A.K. Vainberg, Y. Sniker, I.L. Knunyants, B. Kazmina, U.S. Patent, 3,486,429 (1974).

[14] W.K. Reberts, C.A. Dekker, *J. Med. Chem.* **32**, 816 (1967).

[15] L.W. Hertel, J.S. Kroin, J.W. Misner, J.M. Tustin, *J. Org. Chem.* **53**, 2406 (1988).

[16] R.B. Appell, R.J. Duguid, *Org. Proc. Res, Dev.* **4**, 172 (2000).

[17] R.E. Donaldson, Patent WO1995018815 A1 (1995).

[18] J. Piml, F. Sorm, *Coll. Czech. Chem. Commun.* **29**, 2576 (1979).

[19] M.W. Winkley, R.K. Robbins, *J. Org. Chem.* **35**, 491 (1970).

[20] R.N. Patel, *Curr. Opin. Drug Disc. Dev.* **9**, 744 (2006).

[21] Z. Kazimerczuk, H. Cottam, G. Revankar, R.K. Robbins, *J. Am. Chem. Soc.* **106**, 6379 (1984).

[22] R.T. Lum, J.R. Pfister, S.R. Schow, M.M. Wick, M.G. Nelson, G.F. Schreiner, U.S. Patent, 5,789,416 (1998).

3

Hormone Blocking Anticancer Drugs

3.1 Introduction

Cancer of reproductive system organs comprises the two most common cancers in the United States according to the Center for Disease Control and Prevention. More specifically, these two diseases account for about 225,000 diagnosed cases of cancer of the prostate and 209,000 cases of breast cancer. The list of drugs for treating female cancers significantly exceeds those for treating cancers in males. This is likely due primarily to the more advanced stage of understanding the role of hormones in female reproductive physiology compared to the corresponding area of study in males.

The use of sex hormones and their antagonists is said to have begun near the turn of the nineteenth century with the observation of three cases of advanced breast cancer where removal of women's ovaries had a favorable effects on their tumors. The ovaries are now known to be one of the principal sources of estrogens. In a similar vein, in the mid-twentieth century, a urologist at the University of Chicago reported a major improvement in cancer of the prostate after removal of the testes, a source of androgens. Further developments in the field awaited advances in reproductive physiology and in the chemistry and biochemistry of the sex hormones. Research in the mid-twentieth century identified estrogen receptors; this was followed by the development of assays for those receptors. Those tests revealed that most genital tissues in the female are endowed with receptors for estrogens. Estrogens, research showed, will cause receptor-positive cancers to proliferate. Antiestrogens, such as tamoxifen (**3.4**), were found to bind to estrogen receptors and diminish circulating estrogen levels. This drug and several other estrogen antagonists have been and are still used extensively for treating estrogen receptor-positive breast cancer.

As an aside, post-menopausal osteoporosis is linked to diminished estrogen levels. Most estrogen antagonist drugs exhibit very moderate estrogenic activity in their own right. This

---

Antineoplastic Drugs: Organic Synthesis, First Edition. Daniel Lednicer.
© 2015 John Wiley & Sons, Ltd. Published 2015 by John Wiley & Sons, Ltd.

property has led to their sponsor seeking approval initially for treating osteoporosis. It would likely be much more difficult to obtain approval for treating estrogen receptor-positive cancers.

The relation between androgen receptors and prostate cancer is somewhat more complex. It has however been demonstrated that prostate cancer cells depend on androgen receptor for growth. Specific androgen antagonists have been found to retard the proliferation of cancer of the prostate.

3.2 Estrogen Antagonists

3.2.1 Estrogen Antagonists

3.2.1.1 *Triphenylethylenes*

As not infrequently happens, the field of estrogen antagonists opened with an adventitious discovery. The synthetic estrogen trianisylethylene (TACE, **1.1**) is only sparingly soluble in aqueous media. The analogue with a nitrogen-containing side was arguably prepared in order to provide improve the water solubility of TACE. Pharmacologic studies revealed that this derivative, named **clomiphene (1.2)**, was in fact an impeded estrogen, the term used at that time to denote an agent that inhibited hormonal action. Clomiphene actually consists of a pair of geometrical isomers: the commercial product in turn contains a fixed ratio of those compounds. In addition to its antiestrogenic activity, the drug exerts very weak estrogenic activity. This apparent paradox holds true for all triaryl estrogen antagonist agents.

That adventitious finding pointed the way for finding a means to preparing a compound that would block estrogen stimulation. A synthetic program that was aimed specifically to prepare an antiestrogen that would be used to treat estrogen-stimulated breast cancer led to the antiestrogen tamoxifen (**2.4**).

The treatment of phenol (**2.1**) with 2-chloro-*N*,*N*-dimethylethylamine and base leads to attachment of the basic aminoethyl side chain (**2.2**), a feature that would be found essential for antiestrogenic activity. The reaction of this intermediate with the Grignard reagent from bromobenzene then affords carbinol (**2.3**) largely as the analogue in which the hydroxyl group and a proton are on opposite sides of a central two-carbon unit in the favored rotamer. Toluenesulfonic acid then causes that alcohol to dehydrate. The resulting olefin (**2.4**) is obtained largely as the single isomer, **tamoxifen (2.4)** [1]. The stereochemistry of this last

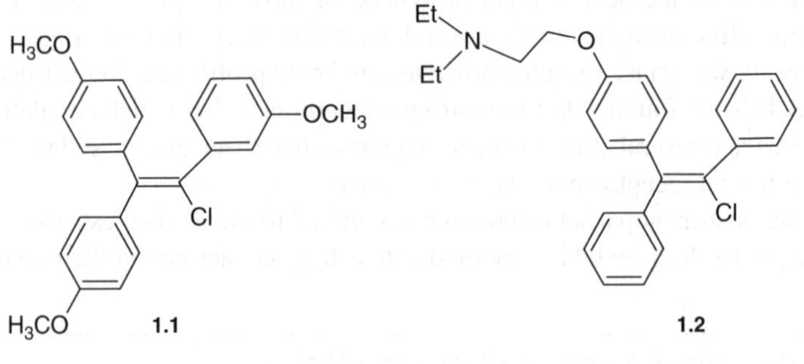

Scheme 3.1 *Clomiphene.*

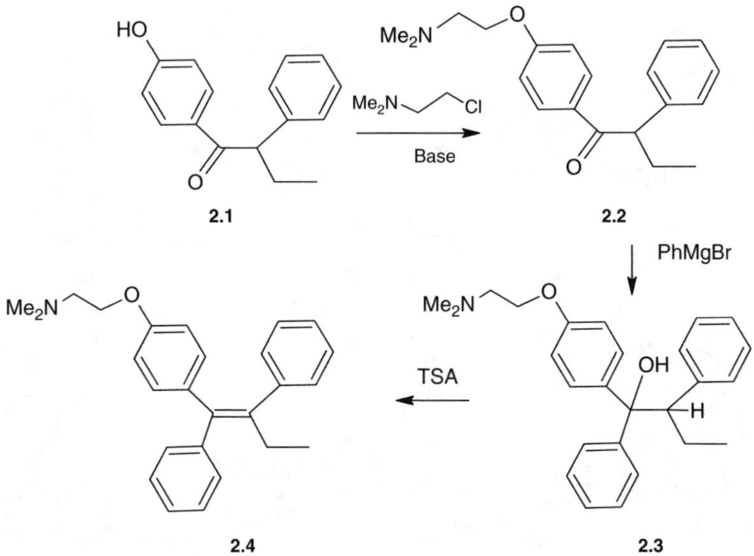

Scheme 3.2   *Tamoxifen.*

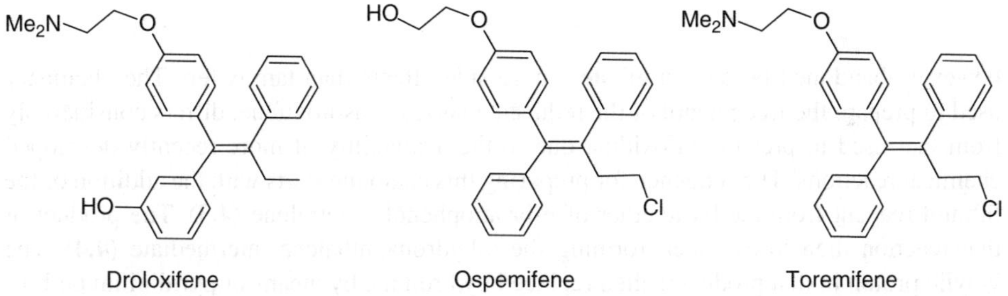

Droloxifene   Ospemifene   Toremifene

Scheme 3.3   *Other "fenes."*

step results from the favored trans relationship of the hydroxyl group and the proton. That in turn traces back to the formation of alcohol (**2.3**) which is determined by the addition of the Grignard reagent to the favored conformation of (**2.2**) [2]. Tamoxifen went on to gain approval for other uses. The most important of these probably comprises delay of the onset of osteoporosis. This effect is attributed to the drugs' weak estrogenic activity. The syntheses of most of the arylethylene follow-on estrogen antagonists (droloxifene [3], ospemifene [4], toremifene [5]) differ in only in detail from the sequence used to prepare tamoxifen.

3.2.1.2   *Cyclized Estrogen Antagonists*

One of the early antiestrogens based on a fused bicyclic nucleus, nafoxidine (**4.7**) [6, 7], was tested in the clinic by the National Cancer Institute. The drug was found to be active against breast cancer with a response rate comparable to that of tamoxifen. Nafoxidne was

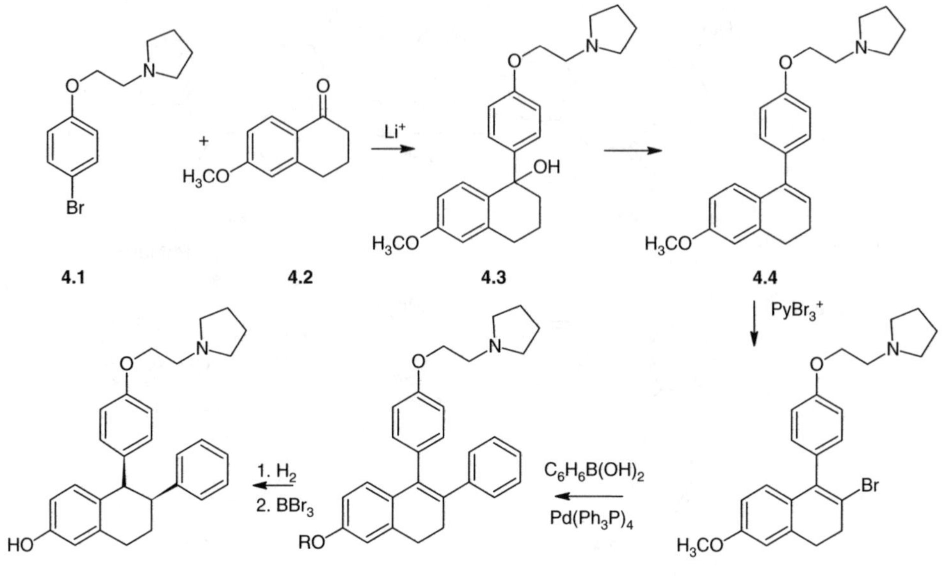

Scheme 3.4 *Lasofoxifene.*

however abandoned because it elicited more side effects than tamoxifen. The chemistry used to prepare the free phenol of the reduced analogue, lasofoxifene, differs considerably from that used to prepare nafoxidine due to the availability of more recently developed chemical reactions. The sequence for preparing this analogue starts with the addition of the lithium reagent from the basic ether of *p*-bromophenol to tetralone (**4.3**). The product of that reaction then loses water, forming the dihydronaphthalene intermediate (**4.4**). The vinylic proton in that product is then replaced by bromine by means of pyridinium perbromide. The required extra benzene ring is added by means of the Suzuki cross-coupling reaction. The intermediate (**4.5**) is accordingly treated with phenylboronic acid in the presence of the palladium complex, Pd(Ph$_3$P)$_4$. Catalytic hydrogenation fully saturates the fused ring. Boron tribromide then cleaves the methoxy function to afford **lasofoxifene (4.8)** [7]. This drug, trademarked Fablyn®, is approved in Europe for treating breast cancer as well as osteoporosis.

The preparation of an analogue in which the central ethylene bond is included in a heterocyclic ring, raloxifene, starts by the displacement of the halogen atom in bromobenzoyl (**5.2**) by sulfur in 3-methoxythiophene (**5.1**). The product from that reaction (**5.3**) is then treated with polyphosphoric acid (PPA). The expected product from that reaction would carry the benzene ring at position 3 in the fused heterocyclic ring. The product instead carries the benzene ring at the position adjacent to sulfur indicating the intermediacy of a rearrangement. The product can be rationalized by assuming that the sulfur inserts into the carbonyl group and the adduct loses water to form an intermediate such as (**5.4**). This product then cyclizes to afford the substituted benzothiophene (**5.5**).

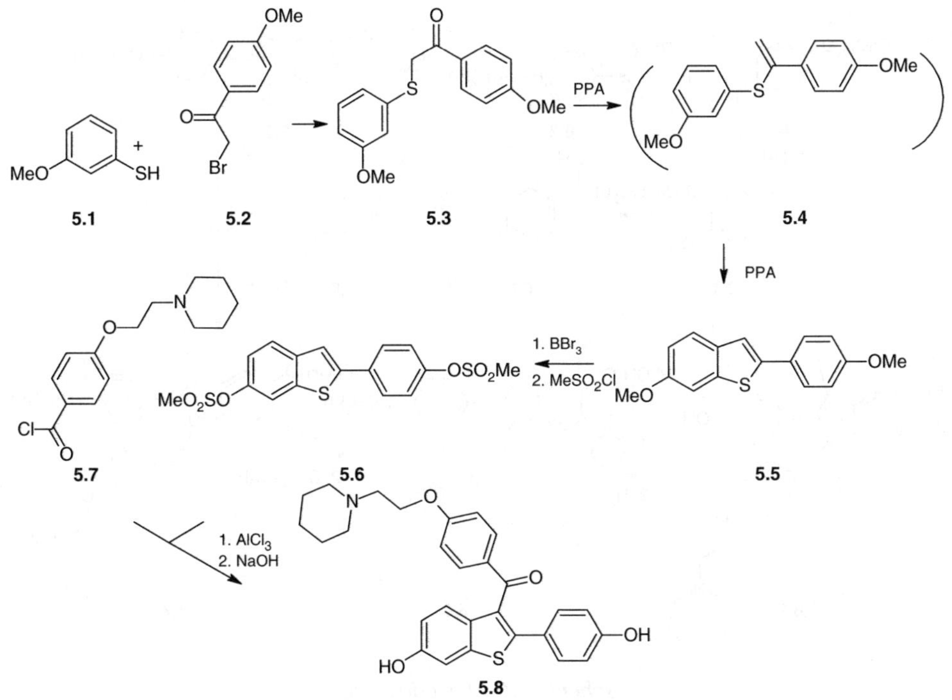

Scheme 3.5 *Raloxifene.*

The methyl ether groups are then cleaved by means of boron tribromide; the resulting phenols are then converted to their methylsulfonyl derivatives (**5.6**) in order to avoid complications in the Friedel–Crafts acylation. The reaction of that intermediate with acid chloride (**5.7**) in the presence of aluminum chloride gives the corresponding acylated derivative. Saponification of the methyl sulfonyl groups affords **raloxifene** (**5.8**) [8]. This compound, Evista®, is approved for essentially the same set of indications as the other triarylethylenes.

An indole nucleus can also serve as the nucleus of a cyclized triaryl estrogen antagonist. The circumstance that one of the aryl groups is attached to the hetero atom illustrates the broad tolerance of the estrogen receptor for structural modifications in the compounds that it will recognize.

This drug too is approved as one of the active ingredients in a drug (Duavee®) intended to fend off menopausal hot flashes. One arm of the convergent synthesis begins with the preparation of the indole nucleus. This involves an alternative method to the Fisher indole synthesis. This procedure is however fairly standard in building other heterocyclic compounds. Thus, the reaction of the protected α-bromoketone with the similarly protected aniline (**6.2**) in the presence of triethylamine affords an indole that incorporates two of the three required benzene rings. The condensation of the carbonyl group with the benzene ring may be facilitated by the high electron density in that ring (**6.3**). The preparation of the other ring involves first building the side chain that will hold that third ring. The alkylation

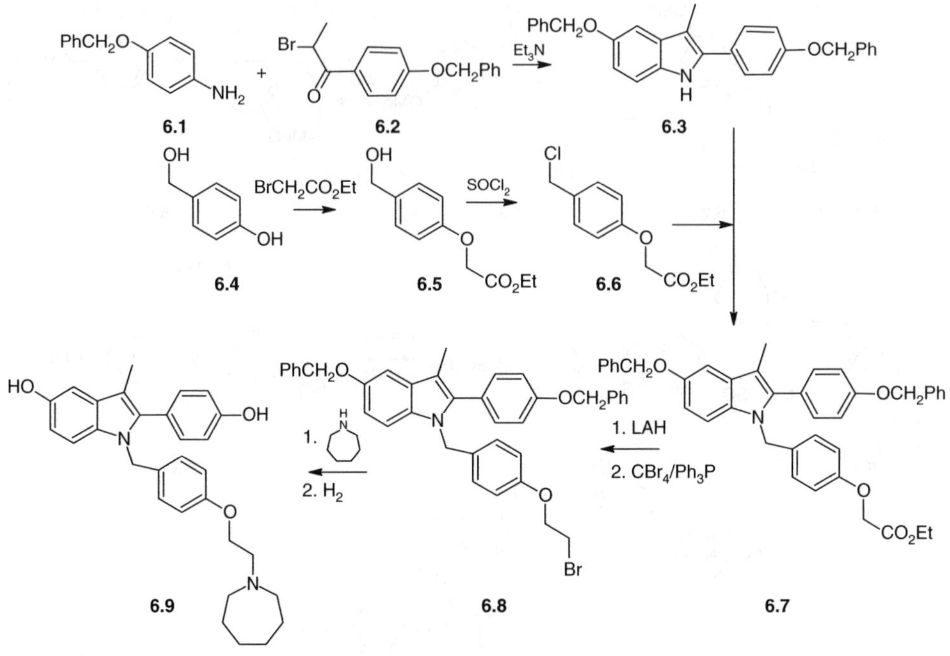

Scheme 3.6   *Bazedoxifene.*

of the phenolic hydroxyl group in (**6.4**) with bromoacetate proceeds selectively on that site. The base used in this reaction will apparently not act on the alkyl hydroxyl. Reaction with thionyl chloride then replaces that hydroxyl by a chloro group. That benzylic halogen is then used to alkylate the indole nitrogen completing the construction of the carbon–nitrogen skeleton (**6.7**). The ester side chain is next reduced to hydroxyl by means of lithium aluminum hydride. The newly introduced hydroxyl is then replaced by halogen by means of the Appel reaction—carbon tetrabromide and triphenyl phosphine. The bromine in the side chain is then replaced by azepine completing the formation of the almost traditional basic ether. Catalytic hydrogenation serves to remove the protecting benzyl ethers yielding **bazedoxifene (6.9)** [9].

3.2.1.3   Steroid

Both classes of antiestrogens discussed this far, the triarylethylenes and the cyclized ana-logues, retained some agonist activity. This has resulted in their application for staving off osteoporosis. That residual estrogenic activity is not particularly helpful when treating breast cancer.

It is perhaps fitting that an estrogen antagonist that is a pure antagonist free of even the weak agonist activity manifested by the drugs discussed previously should itself be a steroid. The compound can be viewed as estradiol equipped with the most unusual side chain. It has been proposed that when this compound occupies a receptor site, the long side chain binds to a proximate function that prevents the steroid moiety from leaving

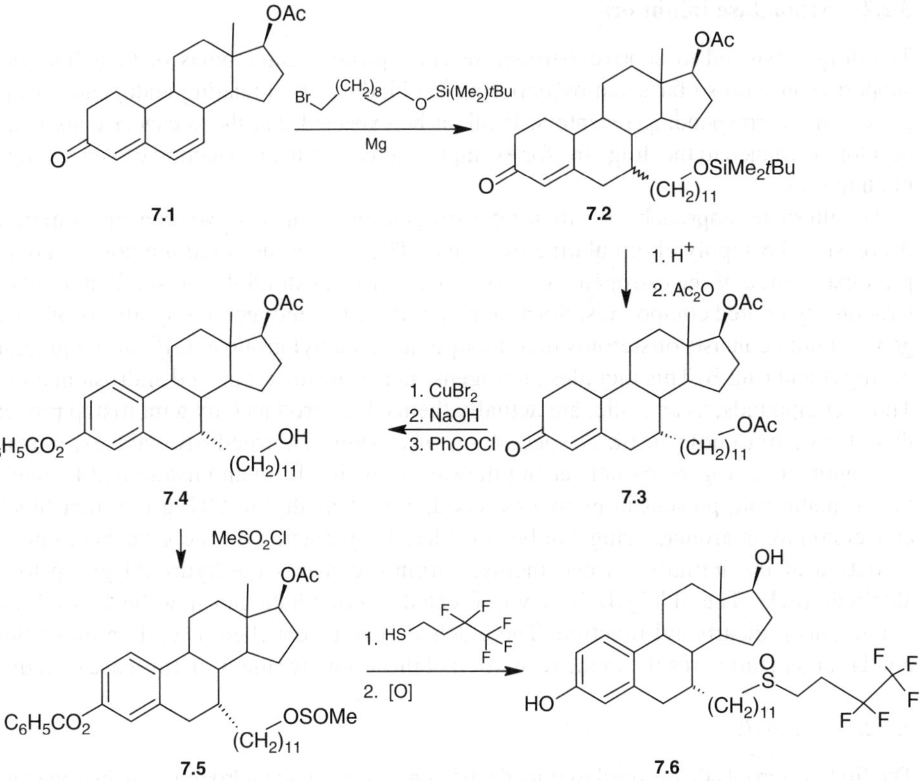

Scheme 3.7 *Fulvestrant.*

the receptor. This in effect removes that receptor from further action. The synthesis opens with the conjugate addition of the Grignard reagent from the dimethyl-*tert*-butylsilyl-protected bromo-11-hydroxyundecanoate to 6,7-dehydro-19-nortestosterone (**7.1**). The product comprises a mixture of α and β C6 isomers. The α-alkyl side chain isomer is then separated from the mixture. The silyl protecting group is then removed by acid hydrolysis and the resulting hydroxyl derivative acylated. Ring A, which differs from phenol by one unsaturated bond phenol, is next aromatized to a benzene ring by oxidation with cuprous bromide. Base-catalyzed saponification is selective for the side chain ester; the acetate at C17 resists opening probably due to steric hindrance about that site. The acylation of the newly formed phenol with benzoyl chloride and an equivalent of sodium hydroxide affords the corresponding ester (**7.4**). The phenoxide ion from the added base increases reactivity at that site, leaving the terminal side chain hydroxyl free. The next few steps are somewhat conjectural. In one way to proceed the terminal side chain hydroxyl would be converted to a leaving group by reaction with methanesulfonyl chloride (**7.5**).

Displacement by sulfur in 3,4-perfluorobutyl mercaptan followed by oxidation would afford **fulvestrant** (**7.6**) [10, 11]. This compound, trade named Faslodex®, is specifically indicated for treating estrogen receptor-positive breast cancer.

3.2.2 Aromatase Inhibitors

The drugs designed to deprive estrogen receptor-positive carcinomas of their hormonal support dealt with so far acted by opposing the binding of circulating endogenous estrogens to the corresponding receptors. It might be expected that those cancer cells would develop resistance to the drugs by, for example, increasing the production of new estrogen binding sites.

   An alternate approach for treating estrogen receptor-positive cancer comprises decreasing the supply of circulating estrogens. The ovaries, as noted previously, are the principal source of the estrogens that comprise estrone, estradiol, and small amounts of structurally related compounds. Starting material for the endogenous synthesis of estrogens actually consists of steroids that incorporate a methyl group at C10, at the juncture of ring A and ring B. This includes androgens such as testosterone and androstenedione. Those compounds, as an aside, are actually themselves produced by a multistep process that starts from cholesterol. A special set of enzymes, termed aromatases, convert androgens to estrogens by converting the steroid A ring from an unsaturated ketone to the aromatic ring present in estrogens. Oxidation of methyl at C10 group that blocks conversion to an aromatic ring can be considered key reaction in the sequence. Further oxidation of the initially formed methyl carbinol converts the hydroxyl group to an aldehyde (**8.3**). The aldehyde in newly created intermediate can now be viewed as a vinylogous β-dicarbonyl function. The aldehyde is thus expelled as carbon monoxide; bond reorganization results in the formation of the aromatic ring A, in this case estradiol.

3.2.2.1 Steroids

The first of steroids that is used to impede the conversion of an androgen to an estrogen has a venerable history as indicated by its six-digit CAS Registry Number: 566-48-3. The compound was in fact patented in 1955. This agent is ironically included in the Drug

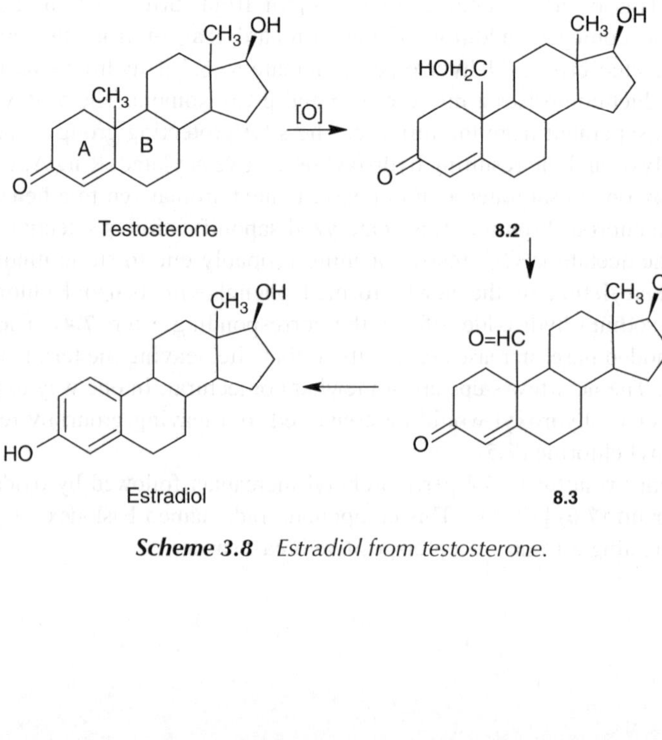

Scheme 3.8   *Estradiol from testosterone.*

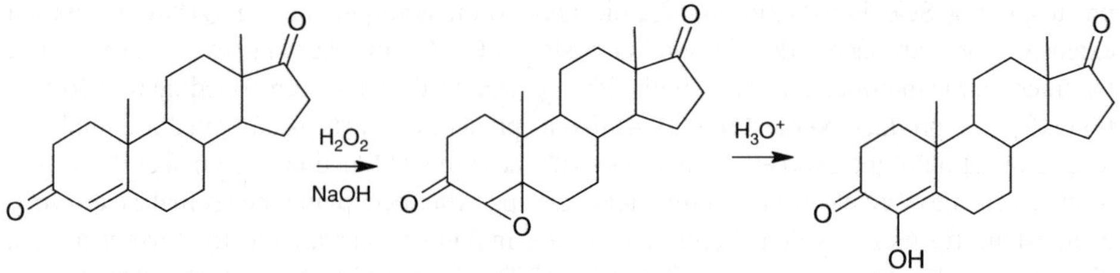

Scheme 3.9 *Formestane.*

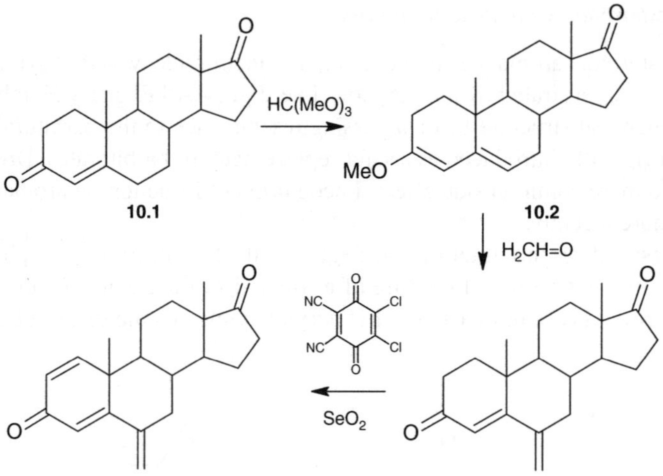

Scheme 3.10 *Exemestane.*

Enforcement Agency (DEA) lists of doping agents because of the possibility that it will be used by athletes taking advantage of the drug's mild anabolic effect. The steroid **formestane (9.3)** can be prepared by a two-step sequence starting with androst-4-ene-3,17-dione **(9.1)**. The steroid is first treated with basic hydrogen peroxide under the special conditions used to oxidize conjugated carbonyl compounds. It is likely that the reaction starts by conjugate addition of peroxide anion. When injected into a patient, the drug binds to the androgen receptor site that starts the process of conversion to estrone catalyzed by the aromatase enzyme. This in effect shuts down the action of that enzyme. This drug is approved in Europe for backup treatment of breast cancer.

A somewhat more complex steroid, exemestane, is a more selective aromatase inhibitor than its predecessor. One preparation for this compound starts with the conversion of androst-4-ene-3,17-dione to its enol ether **(10.2)** with ethyl orthoformate. The crude intermediate is then treated, without prior isolation, with formalin and mild acid; these steps add the desired exomethylene function to position 6 **(10.3)**. The introduction of a double bond at positions 1–2 in steroids such as corticosteroids and androgens generally results in an increase in potency of the precursor drug. Selenium dioxide and 2,2-dicyano-5,6-dichloroquinone (DDQ) have been used extensively to introduce a double bond at positions C1–C2 in other classes of steroids. The treatment of the intermediate (**10.3**) with either reagent leads to the corresponding transformation [12]. The product **exemestane** (**10.4**) like its predecessor, acts as an aromatase inhibitor. The conjugation of the exomethylene group with the carbonyl function at C3 makes that a very reactive center toward electron-rich functional groups. After occupying the aromatase enzyme, exemestane forms a covalent bond at that site, making it unavailable for further action. The drug can thus be termed a "suicide inhibitor." The drug was approved by the FDA in 2005 for use against breast cancer in women whose tumor is estrogen receptor positive. It is now available under the trade name "Aromasin®".

3.2.2.2   Nonsteroidal Aromatase Inhibitors

The first non-steroidal aromatase inhibitor, aminoglutethimide, was first synthesized many decades ago as demonstrated by its' very low, five-digit, CAS Registry Number (77-21-4). This drug's chemical structure bears a passing resemblance to the barbiturates. The drug was actually originally introduced as a safer replacement for barbiturates. Drowsiness is in fact one of the more common side effect. Recognition of its action on aromatase enzymes came much more recently.

The synthesis of aminoglutethimide begins with the nitration of 3-phenylpropion-itrile (**11.1**) with the traditional mixture of nitric and sulfuric acids. Conjugate addition of the anion from (**11.2**) and base to ethyl acrylate yields cyano ester (**11.3**) in which a

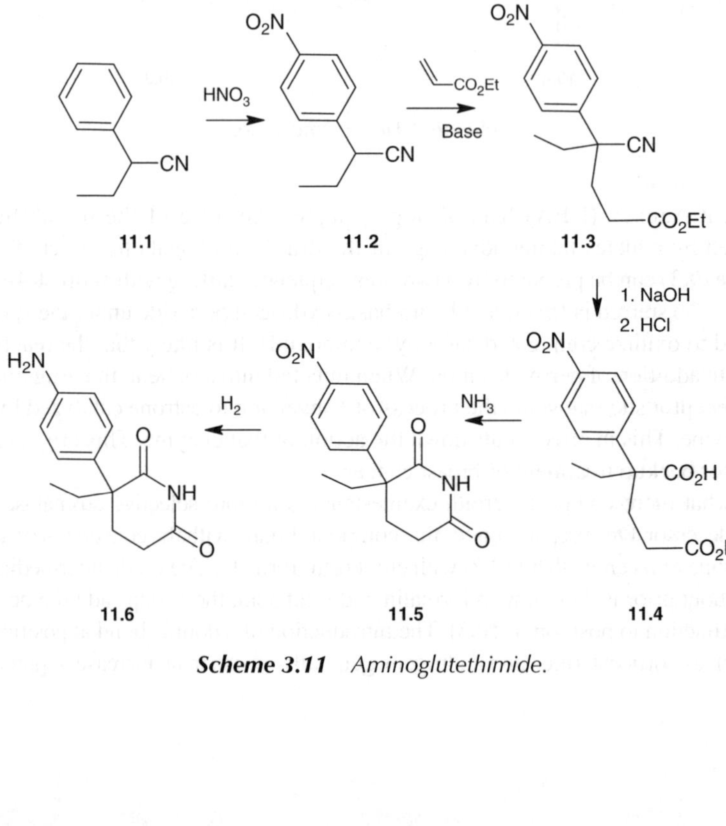

Scheme 3.11   *Aminoglutethimide.*

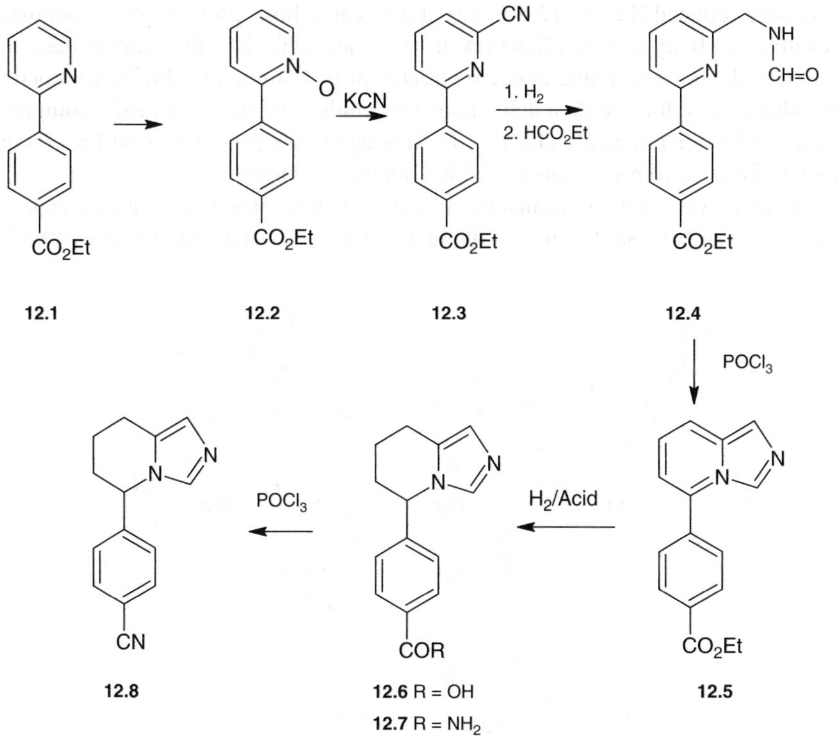

Scheme 3.12   *Fadrozole.*

new side chain has been added. Sequential saponification and acid hydrolysis lead to glutaric acid (**11.4**). Heating that product with a source of ammonia closes the ring to imide (**11.5**). Catalytic hydrogenation reduces the nitro group to an amine, yielding **aminoglutethimide** (**11.6**) [13]. This is currently available under the trade name Cytadren®. The drug is indicated for the treatment of estrogen receptor-positive breast cancer, among other conditions.

The chemical structure of a more recent aromatase inhibitor, fadrozole (**12.8**), is quite different from that of its predecessor but shares a nitrile function with several other quite recent nonsteroidal aromatase inhibitors. It should be noted that nitriles had seldom if ever been considered a pharmacophore prior to the discovery of these compounds. The synthesis of this drug begins with oxidation of nitrogen in the pyridine ring to its N-oxide (**12.1**). This N-oxide is next treated with potassium cyanide in dimethyl sulfoxide. The cyano anion becomes attached to the carbon next to the pyridine nitrogen in a reaction analogous to the insertion of halogen next to ring nitrogen in a heterocyclic molecule, a reaction known as the Polonovski rearrangement (**12.3**). The newly introduced nitrile is next reduced to an aminomethyl group and then formylated by exchange with ethyl formate. The treatment of formamide (**12.4**) with phosphorus oxychloride closes the ring to form the desired imidazopyridine (**12.5**). The remaining steps involve catalytic hydrogenation of an acid solution of that intermediate (**12.5**) to yield the molecule in which the fused pyridine ring has been reduced (**12.5** → **12.6**). The pendant carbethoxy group is next saponified, and the resulting carboxylic acid (**12.6**) taken to an acid chloride; this intermediate is then converted to the corresponding amide by treatment with ammonia (**12.7**). A second round of phosphorus oxychloride then dehydrates the amide. The resulting nitrile comprises the aromatase inhibitor **fadrozole** (**12.8**) [14]. This agent has been approved for second-line treatment of advanced breast cancer and is available as Afema®.

Letrozole is yet another aromatase inhibitor whose structure features benzonitrile moieties. The synthesis starts with the alkylation of 1,2,4-triazole (**13.2**) with 4-cyanobenzyl

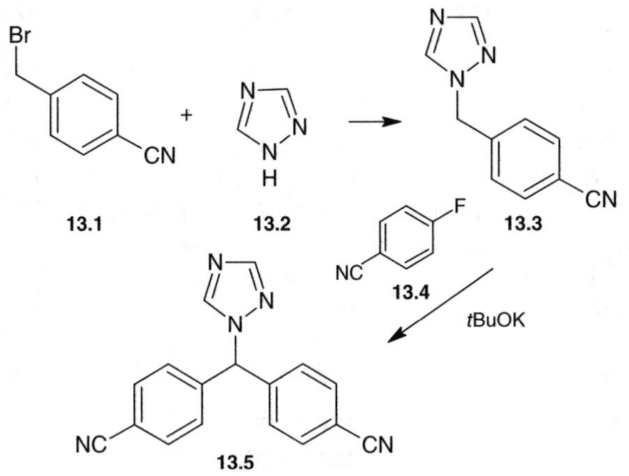

Scheme 3.13   *Letrozole.*

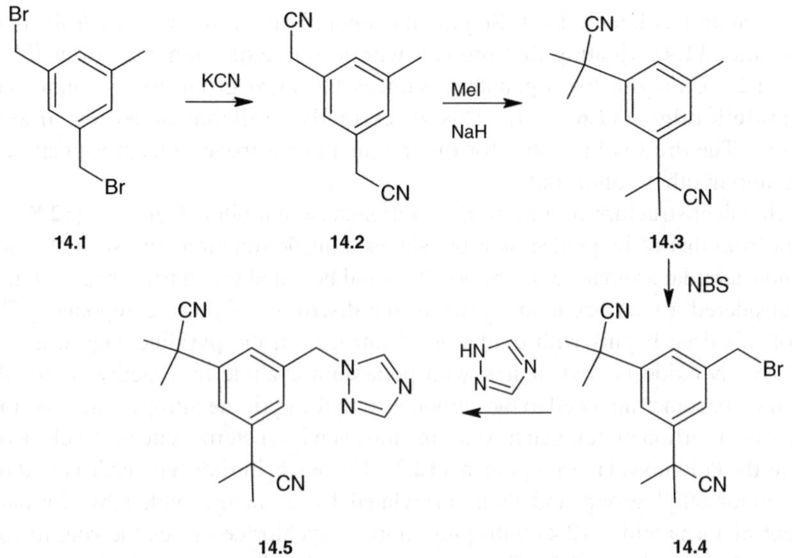

Scheme 3.14   *Anastrozole.*

bromide (**13.1**). The product from that reaction (**13.3**) is then treated with potassium *tert*butoxide to remove one of the protons from the benzylic carbon. The resulting anion is treated with 4-fluorobenzonitrile. Nucleophilic aromatic displacement of fluorine leads to incorporation of a second benzonitrile. The product **letrozole** (**13.5**) [15] is approved by the FDA for treating estrogen receptor-positive breast cancer. Letrozole, trade marked Femara®, is however also prescribed for a considerable number of off-label indications. Many of these rely on the agents' anti-estrogenic activity which is in turn a result of the drug's aromatase inhibition.

The nitrile moieties of yet another aromatase inhibitor are connected to aliphatic rather than aromatic carbon. The relatively straightforward preparation begins with the displacement of bromine from the bis-bromomethyl starting material (**14.1**) by cyanide under phase transfer conditions, owing to its poor aqueous solubility. The protons on the cyanomethyl side chains in this intermediate are rendered more volatile by the electronegative nitriles. The treatment of this intermediate with sodium hydride and methyl iodide replaces those protons by methyl groups (**14.3**). Free radical bromination of the sole remaining methyl group on the benzene ring by *N*-bromosuccinimide catalyzed by benzoyl peroxide gives the bromomethyl derivative (**14.4**). The displacement of that halogen by 1,2,4-triazole gives **anastrozole** (**14.5**) [16, 17]. This drug, trademarked Arimidex® has been approved by the FDA for use in advanced breast cancer: it is also prescribed off-label with much the same indications as the other aromatase inhibitors.

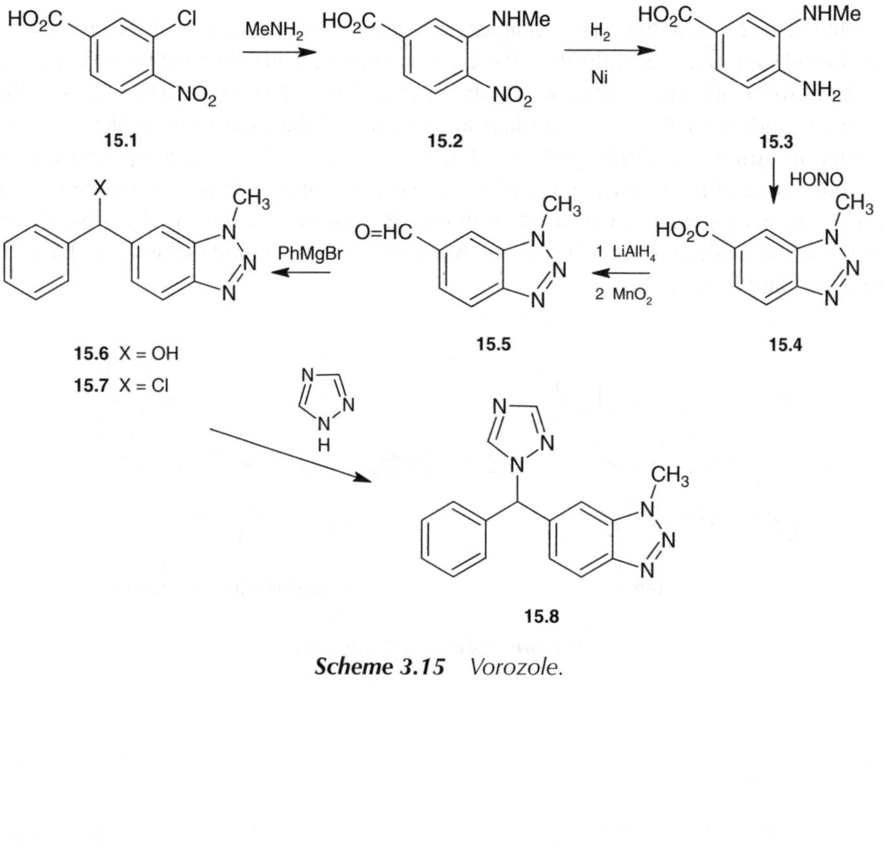

Scheme 3.15   *Vorozole.*

The apparent requirement of one or more nitriles is given lie by the structure of vorozole (**15.8**) which is an effective inhibitor of aromatase. One of the several syntheses of this compound starts with the nucleophilic aromatic displacement of chlorine in the starting compound (**15.1**) by methylamine. This reaction eased by the enhanced reactivity of halogen due to the presence of two electron-withdrawing substituents. Hydrogenation of the intermediate over Raney nickel leads to *ortho* diamine (**15.3**). This last intermediate is then treated with nitrous acid. The reaction can be envisaged by assuming that the reagent converts the primary amine to the corresponding diazonium derivative. The adjacent secondary amine then adds to the diazonium salt to afford benzo-1,2,3-triazole (**15.4**). The carboxylic acid is next converted to a derivative that will accept a new benzene ring. The acid is thus first reduced to its hydroxymethyl derivative by means of lithium aluminum hydride. Manganese dioxide then oxidizes the hydroxymethyl function to aldehyde (**15.5**). The requisite additional benzene is next inserted by the reaction of carboxaldehyde with the Grignard reagent from bromobenzene (**15.6**). The newly formed hydroxyl group is converted to a better leaving group by means of thionyl chloride. The displacement of chloride by 1,2,4-triazole affords the aromatase inhibitor **vorozole** (**15.8**) as a pair of enantiomers [18, 19]. Vorozole, trademarked Rivizor®, in common with the other aromatase inhibitors, is approved for treating estrogen receptor-positive breast cancer. It too is often prescribed off-label for noncancer indications requiring depression of circulating estrogens.

3.3   Androgen Antagonists

As noted in the Introduction to this chapter the drugs for treating cancer of the prostate have been introduced more recently than the corresponding agents for treating breast cancer. The drugs for treating prostate cancer with one exception fall neatly into two classes: these comprise a group of three nonsteroidal agents that inhibit androgen uptake as well as testosterone-stimulated DNA synthesis. The exception consists of a compound that features only three of the four rings found in steroids. It is of note that testosterone proper is not itself an androgen. The unsaturation in ring A must first be reduced by 5-α-reductase enzymes. A sizeable number of highly modified steroids that inhibit that enzyme have been introduced fairly recently.

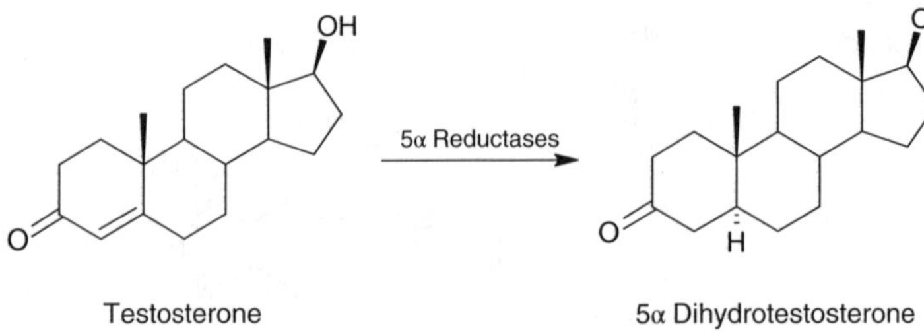

Testosterone   5α Dihydrotestosterone

Scheme 3.16   *5α-reductase.*

3.3.1 Non-steroidal Antian drogens

As is the case with steroid-based estrogen antagonist, most of the steroidal androgen antagonists retain a measure of agonist activity. All but one of the nonsteroidal antagonists that follow are free of such activity, leading some to call these compounds "pure antagonists."

The first of these compounds, **flutamide** (**16.2**), was originally prepared as one of a series of potential bacteriostatic agents [20]. This compound is prepared by treating commercially available substituted aniline (**16.1**) with 2-methylpropionic acid chloride in the presence of triethylamine. Flutamide, which is approved for treating prostatic cancer, is sold under the trade name Eulexin® as well as by its generic name.

Reacting the same starting aniline (**16.1**) with dimethylhydantoin (**16.3**) in the presence of cuprous oxide leads to nucleophilic aromatic displacement of the aniline nitrogen by the

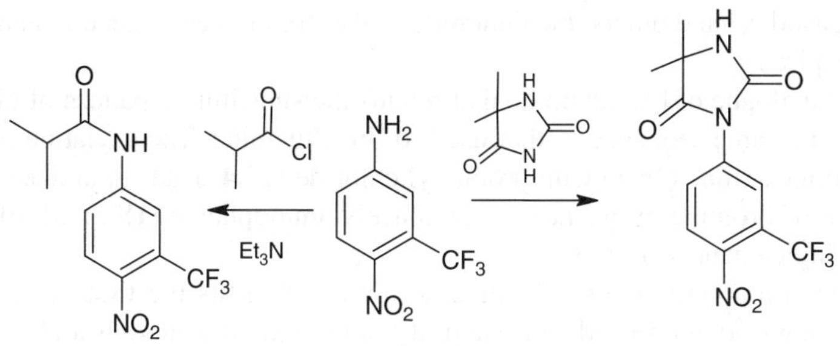

16.2 **16.1** **16.3** **16.4**

Scheme 3.17 *Flutamide and Nilutamide.*

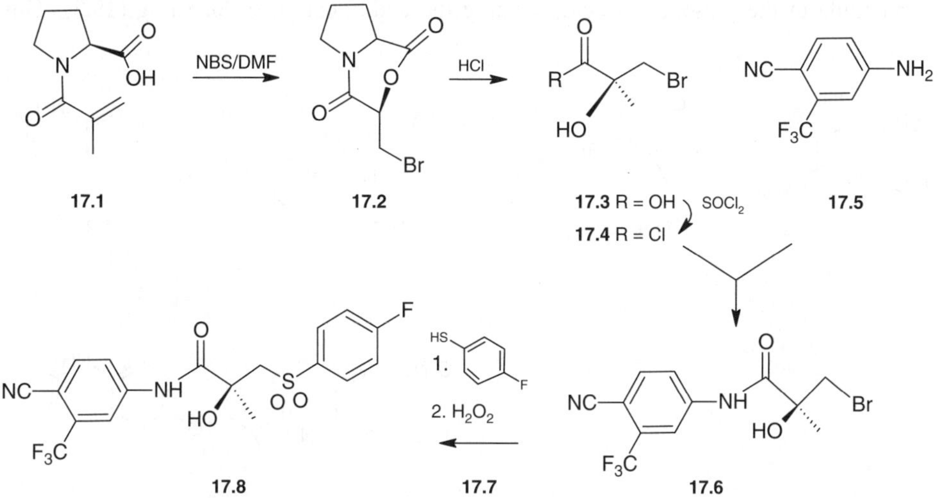

17.1 **17.2** **17.5** **17.8** **17.7** **17.6**

Scheme 3.18 *Bicalutamide.*

acidic hydantoin nitrogen to yield **nilutamide (16.4)** [21]. This androgen antagonist has been approved for the same indication as its predecessor; it is available as Nilandron®.

A somewhat more complex compound, bicalutamide, where the nitro group of the predecessors is replaced by a nitrile is also classed nonsteroidal androgen antagonist. An interesting scheme for preparing the drug in chiral form relies on passing on the chirality of a proline substituent that is later discarded. The enantiospecific synthesis of this compound begins with the preparation of the chiral acid (**17.3**). Bromolactonization of the methacrylamide of S proline (**17.1**) by means of bromine in dimethylformamide (DMF) leads to the fused lactone 17.2. Acid hydrolysis of the product cleaves off the proline to leave behind bromo acid (**17.3**). The acid is next converted to its acid chloride and that is used to acylate the substituted aniline (**17.5**). The resulting intermediate (**17.6**) is next reacted with the anion from mercaptan (**17.7**). The resulting thioether is then oxidized to the corresponding sulfone by means of hydrogen peroxide. The androgen antagonist **bicalutamide (17.6)** is thus obtained [22]. This drug is available under the trade names Casudex®, Casodex® and others; bicalutamide is also sold under its generic name since the patent has expired.

The close analogue of bicalutamide that retains the substitution pattern of nilutamide is prepared by the same sequence as that used for bicaltuamide. The acylation of the seemingly ubiquitous aniline (**16.1**) with propionyl chloride (**17.4**) leads to amide (**18.18**). The displacement of bromine by phenoxide from acetylamidophenol (**18.2**) affords **andarine (18.3)** as a single enantiomer [23].

A tricyclic compound, whose chemical structure suggests the first three rings of an azasteroid, shows potent 5α-reductase activity in both *in vitro* models and *in vivo* experiments. As in the preceding compound, the scheme for preparing bexlosteride is designed to prepare the drug as a single enantiomer. The first step in a synthesis that provides bexlosteride starts by forming the enamine of tetralone (**19.1**) with the chiral amine 2-phenethylamine that will assure that the final product, bexlosteride (**19.4**), will be obtained as a single enantiomer. The resulting enamine (**19.2**) is then condensed with acrylamide in the presence of acid, forming the additional piperidone ring (**19.2**). This

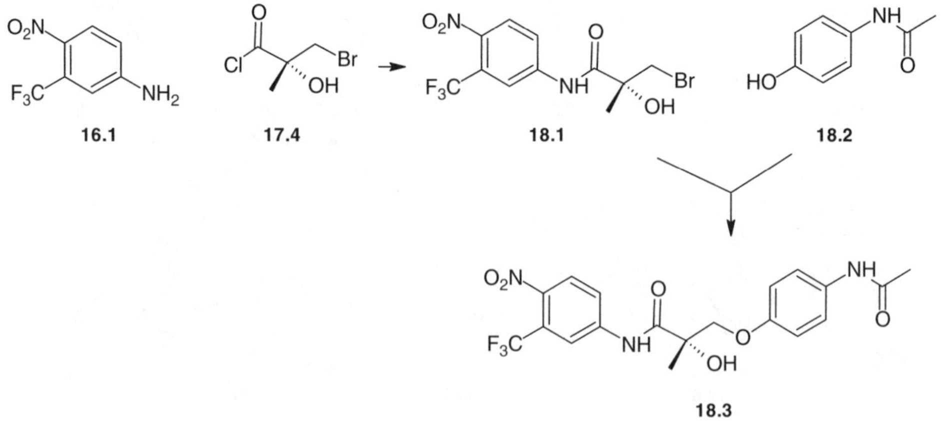

Scheme 3.19 *Andarine.*

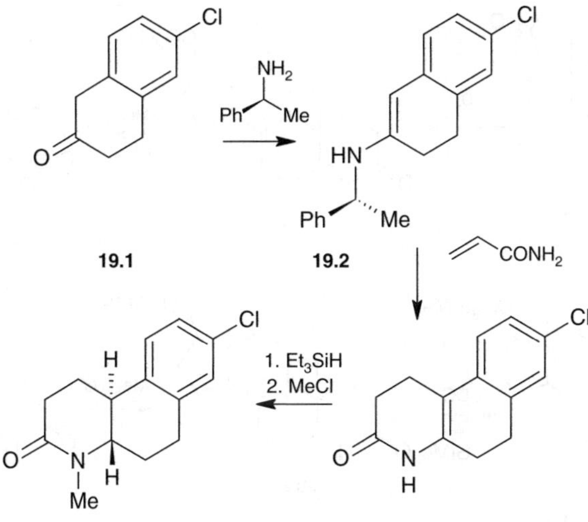

Scheme 3.20   *Bexlosteride.*

transform can be envisaged by assuming the reaction involves first standard enamine chemistry in which the acrylamide adds to the enamine; this is followed by the displacement of enamine nitrogen by that on amide, thus closing ring. The treatment of the cyclized product (**19.3**) with triethylsilane serves to reduce the double bond at the ring fusion. The amide nitrogen is next methylated by means of methyl chloride and base. **Bexlosteride** (**19.4**) is thus obtained [24].

3.3.2   Steroid Androgen Antagonists

3.3.2.1   *Lactam Ring A Compounds*

A group of highly modified steroids exert marked inhibition of 5α-reductase reduction of testosterone to its active metabolite, dihydrotestosterone. These molecules are approved for treating benign prostatic hypertrophy and at a lower dose male pattern hair loss. When first introduced, it had arguably been hoped that inhibition of the reductase would translate to a corresponding effect on prostate cancer. Sizeable clinical trials, however, showed that finasteride (**20.7**) and its analogue, dutasteride (**21.6**), do cause a small statistically significant reduction in the incidence of prostate cancer. The indications for these drugs do not however include treatment of cancer of the prostate. This is an interesting case where the inhibition of a hormone does not translate to an antineoplastic action.

The starting material for the synthesis of finasteride is progesterone (**20.1**), a molecule that incorporates the requisite nucleus with the proper stereochemistry already in place. The initial step comprises removal of one carbon atom from the acetyl side chain, for example, by means of the haloform reaction (potassium hypochlorite). A second oxidation of the product (**20.2**), this time with a mixture of periodate and permanganate, opens ring A with loss of one carbon atom. The reaction arguably involves hydroxylation of the 4–5

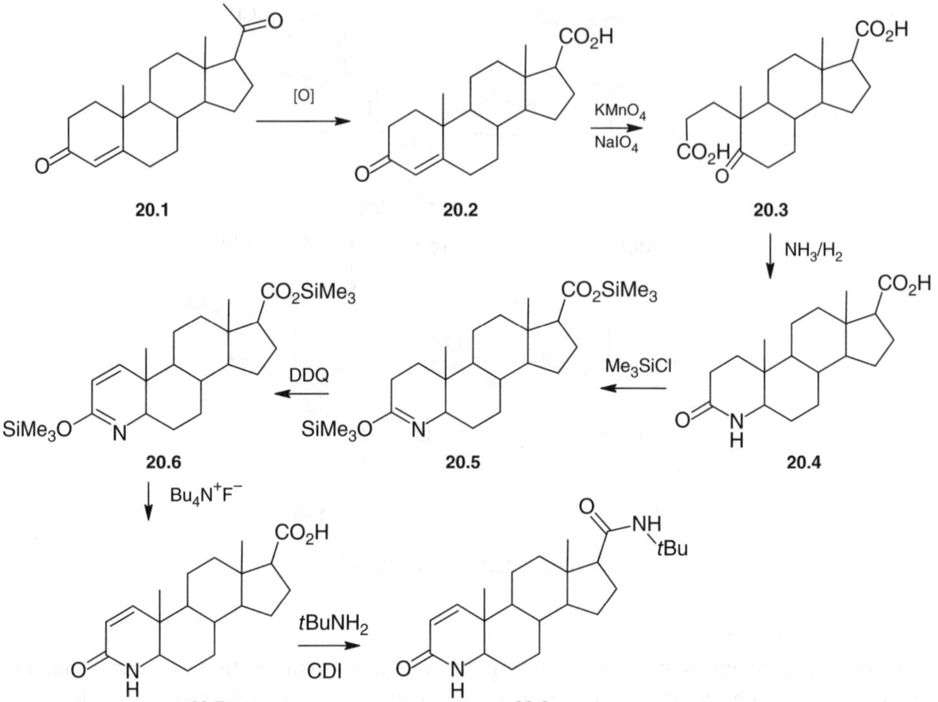

Scheme 3.21 Finasteride.

double bond by permanganate followed by scission of the diol by periodate. Catalytic hydrogenation of that intermediate in the presence of ammonia then leads to lactam (**20.4**). The addition of hydrogen from the more open α-side of the molecule maintains the steroidal configuration of the product. The reaction of the newly formed lactam with trimethylsilyl chloride and imidazole converts the lactam carbonyl to an imino silyl ether; the carboxylic acid is silanated in the process (**20.5**). Resort to one of a steroid chemist's favorite reagents, 2,2-dicyano-5,6-dichloroquinone (DDQ) then introduces a new double bond. This product is not isolated but the reaction mixture treated with tetrabutylammonium fluoride, a reagent specific for removing silyl protecting groups. The product from that reaction is then allowed to react with *tert*-butylamine in the presence of carbonyldiimidazole. The resulting amide, **finasteride (20.8)** [25–27], is a 5α-dehydrogenase inhibitor. The drug is marketed for the treatment of benign prostate hypertrophy under the trade name Propecia®.

As noted earlier, data from a large-scale clinical trial of dutasteride showed that this agent also elicits a small decrease in the incidence of prostate cancer and that it too fails to have an effect on established prostate cancer. The synthesis of this drug differs from its predecessor in that the scheme starts by establishing the amide on the steroid ring D.

The reaction of the same starting material as that used to prepare the preceding lactam (**20.1**) with thionyl chloride affords acid chloride (**21.1**). That crude product is used

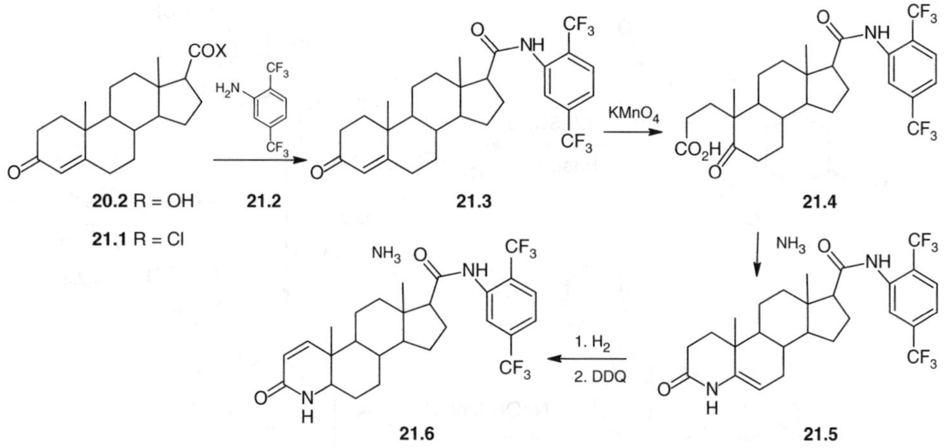

Scheme 3.22 *Dutasteride.*

to acylate 2,4-trifluoromethylaniline (**21.2**). Ring A is next cleaved with loss of the carbon atom at C4 by oxidation with potassium permanganate. Treatment with ammonia presumably converts the ketone in ring B to an imine and the pendant carboxylic acid to an amide. The displacement of imine nitrogen by amide leads to the formation of the cyclic amide (**21.5**). The reduction of the double bond in ring B followed by the treatment of the product with DDQ affords the 4-azasteroid **dutasteride** (**21.6**) [28, 29]. Like its forerunner, this drug, under the trade name Avodart®, is marketed largely for the treatment of benign prostate hypertrophy.

3.3.2.2 Steroids with a Heterocycle at C17

Two steroids bearing a heterocyclic moiety at C17 are potent antiandrogens that are effective against advanced castration-resistant cancers. Unlike the compounds that act via 5α-reductase enzyme, these drugs inhibit several earlier steps in the endogenous synthesis of androgens such as C17 hydroxylation. They also bind to androgen receptors, blocking the action of the access of the androgens that support the tumors.

The synthesis of abiraterone (**22.5**) starts by activating C17 by reaction of trifluoromethylsulfonyl anhydride to afford trifluoromethylsulfonic enol ether (**22.2**). That enol ester is next subjected to the Suzuki cross-coupling reaction. The enol ester (**22.2**) is thus treated with diethyl-3-pyridyldiethylborane (**22.3**) in the presence of quatrotriphenylphosphine palladium. The acetyl ester at C3, which has survived those conditions, is then cleaved by base in methanol. The product of this reaction is **abiraterone** (**22.5**) [30]. This drug is approved for treating metastatic castration-resistant prostate cancer.

A somewhat different strategy was employed for the synthesis of the compound that features a benzimidazole at C17. The sequence begins with a classical Vilsmeier reaction. Thus, the treatment of 3-acetoxy-17-ketoandrostane (**23.1**) with phosphorous oxychloride in dimethylformamide results in the formation of the androstane that incorporates a carboxaldehyde at C16 and an enol chloride at C17. The reaction of that intermediate with

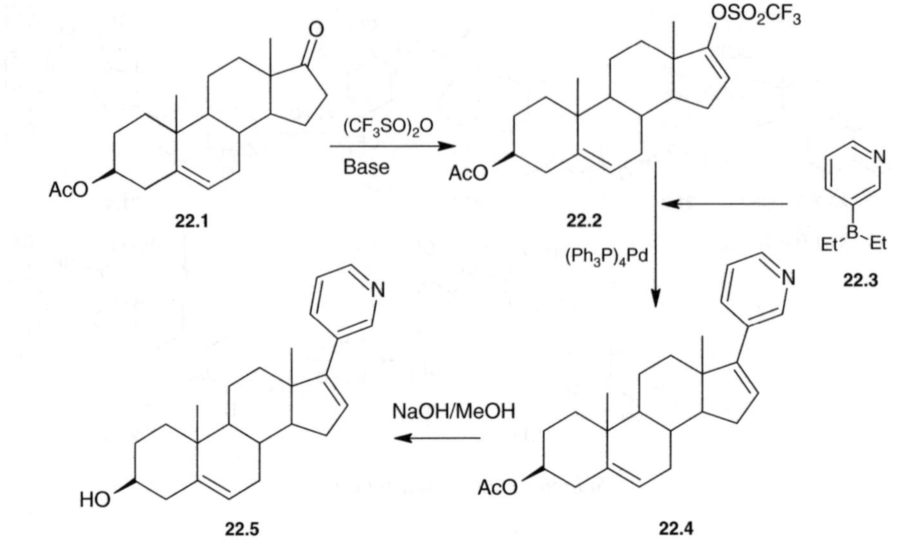

Scheme 3.23 Abiraterone.

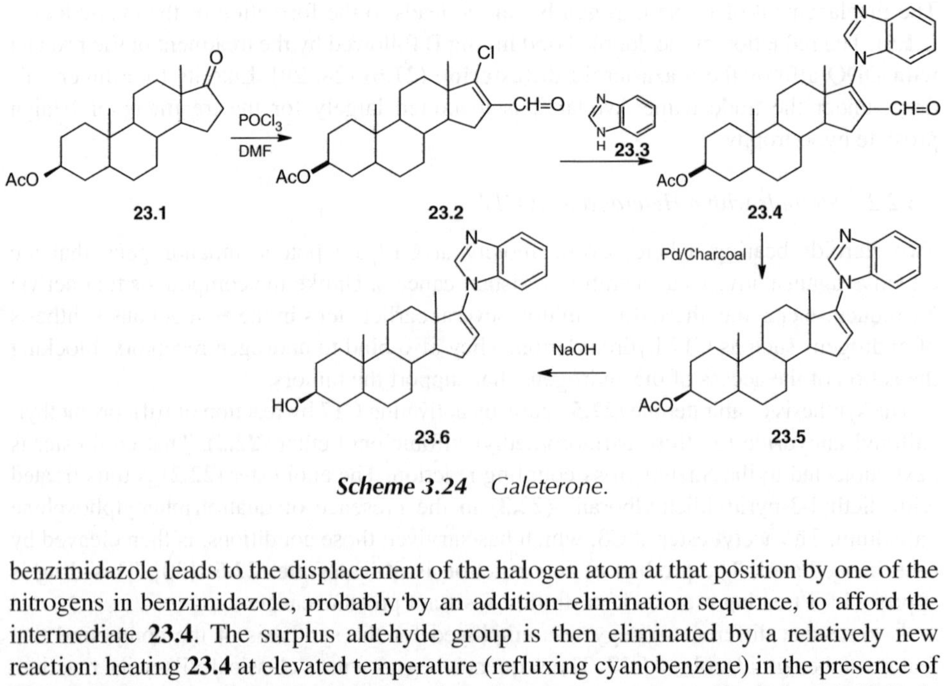

Scheme 3.24 Galeterone.

benzimidazole leads to the displacement of the halogen atom at that position by one of the nitrogens in benzimidazole, probably by an addition–elimination sequence, to afford the intermediate **23.4**. The surplus aldehyde group is then eliminated by a relatively new reaction: heating **23.4** at elevated temperature (refluxing cyanobenzene) in the presence of 10% palladium on charcoal. Saponification cleaves the ester to afford the 17-aliphatic heterocycle-substituted steroid, **galeterone** (**23.6**) [31]. The US FDA designated the drug fast track status for the treatment of advanced castration-resistant prostate cancer.

3.3.2.3   *Miscellaneous Steroids*

There exist a handful of steroids that are used as antineoplastic agents whose mode of action does not fall into one of the foregoing neat categories. One such agent, dromostanolone, for example, is known as a potent anabolic agent that is widely prescribed for that indication. The drug is also used in the treatment of breast cancer in women. This activity relies on the fact that both genders are endowed with androgen receptors. The molecule attaches to androgen receptors where among other activities it inhibits production of prolactin and estrogen receptors.

The synthesis of this venerable steroid begins with the reaction of commercially available androstan-17β-ol-3-one with methyl formate and the strong base sodium methoxide. The newly added formyl function in the product (**24.2**) is shown in the enol form. Catalytic hydrogenation reduces that function to a methyl group. The addition of hydrogen from the bottom face of the molecule leads to the formation of β-methyl isomer where the methyl group occupies the higher-energy axial position. Strong base-induced equilibration of the methyl group leads to the formation of the sterically favored equatorial α-methyl isomer (**24.6**), affording **dromostanolone (24.7)** [32]. The drug is usually prescribed as propionate ester, a derivative available by acylation of the hydroxyl function at position C17 with a reagent such as propionic anhydride.

Insertion of a second methyl group, this time at C7, produces a couple of steroids that significantly increases their potency as anabolic agents. That indication accounts for by far the largest number of prescriptions. The potency of these compounds has landed them on the DEA list of proscribed drugs, as well on athletes' doping websites. Both these compounds, bolasterone and calusterone, have shown modest activity when tested in women with metastatic estrogen receptor-negative carcinoma.

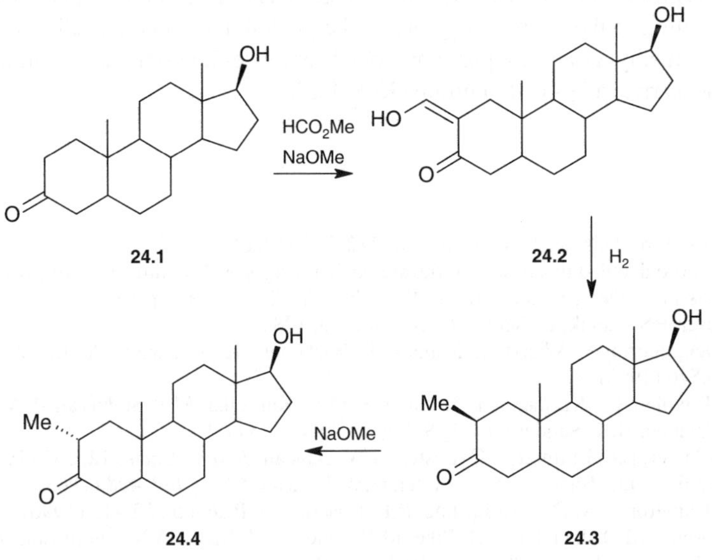

Scheme 3.25   *Dromostanolone.*

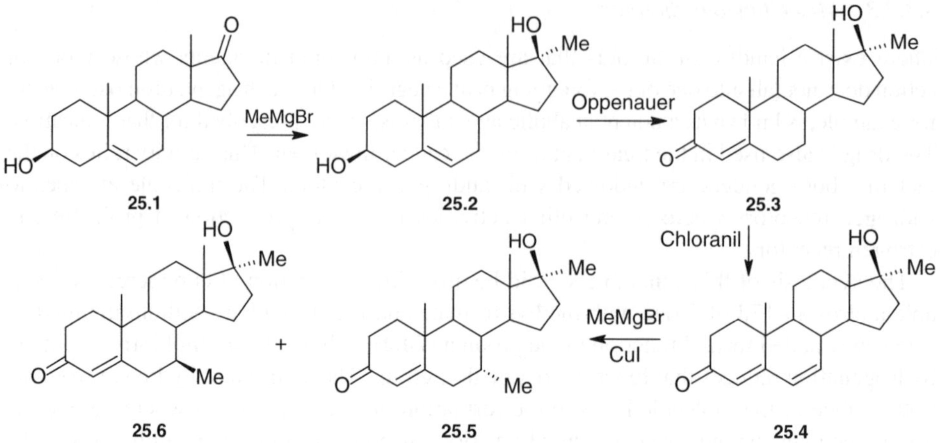

Scheme 3.26   *Bolasterone and calusterone.*

As in the previous example, the preparation of these drugs starts with the addition of a methyl group at C17 via reaction with methylmagnesium bromide. The hydroxyl group at C3 in **25.2** is next oxidized by treatment with cyclohexanone and trisisopropoxy aluminum. This very mild oxidation, called the Oppenauer reaction, in essence comprises the exchange of oxygen between the steroid hydroxyl group and cyclohexanone. The unsaturation is then extended to ring B (**25.3**) by treatment with 2,3,5,6-tetrachloro-1,4-quinone, better known as chloranil. A methyl group is then added via methylmagnesium bromide mediated by cuprous iodide. The latter is added at C7, the terminus of the conjugated system; this regiochemistry is due to the tendency of copper reagents to add to the end of conjugated systems. The product of this last reaction consists of a mixture of the compounds that differ in the orientation of the C7 methyl groups. The predominant product (**25.6**) is that which features an α-methyl group; the pure compound comprises bolasterone. The minor product, **calusterone**, carries a 7β-methyl group (**25.5**) [33].

References

[1]   G.R. Bedford, D.N. Richardson, *Nature* **212**, 733 (1966).
[2]   For a more detailed discussion of the stereochemistry see D. Lednicer, *Strategies for Organic Synthesis and Design*, 2nd Edition, Wiley, New York, NY, 2009, p.198.
[3]   R. Loser, P.S. Janiak, K. Seibel, *DrugsFuture* **9**, 1984.
[4]   M. DeGregorio, V. Wiebe, I. Kangas, P. Harkonen, K. Vaananen, A. Laine, U.S. Patent 5,750,576 (1998).
[5]   R.J. Toivola, A.J. Karjlaainen, T. Gang, K.O.A. Kurkelka, M.J. Sodervall, L.V.M. Kangas, L.G. Blanco, H.K. Sunderquist, U.S. Patent, 4,996,225 (1991).
[6]   D. Lednicer, E.E. Emmert, S.C. Lyster, G.W. Duncan, *J. Med. Chem.* **12**, 881 (1969).
[7]   D. Lednicer, J.C. Babcock, S.C. Lyster, G.W. Duncan, *Chem. Ind.* **498** (1963).
[8]   C.O. Cameron, P.A. Dasilva Jardine, R.L. Rosati, U.S. Patent 5,552,412 (1996).
[9]   C.D. Jones, M.D. Jevnikar, A.J. Pike, M.K. Peters, L.J. Black, A.R. Thompson, J.F. Falcone, J.A. Clemmens, *J. Med. Chem.* **27**, 1057 (1984).
[10] C.P. Miller, M.D. Collini, B.D. Tran, A.A. Santilli, U.S. Patent, 5,998,402 (1999).

[11]   J. Bowler, B.S. Tait, U.S. Patent 4,659,516 (1987).
[12]   J. Bowler, T.J. Lilley, J.D. Pittman, A.E. Wakelin, *Steroids* **54**, 71 (1989).
[13]   V.K. Agrawal, M.K. Singh, A.M. Patel, K. Solanki, U.S. Patent 8,288,571 (2012).
[14]   K. Hoffmann, E. Urech, U.S. Patent 2,848,455 (1958).
[15]   L.J. Browne, U.S. Patent 4,617,307 (1986).
[16]   P. Furet, C. Batzl, A. Bhatnagar, E. Francotte, G. Rhis, M. Lang, *J. Med. Chem.* **33**, 1393 (1993).
[17]   M. Villa, R. Fretta, M. Diulgheroff, M. Pontiroli, U.S. Patent Application 2010/0,099,888 (2010).
[18]   T.G. Gant, S. Sarshar, M.M. Shabbaz, U.S. Patent Application 2010/0,111,901 (2010).
[19]   A.H.M. Raeymaekers, E.J.E. Freyne, J.L.H. Van Gelder, M.G. Vener, European Patent, 0293978 (1988).
[20]   A.G.M. De Knaep, A.M.J. Vandendriessche, D.J.E. Daemen, J.J. Dingenen, *Org. Process Res. Dev.* **4**, 162 (2000).
[21]   J.W. Baker, G.L. Bachman, I. Schmumacher, D.P. Roman, A.L. Tharp, *J. Org. Chem.* **32**, 93 (1967).
[22]   H.Tucker, G.J. Chesterton, *J. Med. Chem.* **31**, 885 (1954).
[23]   H. Tucker, J.W. Crook, G.J. Chesterton, *J. Med. Chem.* **31**, 954 (1988).
[24]   J.T. Dalton, D.D. Miller, Y. He, D. Yin, U.S. Patent 6,996,284 (2006).
[25]   J.E. Audia; L.A. McQuaid, B.L. Neubauer, Blake, V.P. Rocco, U.S. Patent 5,622,962 (1995).
[26]   G.H. Rasmusson, G.F. Reynolds, N.G. Steinberg, E. Walton, G.F. Patel, T. Liang, M. Cacieri, H.H. Chueng, J.R. Brooks, C. Berman, *J. Med. Chem.* **29**, 2298 (1986).
[27]   A. Battacharya, L.M. De Michele, U.H. Dolling, A.W. Douglas, E.J. Grabowski, *J. Am. Chem. Soc.* **110**, 3318 (1988).
[28]   K.W. Batchelor, S.W. Frye, G.F. Dorsey, R.A. Mook, U.S. Patent 5,564,467 (1996).
[29]   S.W. Frye, *Curr. Top. Med. Chem.* **6**, 405 (2006).
[30]   G.A. Potter, S.E. Barrie, M. Jarman, M.G. Rowlands, *J. Med. Chem.* **38**, 2463 (1995).
[31]   R.D. Bruno, T.S. Vasaitis, L.K. Gediya, P. Purushottamachar, A.M. Godbole, Z. Ates-Alagoz, A.M.H. Brodie, V.C.O. Njar, *Steroids* **76**, 1268 (2011).
[32]   H.J. Ringold, E. Batres, O. Halpern, E. Necoechea, *J. Am. Chem. Soc.* **81**, 1513 (1959).
[33]   J.C. Babcock, J.A. Campbell, U.S. Patent 3,341,557 (1967).

4

Topoisomerase Inhibitors

4.1 Introduction

Human DNA comprises a very long microscopically thin molecule that when unwound can measure close two meters end to end. In order to accommodate this extremely long string in a microscopic cell nucleus, that DNA is coiled and the resulting tight package coiled again to achieve a conformation dubbed "supercoiled." The process of assembling new DNA when the cell replicates or creating the new RNA, the ribonucleotide that is directly involved in replicating DNA or that is required for the assembly of amino acids for assembling polypeptides, begins by transferring to RNA the required specific sequence of bases on the genome. This process requires access to the region of DNA that holds those instructions. The necessary sequence may be located deep within the tightly wound genome. In order to achieve that, one or both of the double strands are cut; the region to be read is then passed through the opening. Proteins named topoisomerases are then called upon to temporarily bind covalently to the ends of the cut strand(s) marking the place of the cut. The strands reconnect when the revealed region has been read. Stated simplistically, topoisomerase I comes into play on scission of single strand, while topoisomerase II involves a cut of both DNA strands.

The chemical structures of a number of the older antineoplastic agents, such as the anthracyclines or anthraquinones, consist of linear three- or four-ring flat molecules. The mode of action of those drugs was at first ascribed to intercalation. This process involved insertion between DNA base pairs and thus inhibition of the processes mediated by the DNA. It has more recently been found that intercalation is only the first part of the inhibitory process; the cytotoxic effect is due to the inhibition of topoisomerases. This may involve making permanent the state where a topoisomerase marks the site of a sundered DNA or alternatively blocking the action of the topoisomerase. Either mechanism will disrupt the process whereby specific sections of the genome can be read. This in turn leads to cell death.

Antineoplastic Drugs: Organic Synthesis, First Edition. Daniel Lednicer.
© 2015 John Wiley & Sons, Ltd. Published 2015 by John Wiley & Sons, Ltd.

4.2 Anthracyclines

The regulatory approval, toward the end of the 1970s, of the fermentation products doxorubicin, far better known as Adriamycin® (**1.1**), and its partner lacking the side chain hydroxyl group, daunorubicin (**1.2**) (Cerubidine®), marked the advent of antineoplastic drugs that exhibited some activity against solid tumors. Most analogues of these natural product-based drugs comprised relatively small structural modifications of the natural product such as inverting the stereochemistry of the free hydroxyl group in the sugar (epirubicin, Ellence® (**1.3**)) or modifying the side chain hydroxyl to an ester and the ring nitrogen to an amide (valrubicin, Valstar® (**1.4**)).

Anthracyclines and anthraquinones are not absorbed when administered orally. These older drugs are thus usually given to patients by infusion over a time span. An additional rationale for this method of administration lies in the importance of controlling blood levels of these relatively toxic drugs. These compounds are in addition virtually insoluble in water and require especially designed infusion fluids.

Idarubicin (**2.9**), an anthracycline that lacks the methoxyl group in ring A, comprises an exception since it is relatively well absorbed on oral administration. Since none of the naturally found anthracyclines feature the absence of that substituent, idarubicin is prepared by total synthesis. The commercial, not readily accessible, route to this compound is arguably more direct than the one described in Scheme 4.2.

The somewhat lengthy scheme starts with the addition of lithium acetylide to the carbonyl group in tetralone (**2.1**). The triple bond in ethynylated product (**2.2**) is then in

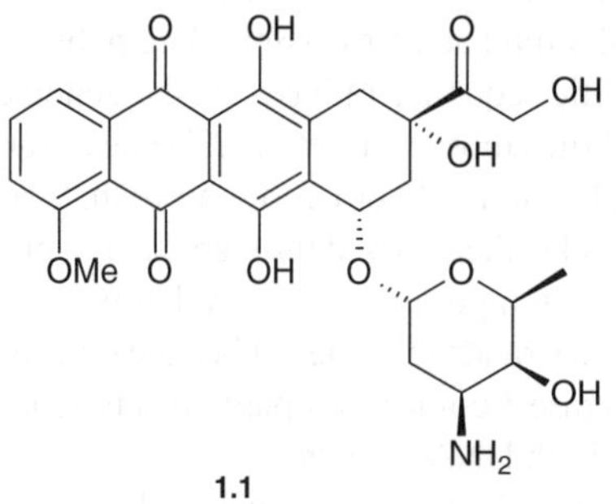

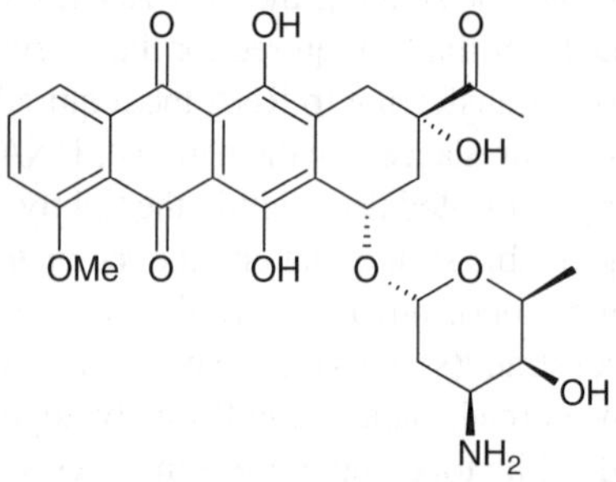

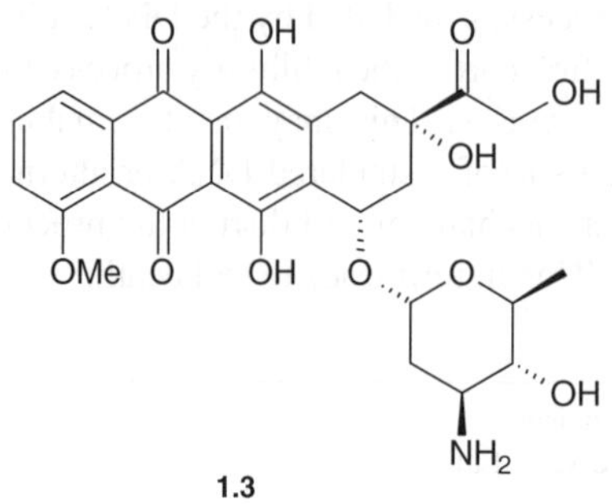

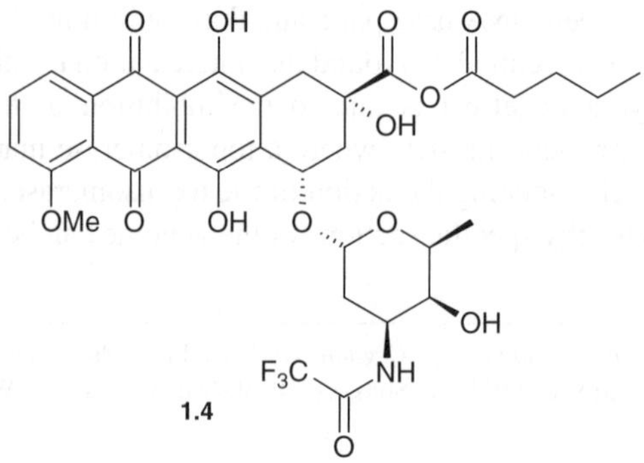

Scheme 4.1 *Rubicin antineoplastic agents.*

effect hydrated by treatment with dilute acid in the presence of yellow mercuric oxide. The first-formed enol rearranges to the keto form under reaction conditions (**2.3**). Aluminum chloride-mediated Friedel–Crafts reaction of hydroxy ketone (**2.3**) with phthalic acid (**2.4**) then yields the anthracycline four-ring skeleton. It now remains to add an additional aliphatic hydroxyl group. The sequence for achieving that begins with the protection of the side chain acetyl methyl group from reagents in the next step (bromine, AIBN) by converting the carbonyl group to an acetal. Reaction with ethylene glycol and in the presence of toluenesulfonic acid catalyst affords acetal (**2.6**). The treatment of this intermediate with bromine in the presence of the AIBN free radical initiator leads to the intermediate brominated the benzylic position. Solvolysis of the unstable bromo derivative in dilute acid replaces the halogen by a hydroxyl group and at the same time hydrolyzes acetal (**2.7**). In order to isolate the intermediate in which both hydroxyl groups in the ring are cis to each other, the crude diol is allowed to react with phenylboronic acid. The cis isomer proceeds to form the cyclic boronate ester (**2.8**); the hydroxyl groups in any trans isomer present are not properly placed to form such a cyclic ester; that isomer thus remains untouched. The difference in chemical properties between the cycloboronic ester and the trans diol facilitates the separation of the desired compound via chromatography. The transesterification of the purified boronic ester with 2-methyl-2,4-pentane diol opens the boronate ester; treatment with boron tribromide then cleaves the methyl ethers to afford the aglycone of **idarubicin** (**2.9**) [1]. The glycoside can in principle be prepared by the reaction of the latter with the silylated doxorubicin sugar.

Amrubicin comprises another anthracycline lacking the 4-methoxyl group. This analogue, too, needs to be prepared by total synthesis. The scheme differs from that used to assemble idarubicin (**2.11**) since amrubicin (**4.8**) also features reversed functionality: the hydroxyl on

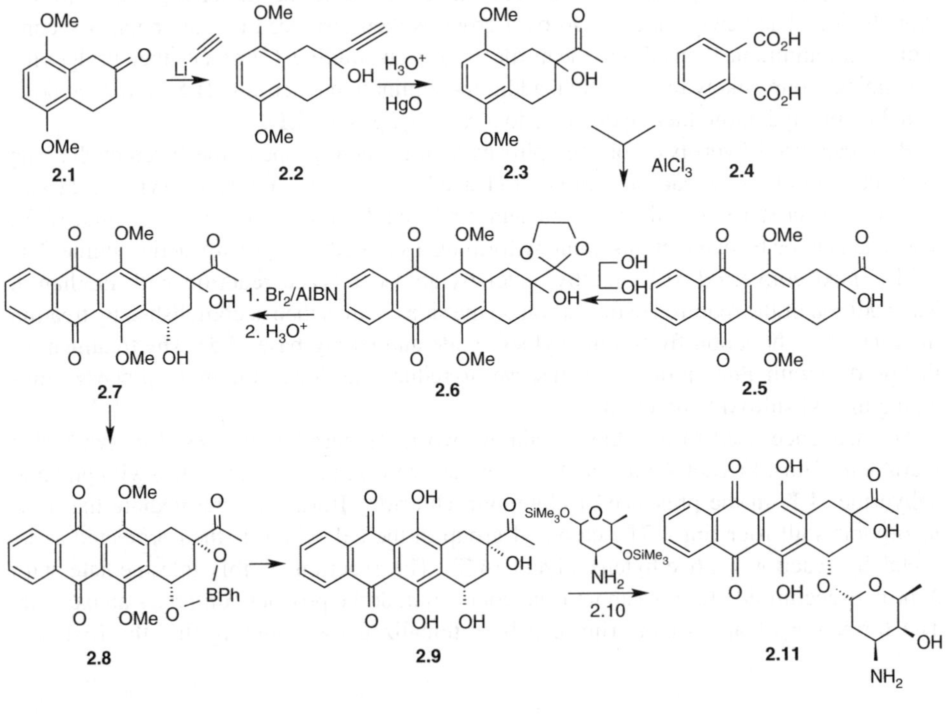

Scheme 4.2 *Idarubicin.*

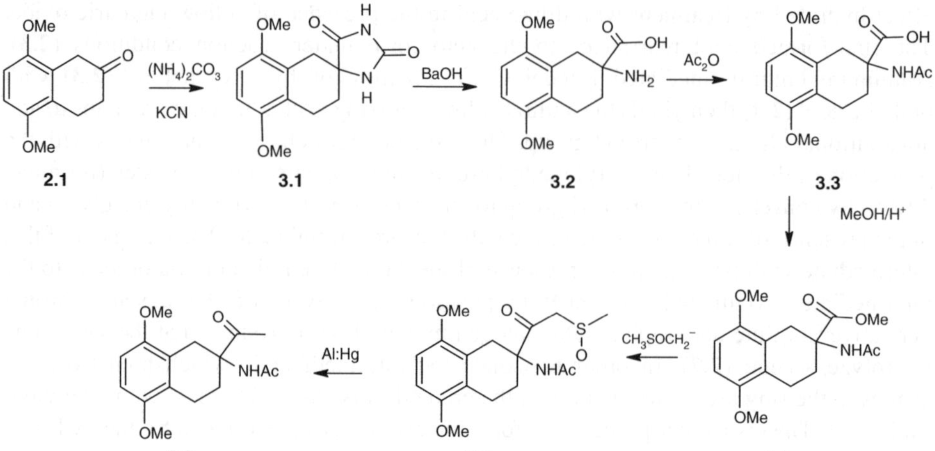

Scheme 4.3   *Amrubicin A.*

the carbon bearing the acetyl group is replaced by an amine, and the amine on the sugar in natural product-based anthracyclines is replaced by a hydroxyl group. The sequence for synthesizing that compound begins with a series of reactions aimed at providing the two-ring unit that will bear the sugar in the final product. The first step comprises the treatment of the same tetralone (**2.1**) as that used to build the corresponding moiety in idarubicin (**2.11**) with potassium cyanide and ammonium carbonate. The formation of the resulting spiro hydantoin involves a complex series of reactions that can be envisaged by assuming that the sequence of steps starts by the addition of cyanide to the carbonyl group to form a cyanohydrin. The newly formed hydroxyl group is then displaced by ammonia. A second mole of the ammonia then adds to the nitrile to form an amidine. Carbonate then bridges the two amine groups likely by sequential addition–elimination reactions. The amidine at position 3 on the hydantoin then hydrolyzes to a carbonyl group (**3.1**).

Base-catalyzed hydrolysis of the spiro hydantoin moiety opens the heterocyclic ring with the ejection of the carbonyl carbon to leave behind the α-aminocarboxylic acid (**3.2**). The amine is next acetylated with acetic anhydride to afford the corresponding amide (**3.3**). The acid may be resolved at this point by forming a salt with an optically active amine. The acid is next converted to its methyl ester by acid-catalyzed reaction with methanol. The addition of the required extra carbon on the side chain is then accomplished by treating the ester with the anion from dimethyl sulfoxide and strong base (**3.5**). The treatment of that product with aluminum amalgam serves to reduce the side chain, in the process eliminating methyl sulfoxide function.

The sequence used to add the remaining two rings parallels that used to synthesize idarubicin. Thus, Friedel–Crafts acylation of the two-ring intermediate (**3.6**) with phthalic anhydride (**4.1**) in the presence of aluminum chloride affords an intermediate that now incorporates all four rings. The carbonyl group in the side chain is then protected as its acetal by reaction with ethylene glycol (**4.3**). The reaction of this intermediate with N-bromosuccinimide follows an unusual course in that the product comprises an oxazine. The formation of an oxazine ring can be rationalized by assuming that the first step

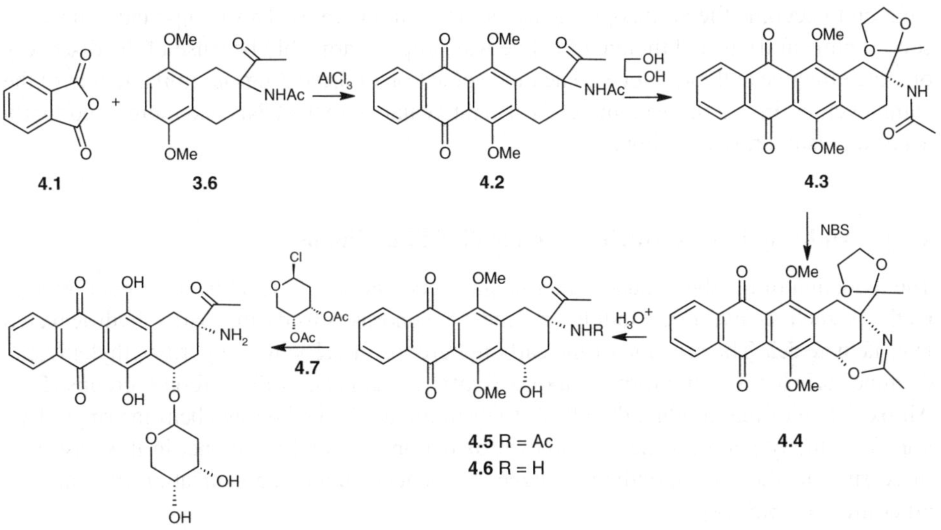

Scheme 4.4 *Amrubicin B.*

comprises free radical bromination on the benzylic position of the only aliphatic ring in the molecule (**4.4**). The oxygen of the enol form carbonyl group then displaces bromine. This cross-ring displacement in effect transfers the stereochemistry from the amide to the new hydroxyl group. Acid hydrolysis then opens the oxazine ring and removes the acetal protecting group. The new hydroxyl group is then glycosidated with the chloro-sugar. Treatment with base then saponifies both the protecting groups on the sugar and that on the amide. Reaction with a reagent such as boron tribromide cleaves the methyl ethers. The anthracycline **amrubicin** (**4.7**) is thus obtained [2, 3]. The compound was granted orphan drug status by the FDA in 2008 for treating small cell lung cancer but has not received approval for other indications as of January 2014. It is available in Japan as Calsed®.

4.3 Anthraquinones and Anthrapyrazoles

It would be tempting to speculate that the class of anthraquinone antineoplastic agents owed their existence to a program designed to simplify the pharmacophore of the antineoplastic anthracyclines—tempting but not in accord with reality. Starting in the mid-1950s, the National Cancer Institute (NCI) instituted and then maintained a high-turnover rodent assay (L-1210 leukemia) for testing compounds from both academic and commercial sources aimed at uncovering new leads for cancer chemotherapy drugs. This program ballooned in 1971 when then-President Nixon declared war on cancer. The NCI at that point instituted an aggressive effort to acquire test samples. One of the first positive hits from the screening program comprised a dark blue dye for ballpoint ink. This compound went on to become the antineoplastic agent ametrantone (**5.3**). This compound underwent several clinical trials and though effective was dropped arguably because of the discovery of more effective anthraquinones. The mode of action of these drugs is similar to that of the anthracyclines in that these compounds also intercalate into the DNA and more importantly act as topoisomerase inhibitors.

4.3.1 Anthraquinones with Two Aminoalkyl Side Chains

The reduction of the dye industry intermediate leucoquinazirin (**5.1**) by any one of several methods such as hydrogen iodide and red phosphorus affords the partly hydrogenated intermediate **5.2**. The reaction of the carbonyl groups in **5.2** with *N*-hydroxyethylethylene diamine (**5.3**) leads to the formation of a Schiff base at each of the carbonyl groups (**5.4**). Air oxidation of that product then leads to the quinone flanked by two benzene rings. This transform likely parallels the well-known oxidation of a hydroquinone, in this case in a fused ring, to the corresponding quinone. Subsequent bond reorganization then affords **ametantrone (5.5)** [4].

A structurally very close analogue of the preceding anthraquinone that incorporates two additional hydroxyl functions on one of the benzene rings proved to have more suitable drug delivery properties and the same or better antineoplastic activity than ametantrone. This drug, mitoxantrone, is indicated for treating leukemia and prostate cancer. It is also surprisingly indicated to halt the progression of multiple sclerosis. The drug is available under its trade name Novantrone® as well as in generic form since the patent expired quite some time ago. The synthesis closely parallels that of its predecessor. The reaction of tetrahydroxy anthraquinone (**6.1**) with the same hydroxyethylene

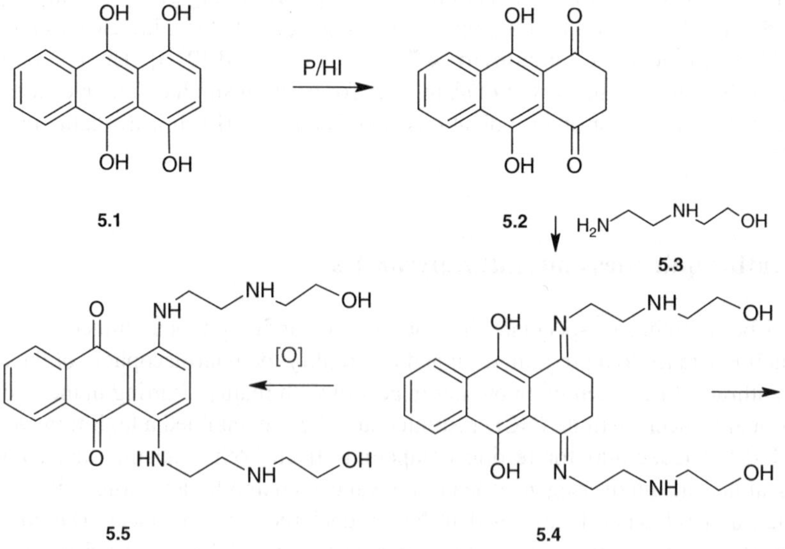

Scheme 4.5 *Ametantrone.*

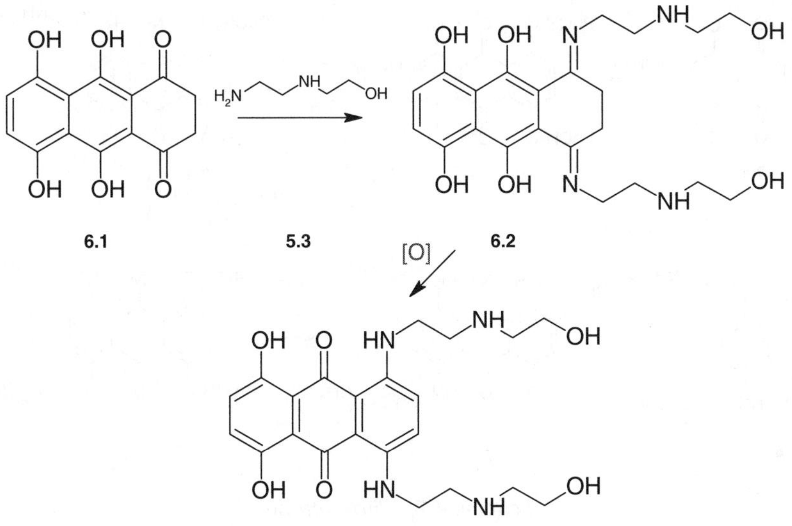

Scheme 4.6 *Mitoxantrone.*

diamine (**5.3**) affords bis imine (**6.2**). Air oxidation followed by bond reorganization results in the formation of an anthraquinone. The antineoplastic agent **mitoxantrone** (**6.3**) is thus obtained [5].

4.3.2 Anthraquinones with a Fused Pyrazole Ring

Several anthrapyrazoles whose structures differ from the preceding anthraquinones by incorporating an additional fused nitrogen-containing ring have been tested in clinical trials. Though they have shown antineoplastic activity in those trials, none seems to have yet been approved by regulatory authorities.

The synthesis of piroxantrone starts with the conversion of the phenolic hydroxyl groups in anthraquinone (**7.1**) to their benzyl ethers (**7.2**) by means of benzyl bromide and a base in order to avoid interference in the formation of the adjacent fused pyrazole ring. The reaction of benzyl-protected anthraquinone (**7.2**) with 1-hydrazino-3-(aminoethoxy) ethane (**7.3**) results in the formation of the desired fused pyrazole ring. This transform involves the formation of an imine between the terminal hydrazine nitrogen and the anthraquinone carbonyl as well as displacement of ring chlorine to form the fused pyrazole; the end product will be the same regardless of the order in which those two steps occur. The relatively easy displacement of chlorine on an aromatic ring can probably be attributed to electron withdrawal by the neighboring carbonyl group (it may not be a stretch to view this group of substituent as a vinylogous acid chloride). The remaining chlorine atom is then again displaced, this time by one of the amines in 1,3-bisaminopropane. Catalytic hydrogenation then cleaves the benzyl ethers to afford **piroxantrone** (**7.7**) [6, 7]. A close analogue (**7.8**) of the latter features a methyl group on the terminal side chain amine.

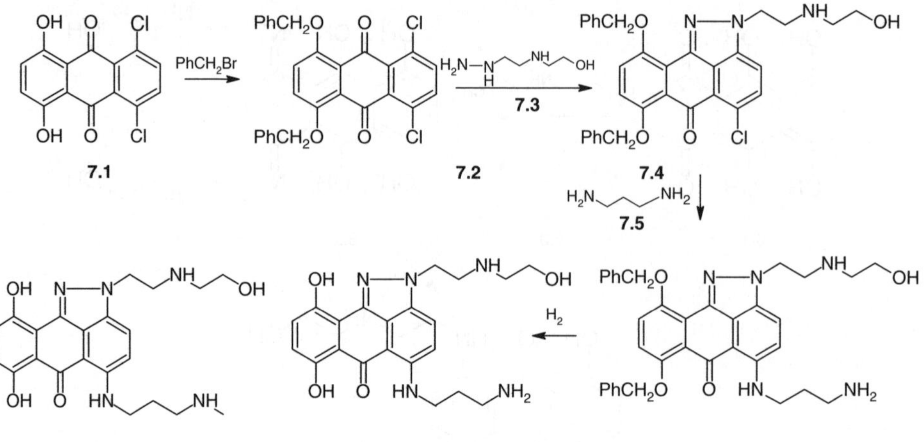

Scheme 4.7   *Piroxantrone.*

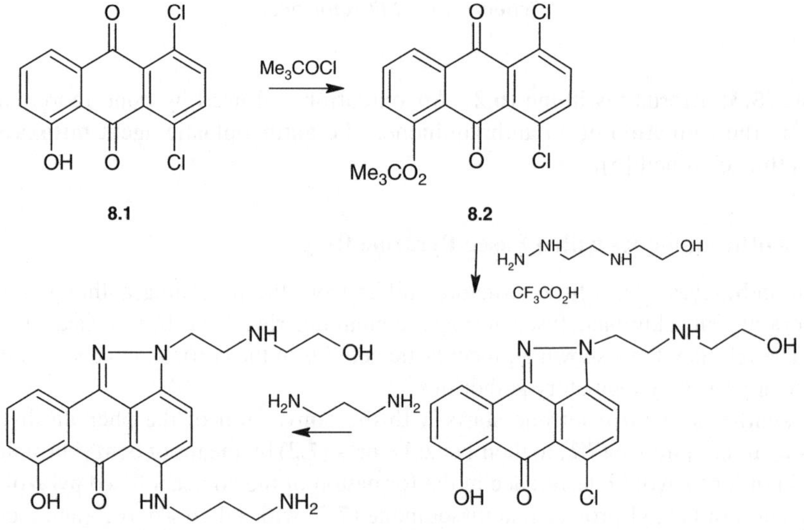

Scheme 4.8   *Losoxantrone.*

The axis of symmetry that slices through the width of the piroxantrone starting material (**7.2**) makes moot the question as to which carbonyl–chlorine pair will react to form the fused pyrazole ring. The closely related starting material (**8.1**) for losoxantrone lacks that symmetry. The scheme for the synthesis of losoxantrone relies on a bulky substituent on the phenolic hydroxyl group to steer pyrazole formation toward the other carbonyl–chlorine pair. The first step in the preparation of losoxantrone thus comprises forming a bulky pivaloyl ester (**8.2**) with the phenol by acylation with pivaloyl chloride. The remainder of the scheme follows that used to prepare piroxantrone. Thus, the reaction of chloroanthraquinone (**8.2**) with the hydrazine-terminated amino alcohol (**7.3**). This reaction builds the fused pyrazole equipped with amino alcohol side chain. The pivaloyl group is then removed by treating the product with trifluoroacetic acid. The remaining chlorine substituent is then displaced by one of the amines in propylene-1,3-diamine to form **losoxantrone** (**8.3**) [6, 7].

4.3.3 Heterocyclic Anthraquinones

Pixantrone comprises an anthraquinone that departs from the usual pattern in that one of the benzene rings that flanks the central quinone is replaced by pyridine and the basic side chains are abbreviated to simple ethylene diamines. The mechanism of action of this drug is basically the same as that of the conventional anthraquinones in that pixantrone is an intracalator that inhibits topoisomerase II. The drug is approved in Europe for use in patients with several episodes of relapsed leukemia.

The synthesis of pixantrone starts with Friedel–Craft acylation of 1,4-difluorobenzene by pyridine-2,3-dicarboxylic anhydride to give a mixture of the two positional isomers created by the reaction of one of the two alternate carboxyl groups.(**9.3**, **9.4**). The treatment of the mixture under more drastic conditions (sulfuric acid spiked with 30% SO_3) gives azaanthraquinone (**9.5**). The next step, displacement of fluorine atoms by one of the amines in ethylene diamine, proceeds under relatively mild conditions due to the reactivity of the carbon–fluorine bond. **Pixantrone** (**9.6**) is thus obtained (Pixuvri®) [8].

Topixantrone incorporates in a single molecule most of the variants that have appeared in the prior anthraquinone antineoplastic agent series. The problem in devising a scheme for preparing this compound, as was the case with losoxantrone, lies in the asymmetry introduced by the pyridyl nitrogen. One of the several routes to topixantrone starts with the same tricyclic intermediate (**9.5**) used in the scheme for producing pixantrone (**9.6**). In the

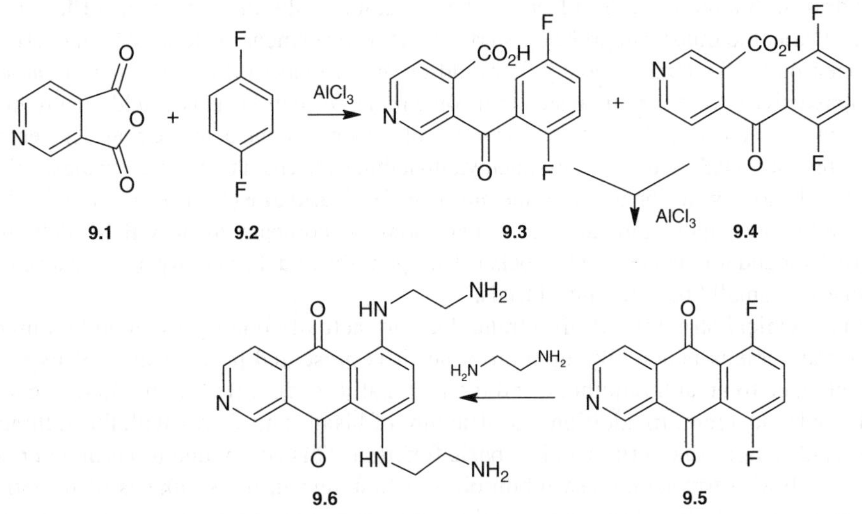

Scheme 4.9 *Pixantrone.*

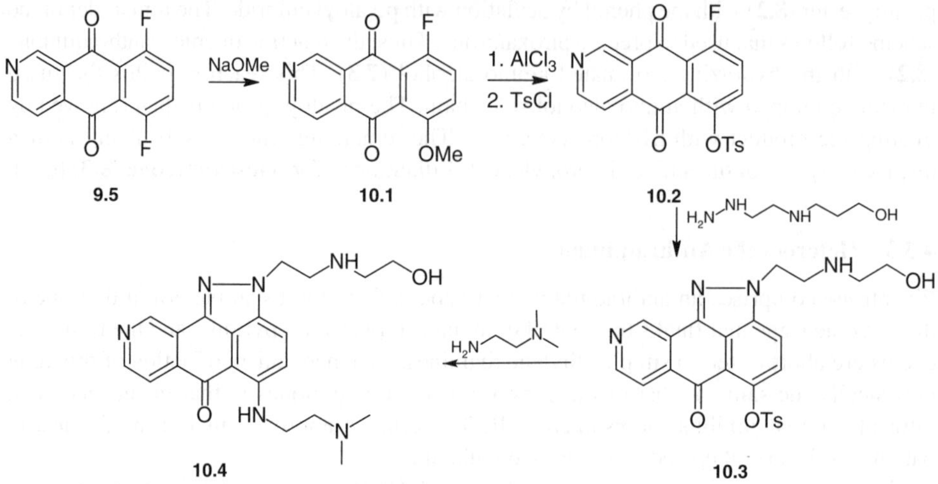

Scheme 4.10   *Topixantrone.*

present case, the first step comprises displacing one of the fluorine atoms. The treatment of (**9.5**) with sodium methoxide from sodium and methanol affords a mixture of two displacement products. The pyridine nitrogen atom is, as might be expected, too distant to effect a difference in reactivity between the two fluorine atoms. The desired isomer is then extracted from the mixture. The methyl ether in (**10.1**) is next cleaved by heating with aluminum chloride. The resulting phenol is then converted to its 4-toluenesulfonate by esterification with 4-toluenesulfonyl chloride. The substituents at two sites that will carry different side chains now differ in reactivity. As was the case in the anthraquinones, whose structures included a fused pyrazole, the reaction of the intermediate (**10.2**) with 1-hydrazino-3-aminoexan-5-ol leads to the formation of the fused pyrazine (**10.3**). In the present case, the transform probably starts by the displacement of fluorine by the terminal nitrogen on the hydrazine moiety. It would require a marked difference the in carbonyl group reactivity were the initial step to involve imine formation. This would require major differences on electron density caused by the substituents on the benzene ring. The reaction of the product (**10.3**) with the *N,N*-dimethylethylamine, the chain to be, leads to the displacement of the toluenesulfonate by amine nitrogen. **Topixantrone (10.4)** is thus obtained [9].

Like its forerunners, topixantrone, better known by company identity BBR 3576, is an intercalator and topoisomerase II blocker. The agent showed the activity on prostate cancer in relatively small Phase II clinical trials.

The chemical structure of bisantrene does not actually qualify as an anthraquinone since the nucleus is actually an anthracene. It is presented here because of its vague resemblance to an anthraquinone and the fact that it is identified in the literature as an antineoplastic agent. In addition, the structure of bisantrene shares with the anthraquinones side chains that terminate in a basic function. This compound too is an intercalator as well as a topoisomerase inhibitor. The first step in the synthesis of bisantrene comprises the Diels–Alder condensation of anthracene (**11.1**) with the rarely used reagent vinylene carbonate (**11.2**) to afford an intermediate where the newly introduced

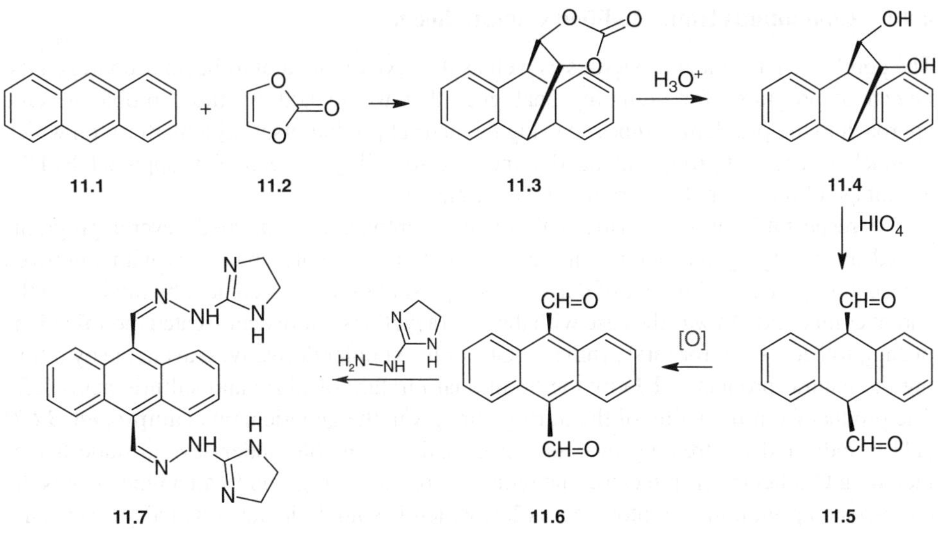

Scheme 4.11   *Bisantrene.*

moiety spans the two central atoms in anthracene (**11.3**). Treatment with acid then hydrolyzes the carbonate ester to leave behind the free glycol (**11.4**). The reaction of that product with hydriodic acid, a standard reagent for cleaving glycols, cuts the glycol in the intermediate (**11.4**), yielding cis bis-aldehyde (**11.5**). Treatment with a mild oxidizing agent restores the double bond that disappeared in the initial Diels–Alder addition (**11.6**). This reaction also restores the flat shape of the molecule. The aldehyde functions are then condensed with 2-hydrazinoimidazoline to afford **bisantrene** (11) [10]. The drug is available in Europe, where it is indicated for the treatment of leukemia, under the trade name Zantrene®.

4.4   Camptothecins

The antineoplastic activity of camptothecin (**12.1**), an alkaloid isolated from the stem and bark of a tree native to China, was discovered in the 1960s. For various reasons, this finding was not exploited until the late 1980s [11]. The very low solubility of this molecule in aqueous media added to the delay in the development of camptothecin as an antineoplastic drug. The concentration of drug in infusion media, for example, was barely sufficient to test the efficacy of the compound in humans. Contrary to expectations, opening of the lactone ring to the respective hydroxy acid resulted in the loss of biological activity; this obviated the use of the acid or hydroxyl group for attaching substituents that would improve solubility. It was established fairly early that camptothecin and its analogues intercalate in DNA and act as topoisomerase inhibitors. The compound and its derivatives comprise one of the few classes of antineoplastics that act on topoisomerase I and thus inhibit the enzyme involved in cutting a single strand of DNA. Camptothecin proper is indicated for use in colorectal cancer and is available as Camptosar®.

4.4.1 Compounds from Modified Camptothecin

The poor pharmaceutical properties as well as the toxicity of camptothecin encouraged the launch of programs for preparing analogues that would address those problems. One approach to the problem comprised using the natural product itself as a platform on which to attach a variety of groups intended to increase solubility in water. This approach had the advantage of retaining the chirality of camptothecin.

The respectable antineoplastic activity of camptothecin generated several programs aimed at modifying the natural product in order to generate analogues with improved physical properties. A number of derivatives represent molecules created by modifying the parent compound. As was the case with the anthracyclines, more deep-seated modifications of camptothecins call for total synthesis. One of the simpler derivatives, available by nitrating the natural product (**12.1**) by the traditional mixture of nitric and sulfuric acids [12]. The product from reduction of the nitro group, given the generic name **camptogen** (**12.2**) [11] is indicated for treating ovarian cancer and is available under several trade names including Orathecin® (rubitecan). The reduction of the nitro group to an amine affords the corresponding 9-aminocamptothecin (**12.4**) which is soluble in aqueous media. This analogue showed promising activity *in vitro* and in laboratory animals and small clinical trials.

A somewhat more complex derivative results in a compound with more suitable aqueous solubility. The synthesis of this agent, irinotecan (**13.5**), starts with camptothecin proper. The scheme used to prepare that compound depends on several somewhat unusual reactions. The synthesis begins by taking advantage of the reactivity toward free radicals of the only free position on the fused pyridine. The scheme starts with a reaction intended to block that reactive position. Thus, the reaction of camptothecin with propionaldehyde in the presence of ferrous sulfate and hydrogen peroxide leads to the introduction of an ethyl group at that position. This transform, also known as the Minisci reaction, involves first the generation of an ethyl radical by the expulsion of carbon monoxide from the aldehyde. That radical then attacks camptothecin at the position that is to be blocked (**13.1**). A second round of hydrogen peroxide converts the quinoline nitrogen to its N-oxide (**13.2**). Irradiating that intermediate leads to the transfer of oxygen from nitrogen to the quinoline benzene ring. In the absence of the blocking ethyl group, this Polonovski-like reaction would probably have delivered oxygen to the blocked position. The phenolic hydroxyl group is then converted to a carbamate by serial reaction with phosgene and then 4-piperidylpiperidine

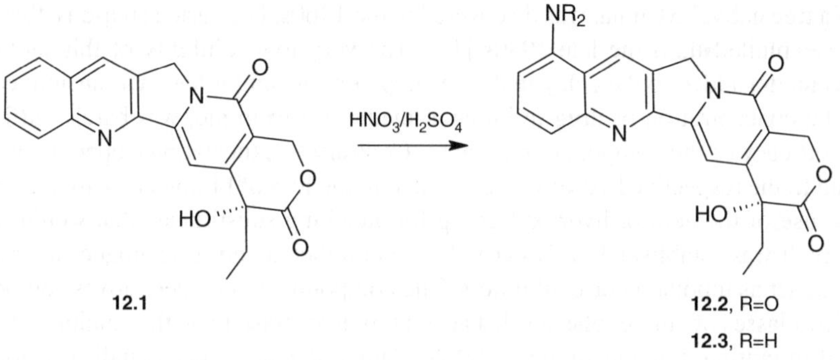

Scheme 4.12 *Camptothecin and camptogen.*

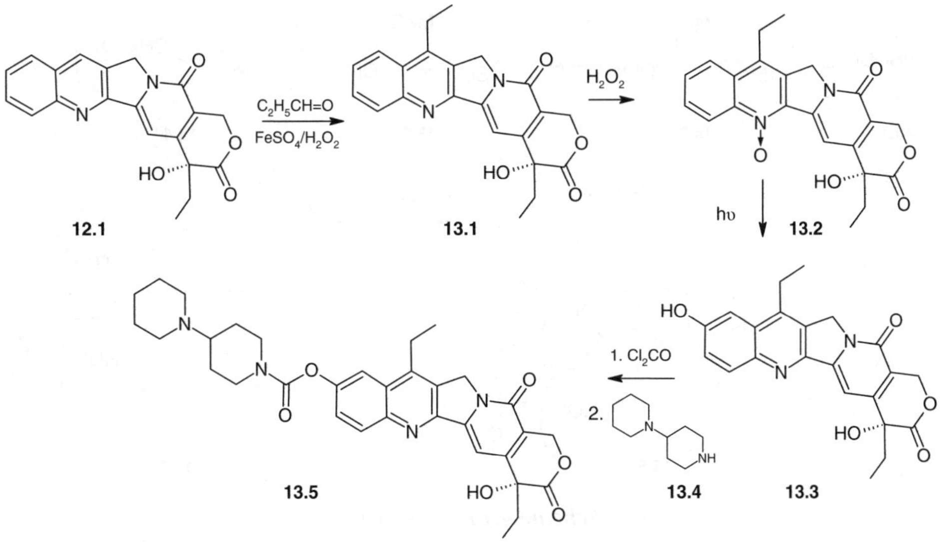

Scheme 4.13 *Irinotecan.*

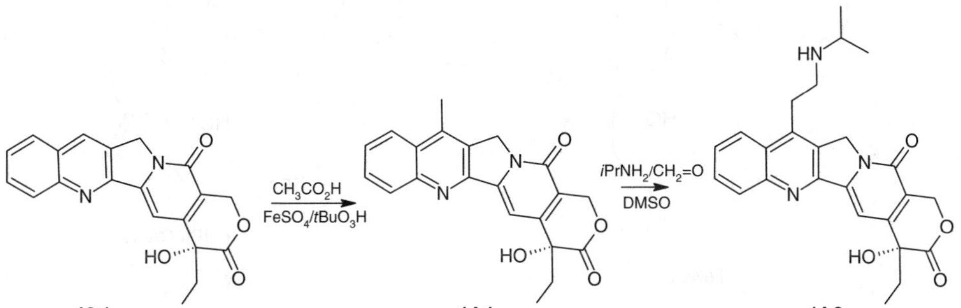

Scheme 4.14 *Belotecan.*

(**13.5**) to afford **irinotecan** (**13.5**) [13]. The polyethylene glycol derivative of this drug, **etirinotecan pegol**, was given fast track designation by the FDA for treating metastatic breast cancer.

The addition of an aminoethyl substituent to the fused pyridine ring of camptothecin improves solubility without losing topoisomerase-inhibiting activity. The preparation of this compound begins with the reaction of camptothecin with acetic acid in the presence of ferrous sulfate and tertiarybutyl peroxide. The methyl radical from loss of carbon dioxide from the acid adds to position 7 on camptothecin (**14.1**). The treatment of that product with isopropylamine and formaldehyde extends the length of the side chain and adds the base to the end of the extended chain. **Belotecan** (**14.2**) is thus obtained [14].

Drugs whose structure includes a covalent carbon to silicon bond can probably be counted on the fingers on one hand. The camptothecin **cositecan** (**15.5**) comprises one of

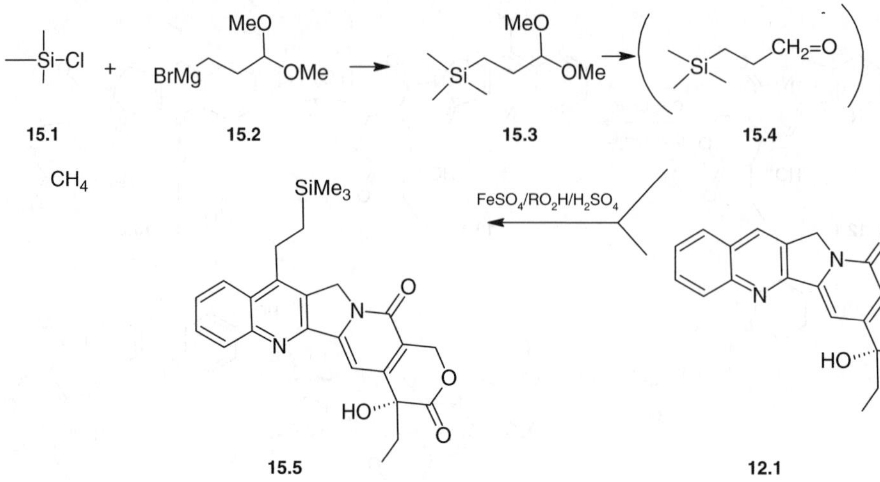

Scheme 4.15   *Cositecan.*

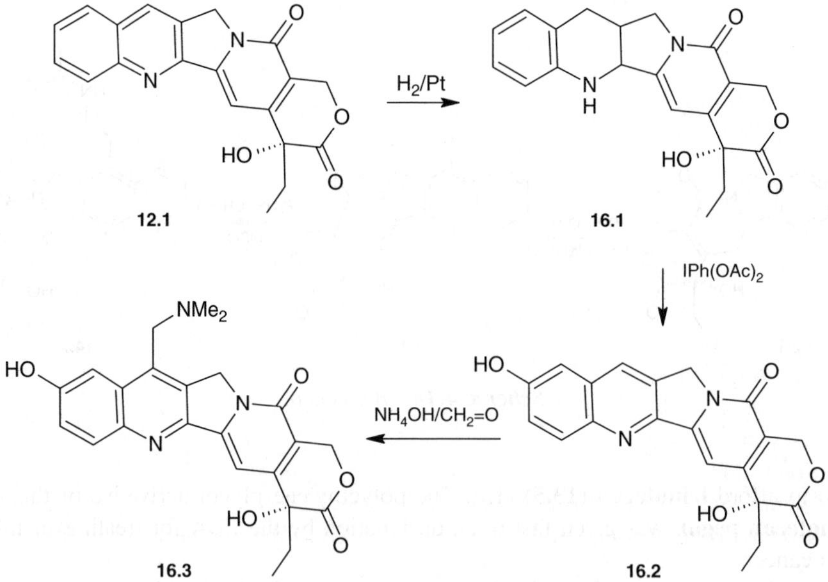

Scheme 4.16   *Topotecan.*

those rare therapeutic agents. This compound too has shown improved pharmacodynamic properties over those of camptothecin. The drug, in common with other campto-
thecin-based antineoplastic agents, acts as an intercalator and as an inhibitor of
topoisomerase I. The antineoplastic activity of the drug has led to a large Phase III
clinical trial. Though not approved as of early 2014, the sponsor has filed the trade name
Karenitecin®.

The synthesis of cositecan starts with the preparation of the intermediate that will be used to introduce the new substituent at position 7. The reaction of the Grignard reagent from 3,3-bismethoxybromopropane (**15.2**) with trimethylsilyl chloride serves to attach silicon to the end of the future side chain. The treatment of camptothecin with the silyl derivative (**15.3**) in the presence of ferrous sulfate, a hydroperoxide, and sulfuric acid probably results first in opening the acetal to the corresponding aldehyde (**15.4**). This then undergoes loss of carbon monoxide to leave behind the substituted ethyl free radical. This last intermediate then goes on to attack position 7 of camptothecin, yielding **cositecan** (**15.5**) [15].

Yet another modified camptothecin analogue, topotecan (**16.3**), like all the other analogues in this series, intercalates in DNA and inhibits topoisomerase I. The drug was approved by the FDA in 2006 for treating cervical cancer in conjunction with cisplatin. This drug is available under the trade name Hycamtin®. The scheme for introducing the hydroxyl group starts with the catalytic hydrogenation of camptothecin to the tetrahydro derivative (**16.1**). The treatment of that reduced derivative with the powerful oxidant (diacetoxy)iodosobenzene in aqueous acetic acid afforded the phenolic derivative (**13.2**). The reagent at the same time restores the aromatic character of the fused pyridine ring [16]. A classic Mannich reaction of that product with ammonium hydroxide and formaldehyde introduced the dimethyaminomethyl analogue. This product comprises **topotecan** (**16.3**) [13].

4.4.2 Camptothecins by Total Synthesis

Syntheses that start from commercially available organic compounds allow the preparation of analogues that can feature structural modification of camptothecin analogues not attainable by modifications of the natural product. The design of the route to be followed for preparing analogues must take into account the fact that camptothecin is a chiral molecule. The target of each of the two total syntheses that follow comprises an analogue of camptothecin that is more soluble in water than the natural product and thus more suitable for formulating infusions.

The convergent scheme devised for the total synthesis of the topoisomerase inhibitor lurtotecan calls for the preparation of the proper enantiomer of fused lactone (**17.7**) and then the synthesis of the quinolone moiety (**18.8**). These fragments are then joined via an alkylation followed by a Hecht vinylation reaction affording lurtotecan (**19.2**).

The synthesis of the moiety that includes lactone (**17.7**) begins with the treatment of 2-methoxypyridine, a compound that can be viewed as the methyl ether of 2-pyridone, with butyl lithium. The resulting anion is then allowed to react with dimethylformamide. This in effect transfers the formyl group to pyridine (**17.2**). A second round of butyl lithium then results in the removal of a proton from the position adjacent to the just-added aldehyde. Treatment with iodine leads to the formation of the iodo intermediate (**17.4**). The reaction of the aldehyde with triethylsilane and methallyl hydroxide leads to reductive alkylation of the formyl group followed by the formation of ether (**17.4**). Palladium acetate-catalyzed Heck vinylation then closes the hydropyran ring. This last intermediate can be visualized by assuming that the unsaturation in the side chain moves up by one carbon; the palladium complex at the vacated position then proceeds to displace iodine and in the process closing the ring (**17.5**). The next reaction introduces

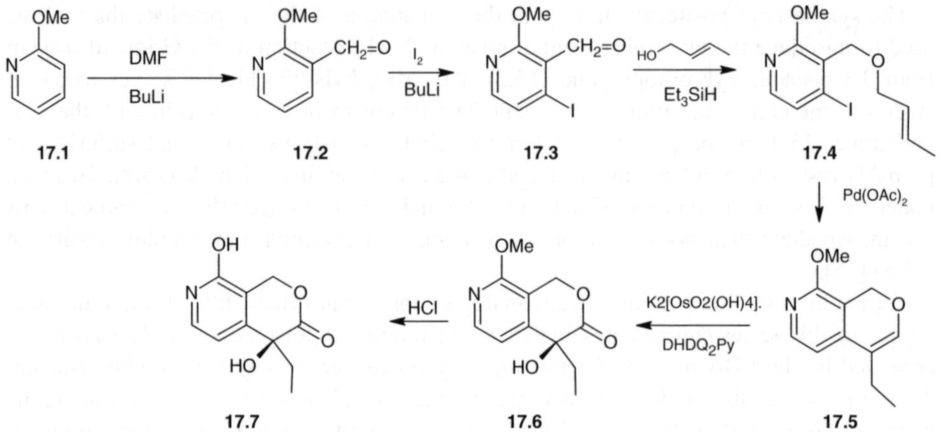

Scheme 4.17   *Lurtotecan A.*

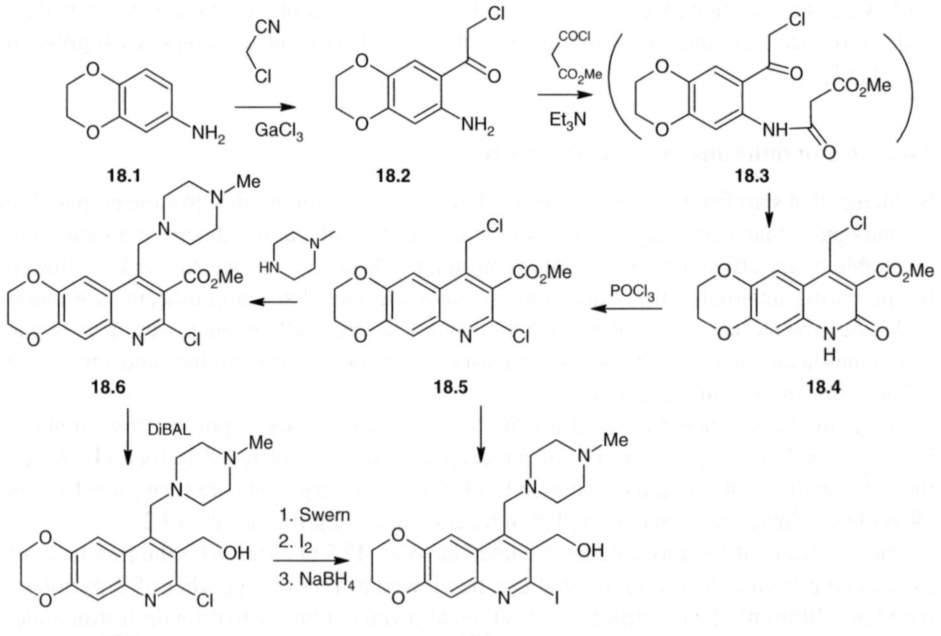

Scheme 4.18   *Lurtotecan B.*

the required chiral center. The stereoselective reaction for introducing the quaternary hydroxyl group starts by applying a chiral version of the Sharpless hydroxylation. The ring unsaturation in (**17.5**) is thus reacted with potassium osmate in the presence of the chiral quinine-based reagent "DHQDPyr." The thus-formed glycol is not isolated since the secondary hydroxyl in the initially formed glycol is oxidized *in situ* by excess

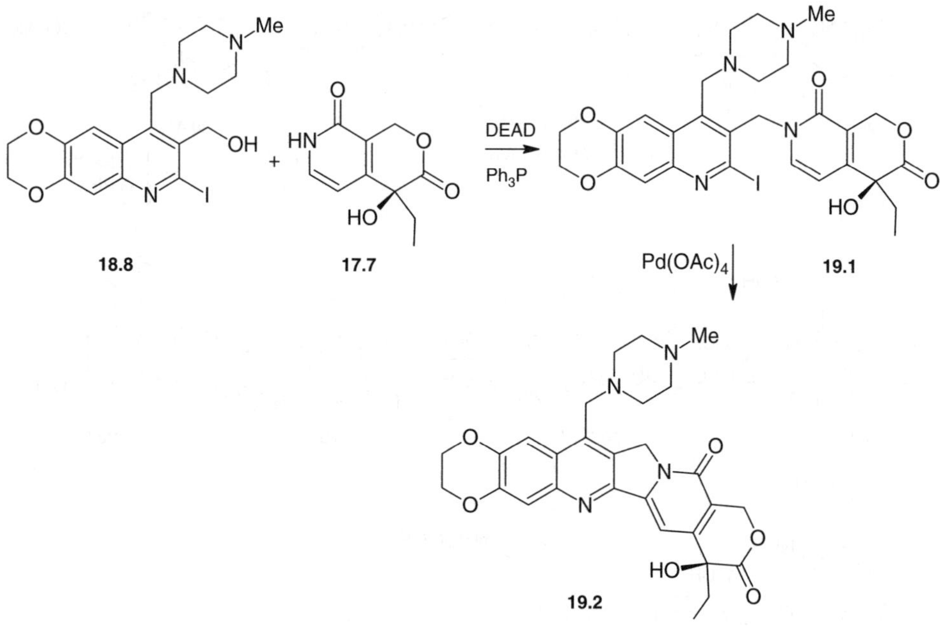

Scheme 4.19 *Lurtotecan C.*

osmate in the reaction mixture. This leads the one of the two moieties that will be united to form lurtotecan (**17.7**).

The construction of the other sizeable parts of lurtotecan starts with Friedel–Crafts acylation of aniline (**18.1**) with chloroacetonitrile. The initially formed product consists of an acyl imine from initial attack on the nitrile. That function hydrolyzes to a carbonyl group (**18.2**) on work-up. A second acylation, this time by methyl chloromethylmalonate onto aniline nitrogen, affords ester (**18.3**). This is not isolated as the base in the reaction medium causes it to undergo Claisen condensation between the methylene group of the malonate and the carbonyl group on the chloroacetyl side chain. Quinolone (**18.4**) is thus obtained. The treatment of this last intermediate with phosphorus oxychloride converts the amide to the corresponding enol chloride (**15.5**). The chlorine atom on the chloromethyl group is then displaced by the secondary amine in *N*-methyl piperidine to afford (**18.6**). The methyl ester on the heterocyclic ring in that last intermediate is next reduced to a carbinol by means of di*iso*butylaluminum hydride (**18.7**). The remaining task involves replacing the ring chlorine by iodine to provide the requisite halide for the coupling reaction that will unite the two moieties that will form the camptothecin analogue. The hydroxyl group in the carbinol is first oxidized to avoid interference in the next step. Halogen interchange then exchanges chlorine by iodine providing the iodomethyl fragment. The aldehyde is then back reduced to the carbinol, providing the other moiety (**18.8**) for building lurtotecan.

The first of the two steps in joining the two previously prepared parts of lurtotecan comprises a relatively straightforward Mitsunobu displacement of the hydroxyl in quinoline (**18.8**) by pyridone nitrogen (**17.7**) in the presence of triphenylphosphine, forming the first linkage between the two moieties. Hecht vinylation of the position adjacent to nitrogen and

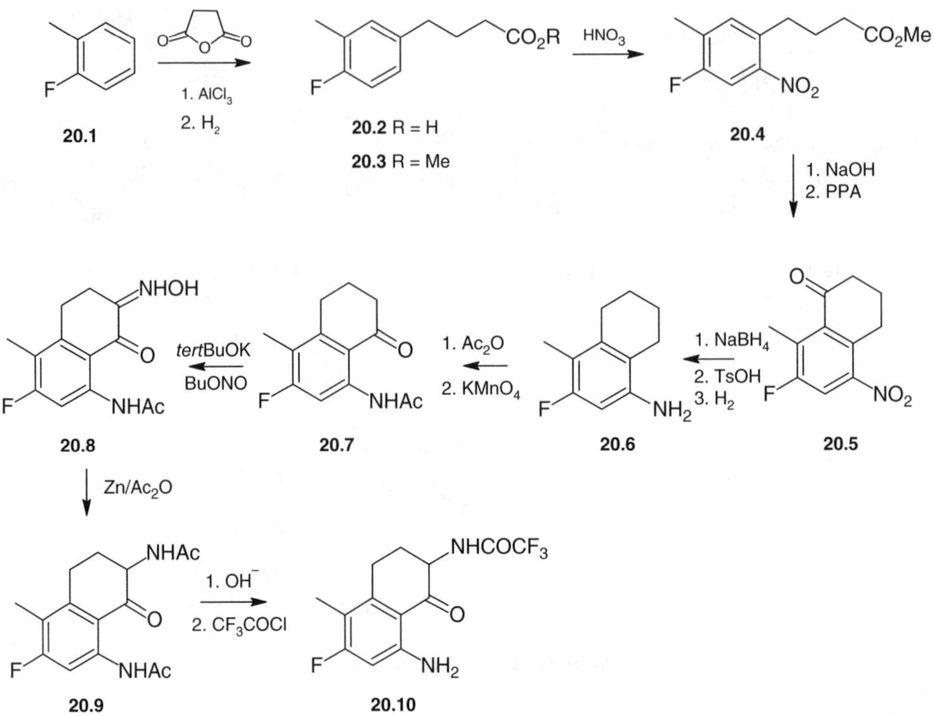

Scheme 4.20   *Exatecan A.*

iodine in the quinoline constituent forms the second link closing the last ring in the process. The topoisomerase inhibitor **lurtotecan** (**19.2**) is thus obtained [17]. The compound essentially failed in Phase II clinical trials conducted a decade or so ago.

The design for the total synthesis of the other relatively water-soluble camptothecin analogue exatecan (**22.1**) also uses a convergent scheme, though the additional fused carbocyclic ring dictates the use of moieties that differ considerably from those used for lurtotecan, as does the route to the individual moieties. The rather lengthy preparation of the fragment that will bear that additional ring starts with Friedel–Crafts acylation of *ortho*-fluorotoluene (**20.1**) with succinic anhydride. Catalytic hydrogenation over palladium of that product serves to reduce the carbonyl to a methylene group (**20.2**). The carboxylic acid is next esterified with methanol (**20.3**). The reaction of that intermediate with nitric acid introduces the nitrogen that will later form the amine in the quinolone ring of the final product. The methyl ester is then saponified and the resulting acid treated with polyphosphoric acid, an alternate catalyst for Friedel–Craft reactions. The next several steps, in essence, move the ring carbonyl group across the carbocyclic ring.

The sequence starts with the reduction of the ketone in (**20.5**) to an alcohol by means of sodium borohydride. Heating that in product the presence of *p*-toluenesulfonic acid cause the alcohol to dehydrate; catalytic hydrogenation then reduces the resulting styrenoid double (**20.6**).

The amine in (**20.6**) is then protected against reagents in subsequent steps by acylation with acetic anhydride. The oxidation of the product with potassium permanganate reintroduces the carbonyl function (**20.7**) across the ring from its former location. (The reason for oxidation at that site is not immediately apparent.) The introduction of the second amine starts by the treatment of the intermediate (**20.7**) with potassium *tert*-butoxide; this introduces an anion at the position adjacent to the ketone. Butyl nitrite then adds to that site in the form of an oxime adjacent to the carbonyl group (**20.8**). That intermediate is then reduce to an amine; acetic anhydride acylates the newly formed amine to an acetamide. Hydrolysis by means of base converts both acetamide groups to revert to primary amines. The treatment of this diamine with trifluoroacetyl chloride proceeds selectively at the more basic aliphatic amine to afford one of the moieties (**20.10**) [18] for building exatecan.

A somewhat briefer sequence is required for the synthesis of the fragment that contains the critical lactone group. The several important gaps in the scheme (indicated by stacked reaction arrows) reflect the difficulty in unearthing relevant references.

The preparation of the second synthon starts with the condensation of pyrimidone (**21.1**) with methyl acrylate. The scheme by which the extra ring is added can be rationalized by assuming that the first step involves the formation of an anion on pyrimidone nitrogen. This charged species would add to the end of the acrylate. The Claisen condensation of the resulting propionate methylene with the ester on the pyrimidone will form the fused ring. The surplus carbethoxy group will then be decarboxylated to form the fused cyclopentanone (**21.2**). The carbonyl group is next protected as an acetal by reaction with ethylene glycol (**21.3**). The reaction of the methyl group on the pyrimidone with ethyl carbonate in the presence of potassium *tert*-butoxide adds a carbethoxy function to the methyl group. The next few steps comprise the conversion of the nitrile to a hydroxymethyl group and

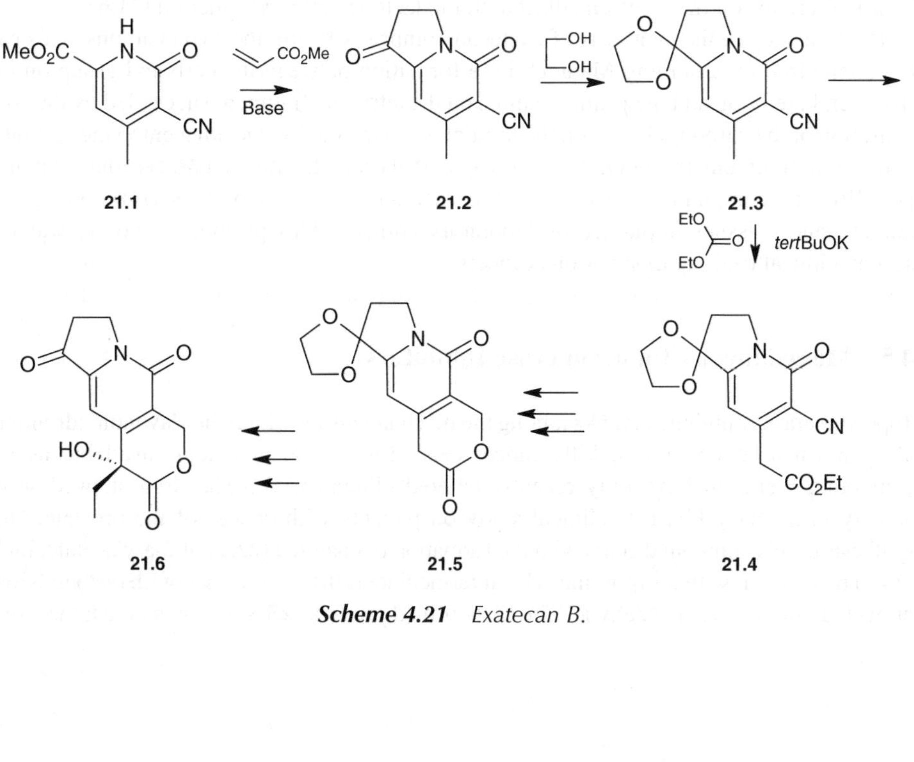

Scheme 4.21 *Exatecan B.*

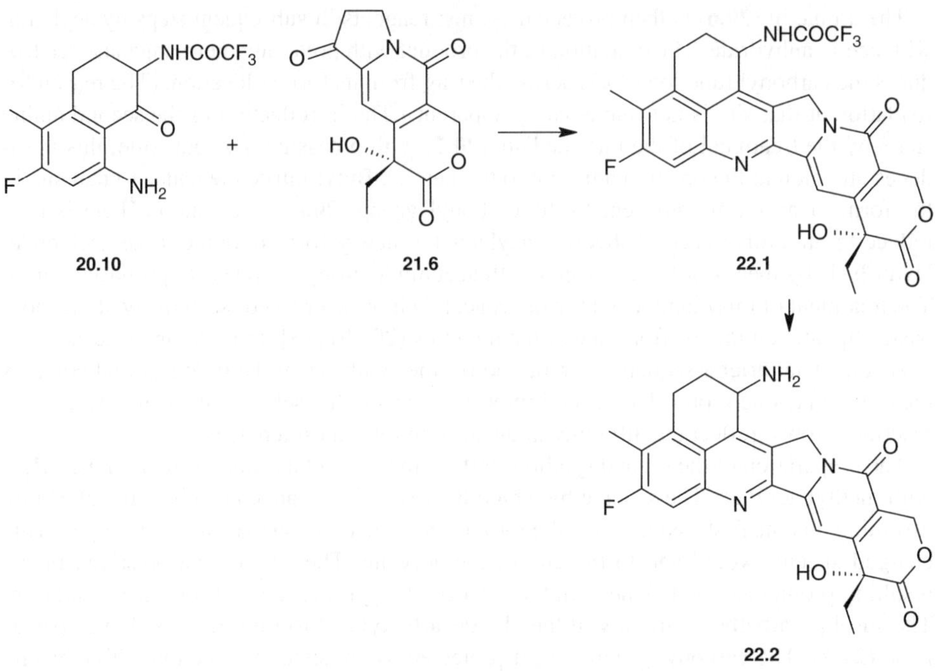

Scheme 4.22   *Exatecan C.*

closing the lactone (**21.5**). This would then be followed by the installation of the ethyl and hydroxyl groups on the position adjacent to the lactone carbonyl function (**21.6**).

The final step in the synthesis of exatecan comprises fusing the two synthons. This can be rationalized by assuming Mannich-base formation between the carbonyl group on the five-membered ring and the primary amine on the tetralone fragment succeeded by the condensation of the carbonyl group on the tetralone and the methylene adjacent to the carbonyl in the other fragment, though not necessarily in that order. **Exatecan (22.2)** is thus obtained [18, 19]. The compound showed potent antineoplastic activity both *in vitro* and against human cancer tissues implanted on laboratory animals. This promising activity was not seen in clinical trials against various cancers.

4.5   Miscellaneous Topoisomerase Inhibitors

Topoisomerase II inhibition ranks among the mechanisms by which the flavonoid idronoxil, also known as phenoxodiol, kills cancer cells. Little clinical data is available as the compound seems to have only recently entered clinical trials. The drug showed some activity in an early Phase II clinical study on patients with cancer of the prostate. One synthesis of this compound starts with the formation of enamine (**23.2**) of 2-arylacetaldehyde (**23.1**) by reaction with morpholine. This intermediate is then condensed with *tert*butylsilyl-protected 5-hydroxyanisaldehyde. The first-formed product (**23.3**) is not isolated; the oxide

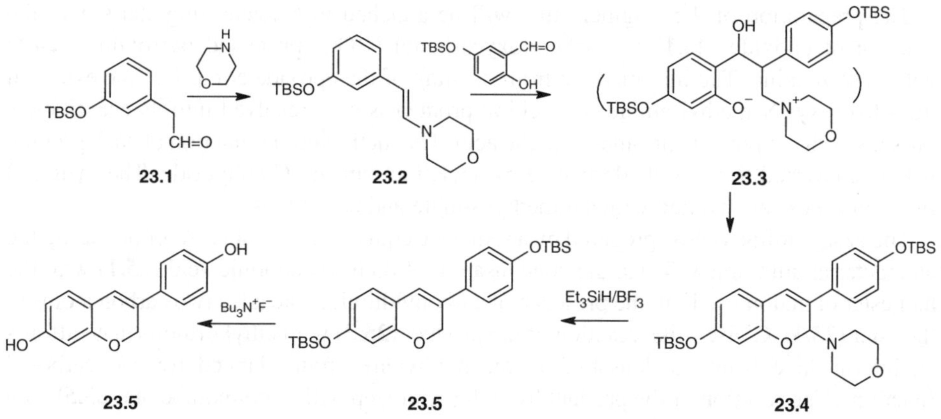

Scheme 4.23 *Idronoxil.*

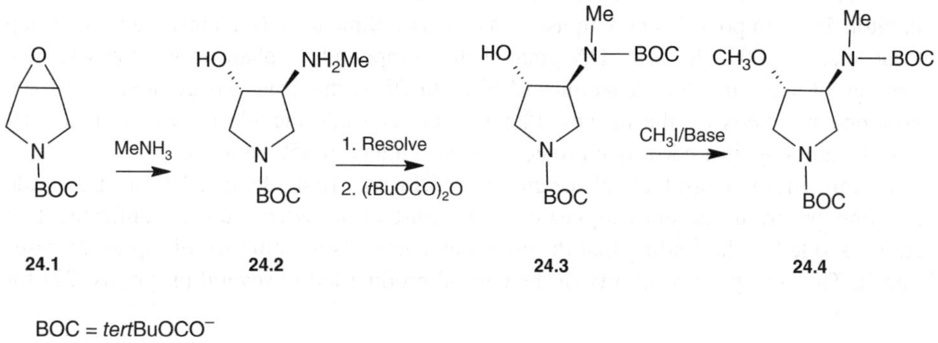

BOC = *tert*BuOCO⁻

Scheme 4.24 *Vosaroxin A.*

anion in the product immediately attacks the methylene group adjacent to the morpholine closing the ring. The hydroxyl group in the initial product dehydrates under reaction conditions so as to build flavonoid (**23.4**). The next step takes advantage of the fact that the carbon bearing the morpholine comprises a hemiaminal. The reaction of the intermediate with triethylsilane in the presence of boron tribromide formally reduces the ring by causing the ejection of the morpholine. The treatment of this last intermediate with tetrabutylammonium fluoride causes the protecting groups to leave, affording **idronoxil** (**23.5**) [20].

It is of interest, in connection with the topoisomerase inhibitor vosaroxin, that bacterial replication is in many ways analogous to that of more complex organisms. The replication of those one-celled organisms is catalyzed by an enzyme called DNA gyrase—a counterpart of the enzyme that serves the same function as do the topoisomerases in more complex organisms. The antibiotic activity of members of the large list of antibacterial quinolones is due to the inhibition of bacterial DNA gyrase. The structure of vosaroxin (**25.7**) could at first sight be mistaken as one of the quinolone antibiotics. The scheme for the synthesis of vosaroxin also involves a convergent design.

The preparation of the fragment that will be attached to benzene ring starts with the reaction of epoxide (**24.1**) in *tert*butyloxycarbonyl (BOC)-protected pyrrolidine (**24.1**) with methylamine. The intermediate from opening of the epoxide comprises the expected trans hydroxyl N-methyl amine. This chiral product is next resolved into its enantiomers via salts with the pure enantiomer of malic acid. The methylamine function of each product is next converted to its BOC derivative by acylation with BOC anhydride. The hydroxyl function is next methylated with a dimethyl sulfate and base (**24.4**).

The azaquinoline is next prepared by a sequence equivalent to that used for preparing the antibacterial quinolones. Thus, the condensation of dichloronicotinic acid (**25.1**) with the half ester of malonic acid in the presence of carbonyldiimidazole adds two carbon atoms to the ester (**25.3**). Claisen-like reaction of the product (**25.3**) with ethyl orthoformate affords an intermediate from condensation at the methylene group flanked by two carbonyl functions. The reaction of the product from the prior step with 1-aminothiazole (**25.5**) then closes the ring (**25.6**). The initial step in this transform may involve either displacement of the ring chlorine or conjugate addition to the formyl group. The displacement of chlorine by the ring nitrogen in the chiral pyrrole (**25.7**) affords the drug **vosaroxin** (**25.7**) [21]. *In vitro* studies showed that the drug intercalates in DNA and also acts as a topoisomerase inhibitor. The compound is now in use in a series of clinical trials against a diverse group of cancers. In 2009, the US FDA granted the compound "orphan drug" status for the treatment of acute myeloid leukemias (AML). In 2011, the drug was granted "fast track development" status by the agency. Planned further trials include a sizeable Phase III clinical studies against AML that projects the enrollment of 450 patients.

The fermentation product rebeccamycin (**26.1**) was first identified in a large-scale screening program for antineoplastic drugs. Further *in vitro* studies confirmed that activity and led to the finding that the compound acted as an inhibitor of topoisomerases I and II. The very poor solubility of the natural product led to several programs [22] for

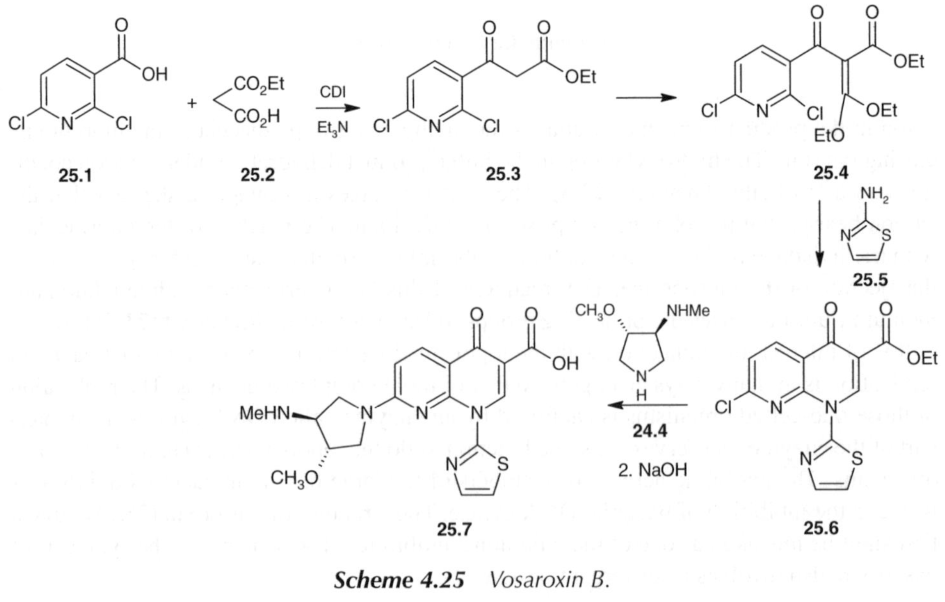

Scheme 4.25 *Vosaroxin B.*

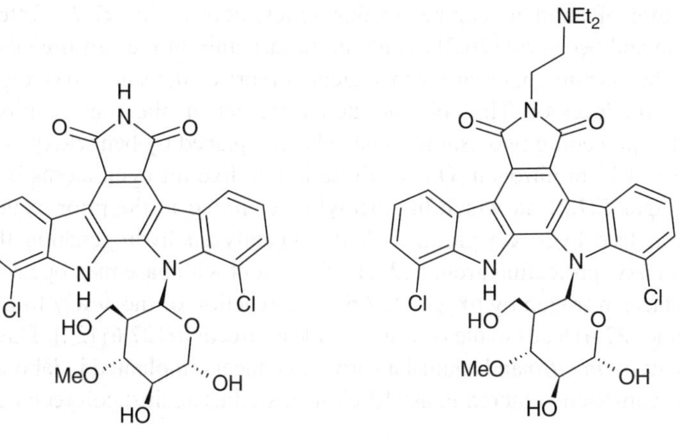

Scheme 4.26 *Becaterin.*

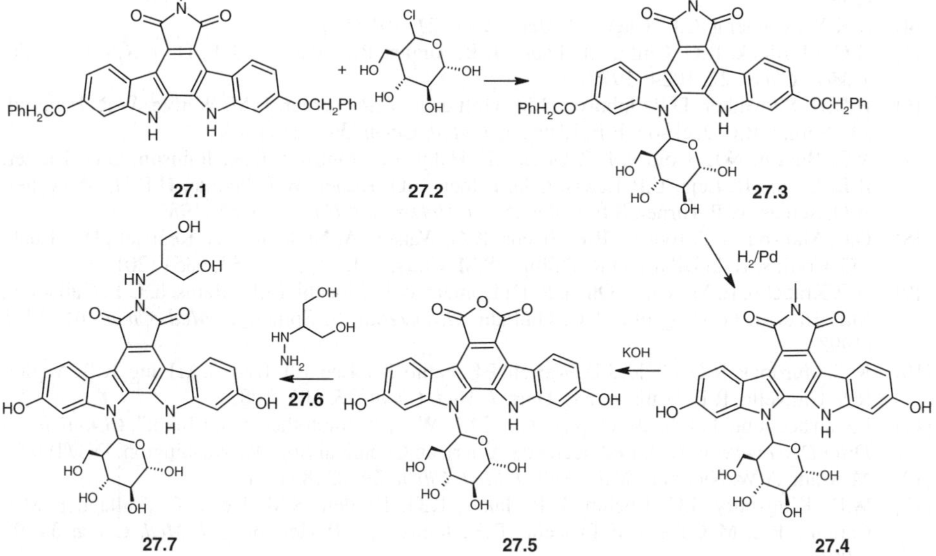

Scheme 4.27 *Edotecarin.*

modifying the structure of the lead in order to identify the derivatives that get around that stumbling block.

The incorporation of an ethyldiethylamino side chain on the imide nitrogen of rebecca-mycin (**26.4**) led to the water-soluble analogue **becaterin**. This drug has been used in Phase III clinical trials. In 2009, the drug was granted the designation of orphan status for the treatment of cancers of the biliary tree.

The structure of another member in this series, edotecarin (**27.7**), differs from both rebeccamycin and becaterin (**26.2**) mainly in the structure of the substituent on the central imide ring. The starting material for this agent comprises the same five-ring system, as is found in the predecessor. The chlorine atoms present on the ring position adjacent to nitrogen in the preceding two examples have been replaced by benzyloxy group one atom further removed from nitrogen. One of those indole-like nitrogen atoms is then glycosidated with sugar (**27.2**), in which the methyl ether found in the prior example has been replaced by a free hydroxyl group. Palladium-catalyzed hydrogenation then serves to remove the benzyl protecting groups (**27.4**). Treatment with base then opens the imide and in effect replaces nitrogen by oxygen (**27.5**). The reaction of the newly formed anhydride with hydrazine (**27.6**) leads to the N-amino imide **edotecarin** (**27.6**) [23]. This drug showed very good activity *in vitro* and against an array of cancers implanted in laboratory animals. In 2008, this compound entered Phase III clinical studies against colorectal cancer.

References

[1] Y.S. Rho, H.K. Ko, H. Sin, D.J. Yoo, *Bull. Korean Chem. Soc.* **20**, 1517 (1999).
[2] K. Ishizumi, N. Ohashi, M. Muramatso, U.S. Patent 4,540,695 (1985).
[3] K. Ishizumi, N. Ohashi, N. Tanno, H. Sato, M. Fukui, S. Morisda, U.S. Patent 4,673,668 (1987).
[4] R.K.Y. Zeecheng, C.C. Cheng, *J. Med. Chem.* **21**, 191 (1978).
[5] K.C. Murdock, R.G. Child, P.F. Fabio, R.B. Angier, R.E, Wallace, F.E. Durr, R.V. Citavella, *J. Med. Chem.* **22**, 1024 (1979).
[6] H.D.H. Showalter, D.R. Johnson, J.M. Hoftiezer, W.R. Turner, L.M. Werble, W.R. Leopold, J.L. Shillis, R.C. Jackson, E.E. Elslager, *J. Med. Chem.* **30**, 121 (1987).
[7] V.G. Beylin, N.L. Colbry, L.P. Goel, J.E. Haky, J.E. Johnson, D.R. Johnson, G.D. Kanter, R.L. Leeds, B. Leja, E.P. Lewis, C.D. Ritner, C.D. Ritner, W.T. People, H.D.H. Showalter, A.O. Serces, W.P. Turner, S.E. Uhlendorf, *J. Heterocycl. Chem.* **26**, 85 (1989).
[8] O.C. Mansour, B.J. Evison, B.E. Sleebs, K.G. Watson, A. Nudelman, A. Rephaeli, D.P. Buck, J.G. Collins, R.A. Bilard, D.R. Phillips, S.M. Cutts, *J. Med. Chem.* **53**, 6851 (2010).
[9] A.P. Krapcho, E. Menta, A. Oliva, R. Di Domenico, L. Fiocchi, M.E. Maresch, C.E. Gallagher, M.P. Hacker, G. Beggiolin, F.C. Giuliani, G. Pezzoni, S. Spinell, *J. Med. Chem.* **41**, 5431 (1998).
[10] K.C. Murdock, R.G. Child, J.D. Warren, P.E. Fabio, V.J. Lee, P.T. Izzo, S.A. Lang, R.B. Angier, R.V. Citavella, R.E. Wallace, F.E. Durr, *J. Med. Chem.* **25**, 505 (1982).
[11] For an accounted of the development *see* M.E. Wall, "Camptothecin and Taxol", *Chronicles of Drug Discovery, V. 3*, D. Lednicer, ed., American Chemical Society, Washington, DC (1993).
[12] M. Wani, A.W. Nicholas, M.E. Wall, *J. Med. Chem.* **29**, 2358 (1986).
[13] W.D. Kingsbury, J.C. Boehm, D.R. Jakas, K.G. Holden, S.M. Hecht G. Gallagher, M.J. Caranfa, F.L. McCabe, L.F. Faucette, R.K. Johnson, R.P. Hertzberg, *J. Med. Chem.* **34**, 98 (1991).
[14] S. Hegde, M. Schmidt, *Ann. Reports Med. Chem. Ann. Reports Med. Chem.* **40**, 577 (2004).
[15] K. Narkunan, X. Chen, H. Kochat, F. Hausheer, U.S. Patent 7,687,496 (2010).
[16] J.L. Wood, J.M. Fortunak, A.R. Mastrocola, M. Mellinger, P.L. Burk, *J. Org. Chem.* **60**, 5739 (1995).
[17] F.G. Fang, D.D. Bankston, E.M. Huie, M.R. Johnson, C.S. LeHoulier, C. Lewis, C. Lovelace, M.W. Lowery, *Tetrahedron* **53**, 10953 (1997).
[18] M. Sugimori, A. Ejima, S. Ohsuki, K. Uoto, I. Mitsui, K. Matsumoto, Y. Kawato, M. Kasuoka, K. Sato, H. Tagawa, H. Terasawa, *J. Med. Chem.* **37**, 3038 (1954).
[19] M. Sugimori, A. Ejima, S. Ohsuki, K. Uoto, I. Mitsui, Y. Kawato, Y. Hirota, K. Sato, H. Terasawa, *J. Med. Chem.* **41**, 2308 (1998).

[20] A.J. Tilleya, S.D. Zanattaa, C.X. Qinb, I.-K. Kimc, Y.-M. Seokc, A. Stewart, O.L. Woodmand, S.J. Williams, *Bioorg. Med. Chem.* **20**, 2353 (2012).

[21] A. Sudhaka, T. Subramani, M. Sheik, M.M. Rahuman, R. Subbiah, U.S. Patent Application 2010/0,203,162 (2010).

[22] *See also* E.R. Pereira, L. Belin, M. Sancelme, M. Prudhomme, M. Ollier, M.R.D. Sevère, J.-F. Riou, D. Fabbro, T. Meyer, *J. Med. Chem.* **22**, 4471 (1996) and F. Anizon, P. Moreau, M. Sancelme, W. Laine, C. Bailly, M. Prudhomme, *Bioorg. Med. Chem.* **11**, 3709 (2003).

[23] M. Ohkubo, T. Nishimura, H. Kawamoto, M. Nakano,T. Honma, T. Yoshinari, H. Arakawa, H. Suda, H. Morishima, *Bioorg. Med. Chem. Lett.* **10**, 419 (2000).

5

Mitotic Inhibitors

5.1 Introduction

Cell division is the essential process for the survival of the individual and in a longer view persistence of the species. This process has been the focus of many studies leading to today's detailed image of cell division. The following discussion only skims the surface of this crucial life process. Every cell in a live organism is provided with a microskeleton of fine tubular strands composed of the polypeptides tubulin α and tubulin β. These microtubules are responsible for all movements within a cell. Movement is particularly important at the time when the cell is about to replicate by splitting into two identical halves, sometimes called daughter cells. After the genes have duplicated, microtubules radiate from structures known as mitotic spindles. These attach to one each of the by-now duplicated genes. The tubules then contract physically drawing each of the genes toward one of the pair of spindles from which the microtubules originate. The microtubules then relax and break their bond to the genes. Antineoplastic inhibitors interfere with this process and bring mitosis and thus replication to a halt. Tubulin inhibitors interfere with this process by several different mechanisms that depend on which receptors are affected. Paclitaxel more familiarly known as Taxol® binds to a special receptor. On binding, taxanes stabilize the microtubule-attached components of mitosis. This freezes the process short of normal completion, halting cell division. Most of the other mitotic inhibitors bind to receptors characterized as either vinca alkaloid sites or colchicine sites. A detailed discussion of consequences of binding to one or the other is beyond the scope of this work. The vinca alkaloids, which are said to be the second most widely used class of anticancer drugs, are given short shrift in these pages. These drugs, vincristine and vinblastine, are natural products with complex chemical structures. There has consequently been relatively little if any synthetic activity reported in connection with these drugs.

Antineoplastic Drugs: Organic Synthesis, First Edition. Daniel Lednicer.
© 2015 John Wiley & Sons, Ltd. Published 2015 by John Wiley & Sons, Ltd.

5.2   Taxanes

Monroe Wall, the scientist who discovered camptothecin, followed that achievement by the discovery of the drug paclitaxel, more widely known among chemists as Taxol® (**1.6**). This compound like camptothecin lay fallow until it was found to kill neoplastic cells by freezing the tubulin-mediated cell replication short of completion. The occurrence of the original natural product, named Taxol at the time, seemed to be confined to the bark of the Pacific yew, *Taxus brevifolia*, leading to half-serious concerns about the survival of that plant species. A taxane lacking the long side chain, baccatin III (**1.1**), was found to occur in good supply in needles of some of the more common yews including those used to adorn many dwellings. The chemistry used to convert baccatin to paclitaxel opened the way for the preparation of analogues. The complex structure of the taxanes with its multiplicity of chiral centers is in some ways reminiscent of corticosteroids. As is the case in steroids, modified analogues are built on the framework of taxanes obtained from yews. All analogues that have been accorded nonproprietary names comprise modifications on the isoserine side chain. The first three members of the quartet have been approved for treating cancer by regulatory agencies.

A converging synthesis is used to carry out the conversion of baccatin III to paclitaxel. The principal structural change required for that conversion comprises the incorporation of the isoserine side chain at position 13 in baccatin III (**1.1**).

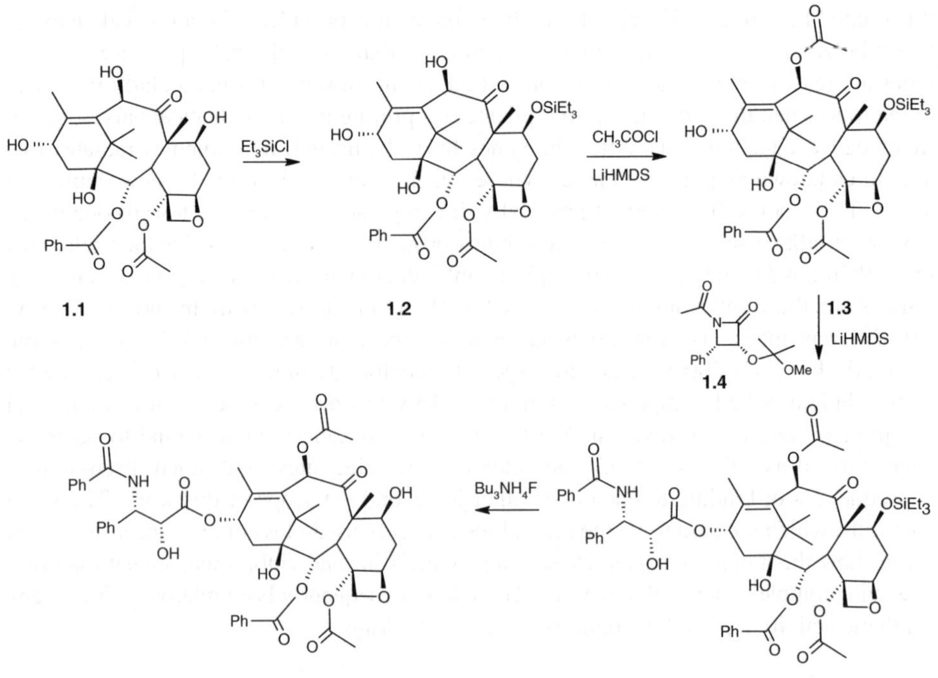

Scheme 5.1   *Paclitaxel from baccatin III.*

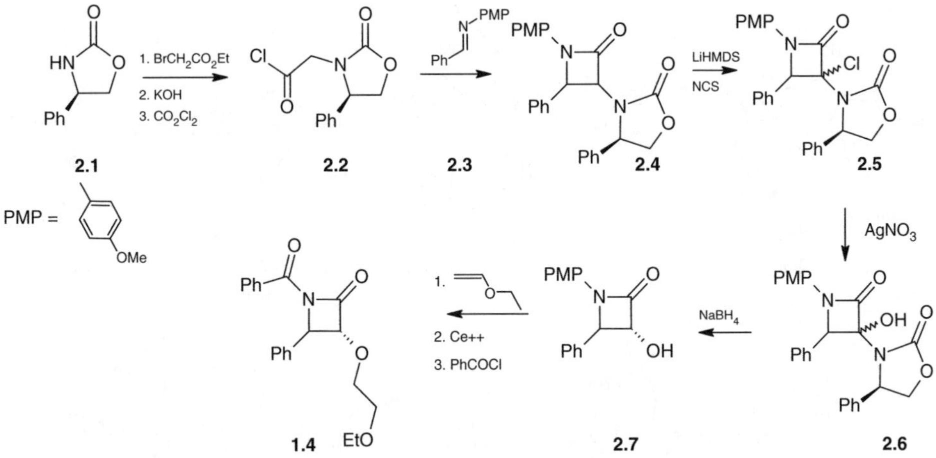

Scheme 5.2 *Synthesis of β-lactam.*

The scheme for preparing the proper enantiomer of all-important β-lactam (**1.4**) depends on carrying a chiral auxiliary better than halfway through the synthesis. The scheme starts with the alkylation of the chiral oxazolidone (**2.1**) with ethyl bromoacetate; the ester is then saponified, and the resulting acid converted to the acid chloride by means of oxalyl chloride (**2.2**). That acid chloride is then converted to the corresponding ketene by treatment with triethylamine. The condensation of that intermediate with the p-methoxy phenyl (PMP) protected imine from benzaldehyde (**2.3**) leads to the formation of β-lactam by a 2+2 cyclo-addition. The anion from the reaction of the product (**2.4**) with the strong base lithium hexa-methyldisilazide (LiHMDS) is then treated with N-chlorosuccinimide affording a mixture of the halogenated derivatives (**2.5**). Silver nitrate then replaces chlorine with a hydroxyl group; that carbon has been converted to a hemiaminal in the process. The latent carbonyl group is next reduced to an alcohol with sodium borohydride. That newly formed hydroxyl group is then protected as an ethoxyethyl ether by reaction with ethoxy vinyl ether [1].

The synthesis of paclitaxel proper begins by covering the most reactive hydroxyl at position 7. Thus, the treatment of baccatin with triethylsilyl chloride affords the corresponding triethylsilyl ether (**1.2**). The acetylation of the alkoxide from the reaction of the last product with LiHMDS followed by acetyl chloride affords the acetate of the next most reactive hydroxyl group (**1.3**).

The addition of the important isoserine fragment starts with the removal of the proton on the alcohol at C17 with strong base; the reaction of the resulting alkoxide with β-lactam (**1.4**) results first in addition to the carbonyl lactam. The lactam ring opens effectively incorporating the isoserine side chain; ring strain facilitates the opening of the amide bond. The removal of the silyl ether by fluoride ion completes the conversion to **paclitaxel** (**1.6**) [2].

Docetaxel is a well-established tubulin inhibitor that acts by binding to tubules. This antineoplastic compound has been and is still used for treating a selection of solid tumors including breast, ovarian, and non-small cell lung cancer. The patent having expired, that drug is now generic and marketed under several names including Docetaxel®, Taxotere®, and Docefrez®.

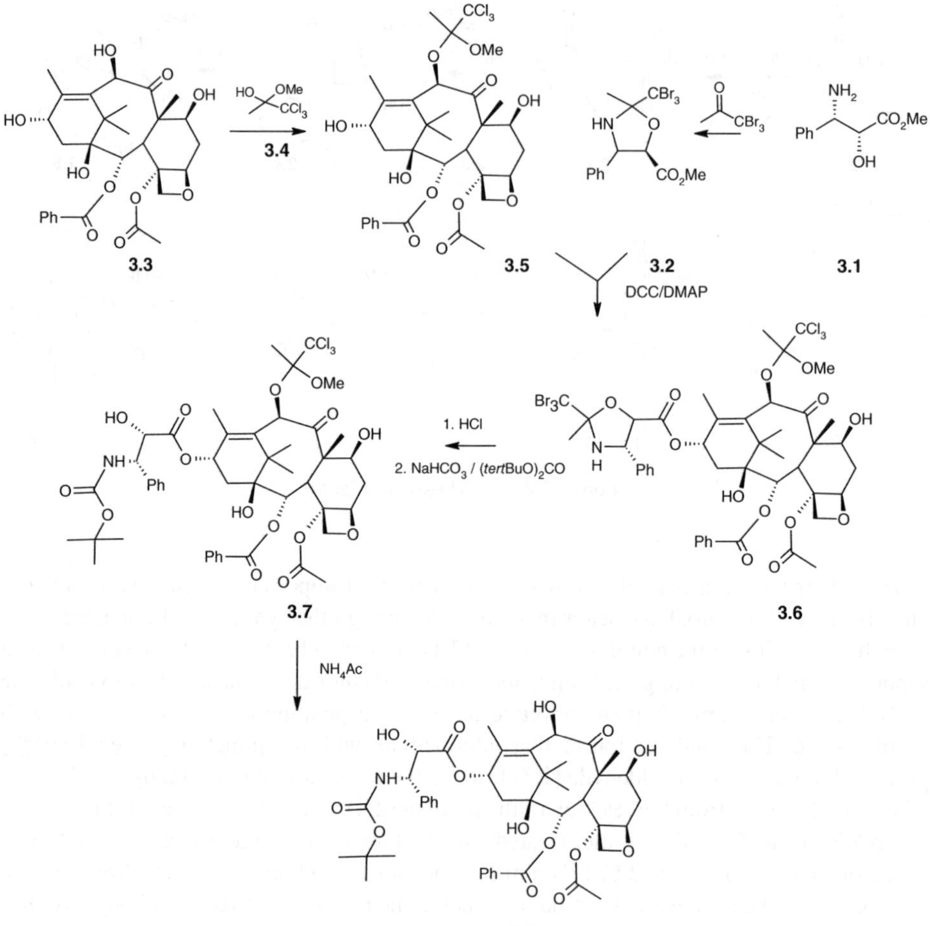

Scheme 5.3   *Docetaxel.*

The preparation of docetaxel on commercial scale from desacetyl baccatin III requires a rather lengthy convergent scheme. The synthesis starts with the preparation of a protected version of the larger part of the fragment that will be part of the isoserine side chain at C13. The reaction of a single enantiomer of aminoalcohol (**3.1**) with tri-bromoacetone affords oxazoline (**3.2**). The next step comprises masking the reactive hydroxyl at C7 in desacetyl baccatin III. That compound (**3.3**) is thus treated with the methyl acetal of trichloroacetone (**3.4**); hemiacetal interchange results in the transfer of the chloroacetone to the taxane affording the hemiacetal (**3.5**). Dicyclohexylcarbodiimide (DCC)-mediated condensation in the presence of dimethylaminopyridine (DMAP) adds the previously prepared carbethoxy oxazoline that will constitute the side chain at C13 (**3.6**). The reaction of that intermediate with acid opens the oxazoline and expels the bromoacetone. The crude product from that reaction is treated in turn with sodium bicarbonate and then bis(*tert*-butoxy)carbonate to yield the tributyloxy derivative (**3.7**).

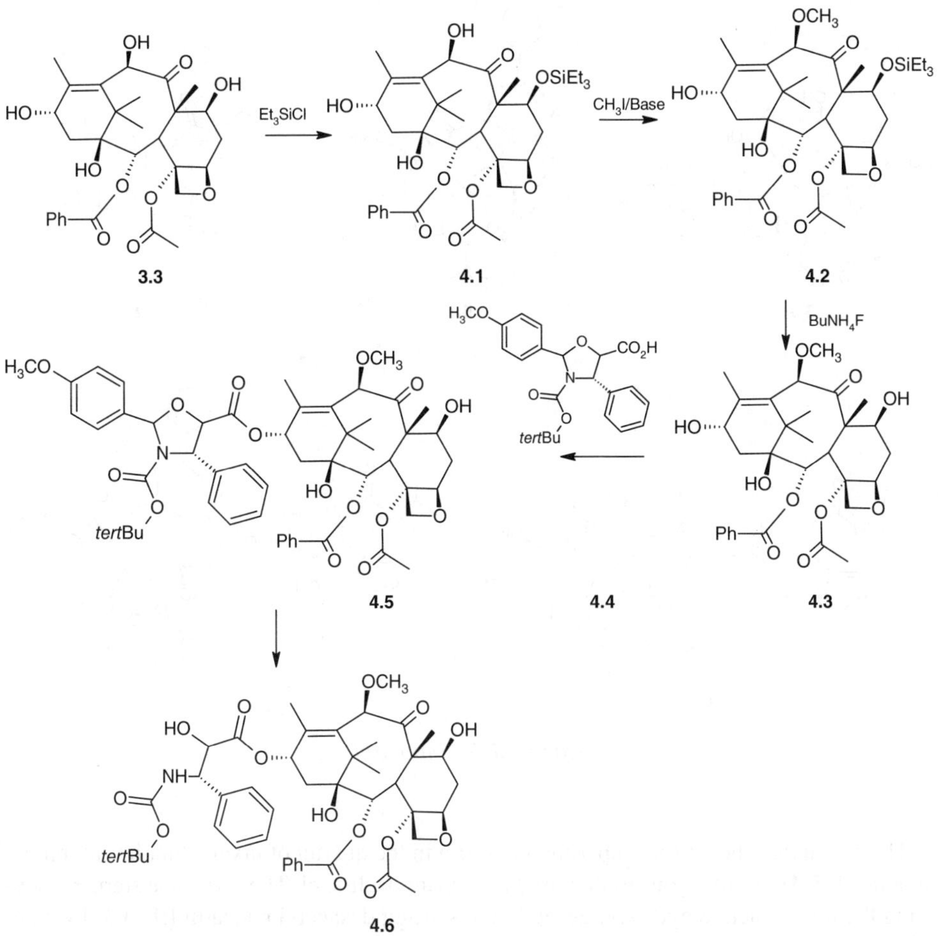

Scheme 5.4 *Cabazitaxel.*

The last protective group, the oxaminal linkage at C7, is next removed by treatment with ammonium acetate, extruding benzaldehyde in the process. The taxane **docetaxel** (**3.8**) is thus obtained [3].

The synthesis of a related taxane, cabazitaxel (**4.5**), also uses desacetyl baccatin III as starting material. The preparation of this analogue starts by masking the most reactive hydroxyl group by reaction with triethylsilyl chloride. The treatment of the product with base and methyl iodide or another source of a methyl group converts the hydroxyl group at C10 to its methyl ether (**4.2**). The reaction of that intermediate with a source of fluoride removes the masking silyl group. The addition of the side chain proceeds much as in the case of docetaxel by coupling the isoserine-like fragment (**4.4**) in the presence of DCC and DMAP. The treatment of this intermediate with ammonium acetate breaks up the oxaminal linkage extruding 4-methoxybenzaldehyde. This then yields the antineoplastic agent **cabazitaxel** (**4.6**) [4], under the trade name Jevtana®.

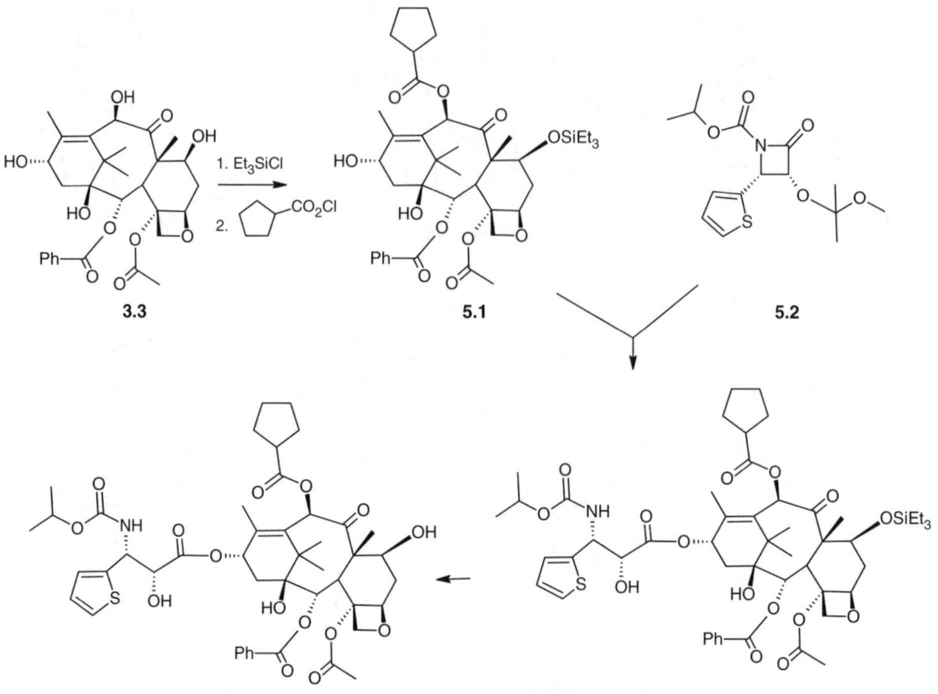

Scheme 5.5   *Simotaxel.*

The synthetic scheme for preparing the fourth in the quartet of taxane tubulin inhibitors, simotaxel (**5.4**), is analogous to that used to prepare paclitaxel. The sequence starts by covering the most reactive hydroxyl group in the starting desacetyl baccatin III (**3.3**) by reaction with triethylsilyl chloride. The acylation with cyclopentylcarbonyl chloride affords the corresponding ester (**5.1**). The β-lactam in which thiophene (**5.2**) replaces the phenyl group found in the paclitaxel synthesis is prepared in a convergent set of reactions. The treatment of taxane (**5.1**) with strong base leads to the formation of the alkoxide at the remaining open hydroxyl group. The reaction of that intermediate with β-lactam (**5.2**) then attaches the modified isoserine side chain. Cleavage of the silyl ether with fluoride ion affords **simotaxel** (**5.4**) [4]. Clinical trials are said to have been halted.

5.3   Wholly Synthetic Compounds

A sizeable group of compounds with widely varying chemical structures owe their antineoplastic activity to the inhibition of the polymerization of tubulin for building microtubules. Most of these drugs bind to tubulin and thus inactivate the substance. The end point of that process may comprise inhibition of the creation of new blood vessels in solid tumors or halting the growth of ovarian cancer cells.

5.3.1 Carbocyclic Compounds

A structurally rather simple compound, batabulin proved to bind covalently to tubulin, disrupting its polymerization. This ends in the collapse of the cytoskeleton killing the cell. The reaction of pentafluorosulfanyl chloride with the substituted aniline (**6.2**) affords **batabulin** (**6.3**) [5].

Combretastatins are a series of naturally occurring compounds found in the African willow bush. One of those, combretastatin A4 (**7.4**), is an especially active member of the series that binds to tubulin and is markedly cytotoxic. The phosphate group attached to the free phenol group improves the pharmacodynamic properties of this compound. In 2009, the FDA granted that drug fast track designation. The as yet to be approved agent bears the trade name Zybrestat®. The preparation of this compound opens with the conversion of benzylic bromide (**7.1**) to the phosphonium salt by reaction with triphenylphosphine. This salt is then converted to its ylide by treatment with strong base. That transient intermediate is then allowed to react with the benzaldehyde under the low salt conditions that favor the formation of *cis* double bonds. The product from that reaction (**7.4**) is next converted to its covalent phosphate by reaction with bis-butoxy phosphate. The butoxy group cleaves off during work-up to afford **fosbretabulin** (**7.5**) [6, 7].

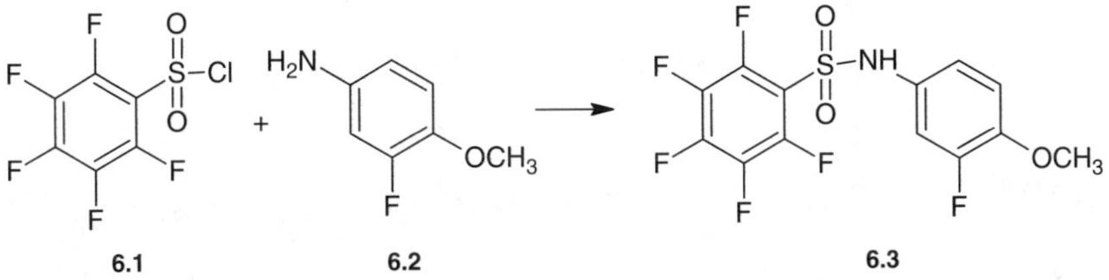

Scheme 5.6 *Batabulin.*

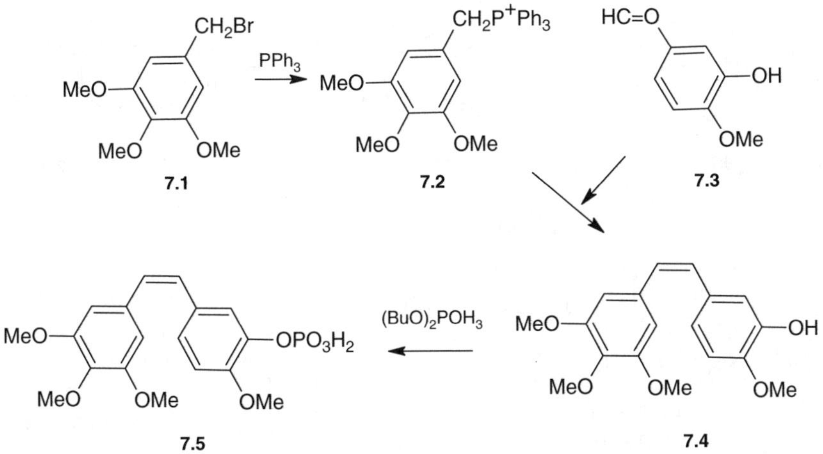

Scheme 5.7 *Fosbretabulin.*

5.3.2 Peptide-Like Inhibitors

The dolastatins are a group of cytotoxic agents isolated from marine sea hares. Several members of this group, such as dolastatin 15, display potent antimitotic activity. It has been demonstrated that the cytotoxic activity of dolastatins traces back to the circumstance that these agents bind to tubulin. Bound tubulin cannot polymerize to form the microfibrils that are crucial for cell division.

Tasidotin (**5.8**) is a synthetic analogue of one of the marine natural products, dolastatin 15. This agent is prepared by the use of solid-phase synthesis. A typical synthesis begins by attaching the last member of the target chain—in this case a peptide—to a solid support by means of a covalent bond. The bonds should be inert to the reaction condition down the line but scissile to a unique set of reaction conditions. The supported fragment is then allowed to react in turn to each amino in the sequence with washes between steps. Functional groups that might interfere with the chain-lengthening reactions are covered as needed by protecting groups. The product is then released from the

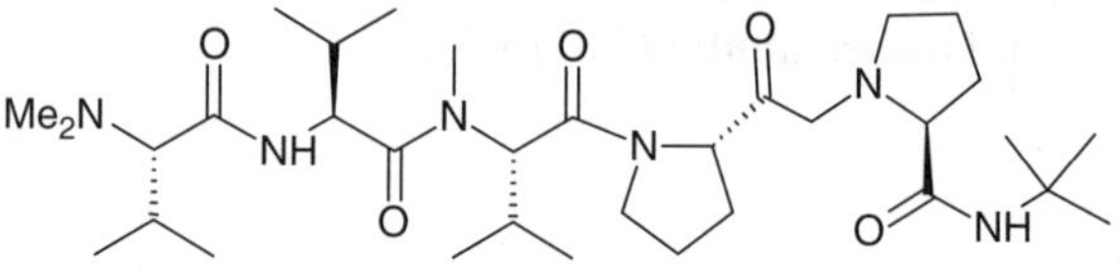

Scheme 5.8 *Tasidotin.*

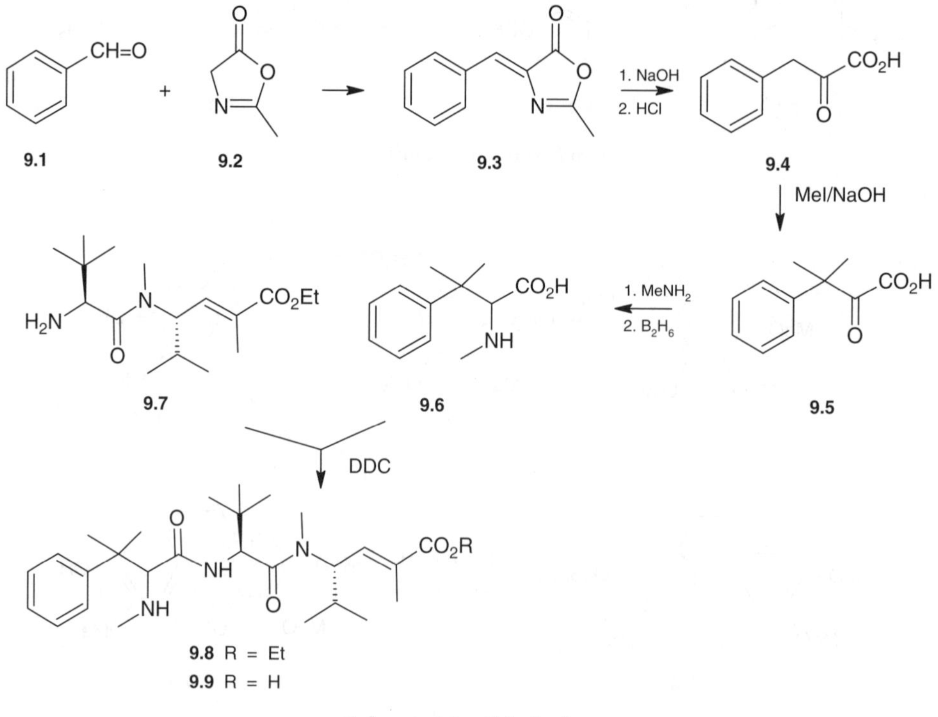

Scheme 5.9 *Taltobulin.*

support and purified. In the case at hand, this last step yields **Tasidotin** (**5.8**). This tubulin-binding agent has shown *in vitro* activity against a wide selection of cancers. The agent entered clinical trials in 2006.

Taltobulin is yet another peptide that bears a distant relation to the dolastatins. This inhibitor of tubulin polymerization has been granted orphan drug status by both the US FDA and the European Community for treating soft tissue sarcomas and ovarian cancer. The first step in the synthesis comprises base-catalyzed condensation of benzaldehyde with azlactone (**9.2**). Consecutive treatment with acid and base opens the azlactone ring to yield ketoacid (**9.4**). Exhaustive methylation with methyl iodide and base leads to the corresponding gem-dimethyl derivative (**9.5**). The reaction of the carbonyl group in this intermediate with methylamine leads to the Mannich base; this is reduced to amine (**9.6**) as a pair of racemates. The reaction of this last product with chirally pure dipeptide (**9.7**) in the presence of dicyclohexylcarbodiimide affords amide (**9.8**). Saponification then affords **taltobulin** (**9.9**) as a pair of enantiomers [9].

5.3.3 Monocyclic Heterocyclic Inhibitors

The acetal **tubulazole** (**10.1**) was one of the earlier synthetic molecules selected for further testing on the basis of tubulin-inhibiting activity. More specifically, this agent blocked the polymerization of tubulin required for forming the cellular microskeleton as well as the microtubules involved in mitosis. The presence of the 1,4-dichlorobenzene ring as well as the "azole" ending of the nonproprietary name suggests that the compound may have originated in a program intended to discover new antifungal agents.

The analogue in which the halogenated benzene ring is replaced by an anisole has much the same tubulin-inhibiting activity as the forerunner. In addition to that effect on tubulin, erbuzole also acts as a radiosensitizer for X-ray antitumor therapy.

The straightforward scheme for preparing erbuzole begins with the formation of the starting material by alkylating imidazole an anisyl bromobenzoate. This is followed by reaction with glycerol to form the glyceryl 1,3-acetal from benzoyl ketone (**11.1**). It should be noted that the product will consist of a pair of diastereomers since this reaction introduces a pair of new asymmetric centers. In order to separate the resulting enantiomers, the mixture of products from acetal-forming reaction is converted to the corresponding benzoyl esters. The desired still-racemic enantiomer is next separated from the mixture. The benzoyl

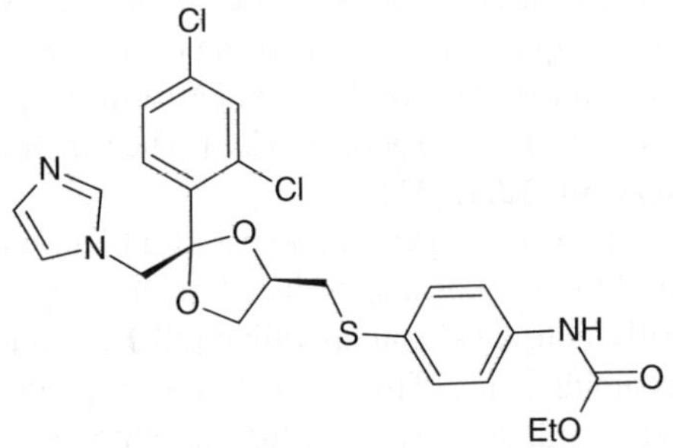

10.1

Scheme 5.10 *Tubulazole.*

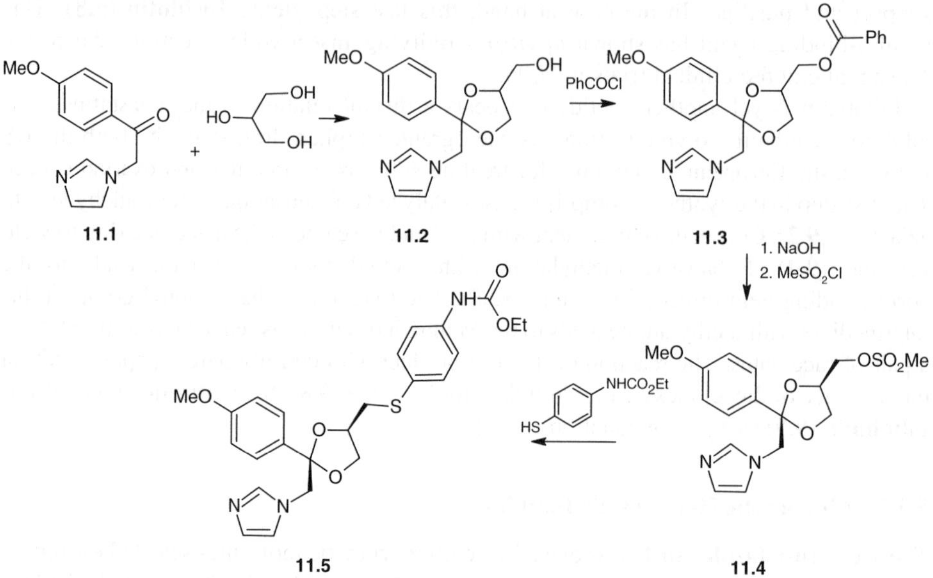

Scheme 5.11 *Erbuzole.*

ester is then replaced by a better leaving group. Thus, saponification of the now-separated benzoyl ester followed by reaction of the resulting carbinol with methanesulfonyl chloride affords mesylate (**11.4**). The treatment of this intermediate with the 4-carbamoyl ester of *p*-aminothiophenol affords **erbuzole** (**11.2**) [10].

Two quite diverse six-membered heterocyclic rings, a piperazine and a pyrimidine, form the bases for a pair of tubulin inhibitors. The structure of first of that pair, plinabulin, is said to be based on the activity in an *in vitro* screen for antineoplastic agents of an extract from the marine fungus halimide. The purified compound binds to tubulin consequently inhibiting its polymerization. At the whole organism level, plinabulin acts as a vascular-disrupting agent that acts selectively on blood vessels in solid tumors.

The acylation of piperazino-2,5-dione (**12.1**) with acetic anhydride affords the corresponding *N,N′*-diacetate (**12.2**). The reaction of that product with the substituted imidazole aldehyde in the presence of base affords the condensation product (**12.4**). The reaction interestingly stops at the addition of a single imidazole. A stronger base, lithium diisopropylamide (LDA), is required for effecting the condensation of a second aldehyde via an aldol reaction. The condensation of the product from the prior reaction (**12.4**) with benzaldehyde mediated by LDA leads to adduct (**12.5**). The saponification of the acetamide groups then affords **plinabulin** (**12.6**) [11].

The very concise preparation of the pyrimidine-based tubulin inhibitor involves first displacement of one of the chlorine substituents in 2,5-dichloropyrimidine (**13.2**) by amine nitrogen in enantiomerically pure alkylaminopyridine (**13.1**). Palladium-catalyzed coupling between chlorine on the pyrimidine ring of the product from the preceding reaction and boronic acid (**13.4**) of the substituted aniline leads to the tubulin inhibitor **lexibulin** (**13.5**) [12]. The biological properties of the drug are quite similar to those of plinabulin. Lexibulin in contrast to many other chemotherapy drugs is well absorbed when taken by mouth.

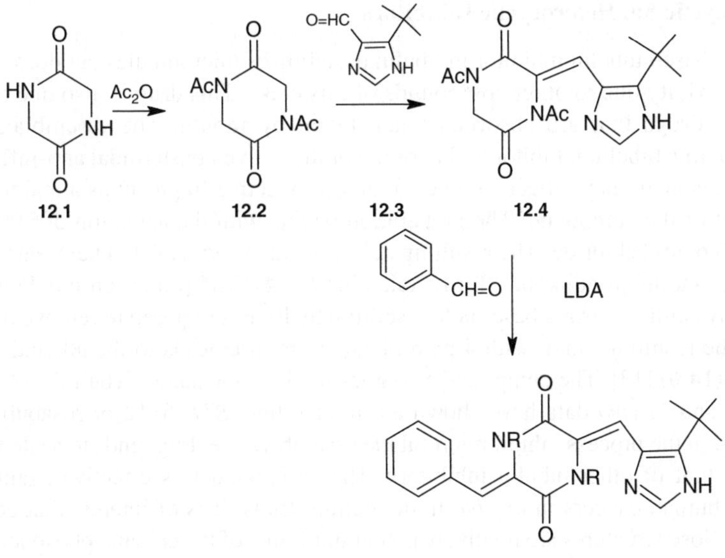

Scheme 5.12 Plinabulin.

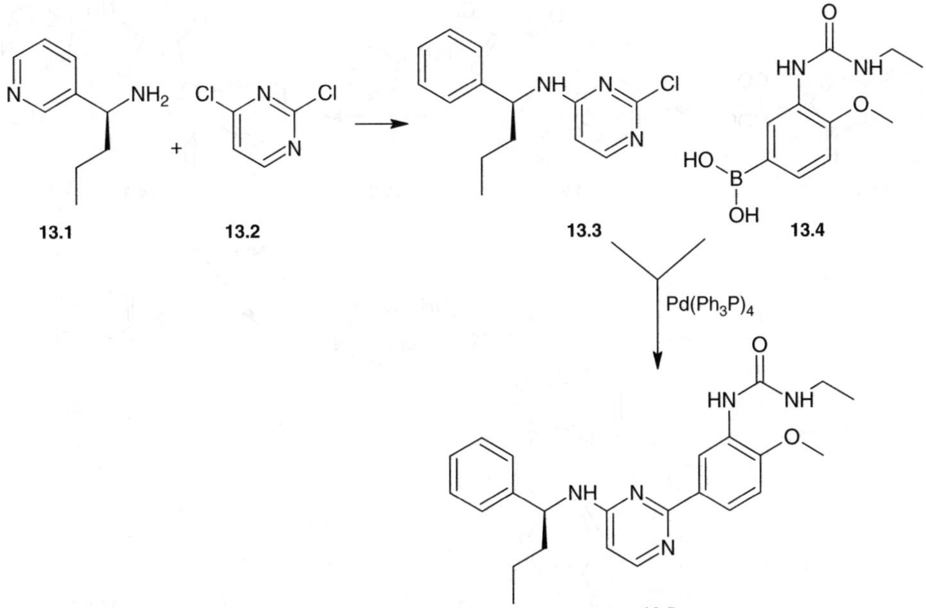

Scheme 5.13 Lexibulin.

5.3.4 Bicyclic 5:6 Heterocyclic Inhibitors

The indole-based tubulin inhibitor indibulin destabilizes microtubules but does not exhibit the neurotoxicity due to other compounds of this class. This drug is also one of the chemotherapy drugs that are absorbed when taken by mouth. The resemblance of the structure of this tubulin inhibitor to that of the orally active nonsteroidal anti-inflammatory drug indomethacin may reflect a choice of an orally active fragment as a major structural component of the compound. The preparation begins with the acylation of indole proper (**14.1**) with oxalyl chloride. The resulting acid chloride is not isolated but treated immediately with 4-aminopyridine to afford oxalamide (**14.4**). The proton on indole nitrogen is surprisingly acidic; a strong base such as sodium hydride is required to remove that proton. Treating the resulting anion with 4-chlorobenzyl bromide leads to the alkylation product **indibulin** (**14.6**) [13]. The compound also goes by the trade name Zybulin®.

In vitro and *in vivo* data have shown an indolizidine, STA-5312 or rosabulin, to bind tubulin and in the process inhibit microtubule assembly. The drug binds to a site that is distinct from that of other tubulin inhibitors. The compound was effective against a wide variety of human cancers in *ex vivo* tissue culture trials. It is of interest that compounds with quite close structures are relatively potent inhibitors of the enzyme phosphodiesterase. The preparation of this indolizidine begins with the bromination of 4-cyanoacetophenone (**15.1 → 15.2**). This very reactive halogenated ketone is then used to alkylate the nitrogen atom in 1-methylpyridine. The key reaction in the synthesis comprises addition of an extra carbon atom followed by cyclization to a fused five-membered ring. In actual practice, the charged species (**15.4**) is treated with a mixture of dimethylformamide and dimethyl sulfate

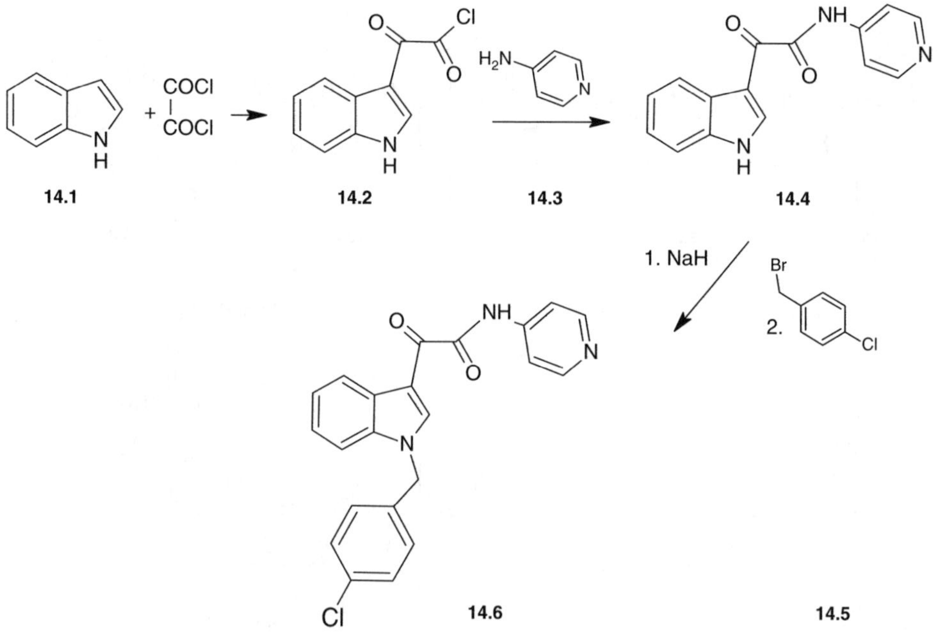

Scheme 5.14 *Indibulin.*

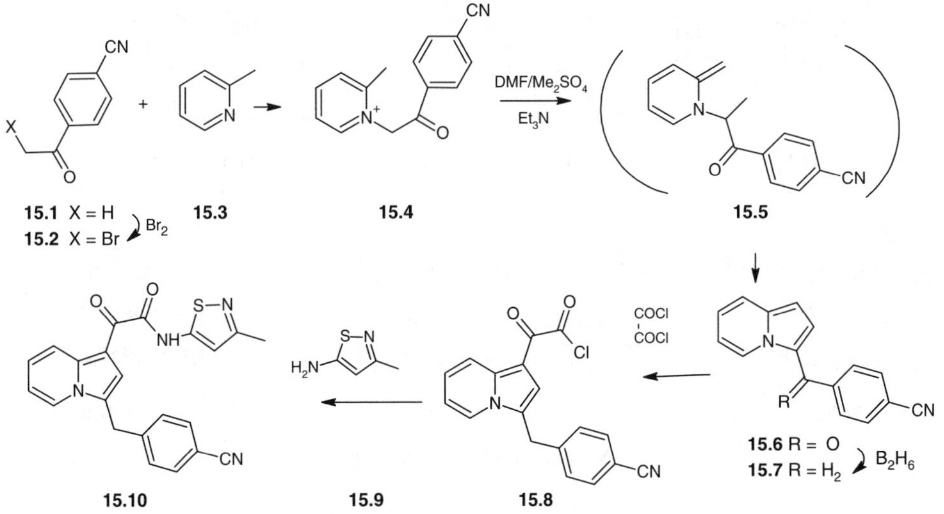

15.1 X = H
15.2 X = Br ⟍ Br₂   **15.3**   **15.4**   **15.5**

15.10   **15.9**   **15.8**

15.6 R = O
15.7 R = H₂ ⟍ B₂H₆

Scheme 5.15   Rosabulin.

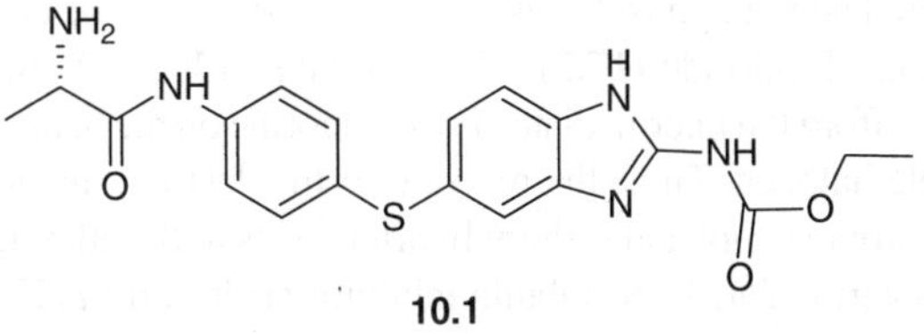

10.1

Scheme 5.16   Denibulin.

in the presence of triethylamine. The product of this reaction can be visualized by assuming that the combination of reagents forms a reactive methylene species; that carbon atom adds to the position adjacent to the carbonyl group (**15.5**). That then closes to the five-membered ring. Exhaustive reduction by means of diborane converts to carbonyl carbon to a methylene group (**15.7**). Friedel–Crafts acylation with oxalyl chloride leads to the corresponding acid chloride (**15.7**). The treatment of that intermediate with isothiazolamine (**15.9**) then affords the amide and thus **rosabulin** (**15.10**) [14].

The benzimidazole **denibulin** irreversibly binds to tubulin and inhibits microtubule assembly. This results in the disruption of the tumor cells microskeleton. Clinical trials of the drug have focused on disruption of the tumor cell vasculature.

Most of the wholly synthetic tubulin inhibitors bind to the target molecule and thus inhibit its polymerization to form microtubules. The mechanism by which the triazolopyrimidine cevipabulin inhibits tubulin is closer to that of the taxanes rather than the synthetic compounds. The synthesis of this tubulin inhibitor starts with the displacement of bromine in bromo-2,4,6-trifluorobenzene by the anion from the reaction of ethyl malonate with a base such as sodium ethoxide. The reaction of the malonate with 2-amino-1,2,4-triazole results in fusing a pyrimidine onto the molecule by a sequence characteristic of malonates. The overall result can be

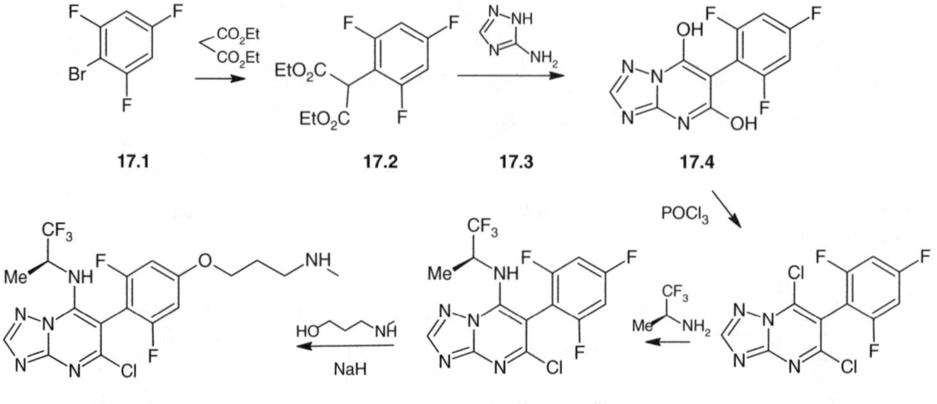

Scheme 5.17   *Cevipabulin.*

visualized by assuming that the primary amine in the triazole adds to one of the ester carbonyl groups. Addition of a ring nitrogen to the other malonate ester completes the ring. The carbonyl groups then revert to the enol form. In a reaction used extensively in heterocyclic chemistry, the treatment of the bis-enolate with phosphorus oxychloride converts the enol hydroxyl groups to chlorides. The reaction of chloride (**17.5**) with resolved trifluoro-2-aminopropane displaces one of the chlorines to afford the intermediate (**17.6**). Fluorine on the benzene ring constitutes the next most displaceable halogen. Thus, the reaction of this last intermediate with the alkoxide from *N*-methylaminopropran-3-ol and sodium hydride leads to the ether from the displacement of fluorine. This last intermediate is the tubulin inhibitor **cevipabulin** (**17.6**) [15, 16].

5.3.5   Bicyclic 6:6 Heterocyclic Tubulin Inhibitors

The structurally relatively simple quinazoline verubulin (**18.6**) binds to tubulin and prevents the polymerization of that substance required for building microtubules. This compound too is classed as a vascular-disrupting agent. A particularly promising feature of this compound lies in the fact that it achieves high levels in the brain as demonstrated in Phase I clinical trials. The first step in preparing this compound comprises treating anthranilic acid amide with acetic anhydride. The reaction of the cyclized product (**18.2**) with ammonia replaces oxygen by nitrogen to afford quinazolone (**18.3**). As in the preceding example, phosphorus oxychloride serves to replace the carbonyl group by an enol chloride. The newly introduced halogen is displaced by nitrogen in 4-methoxyphenylmethylamine (**18.5**) to afford **verubulin** (**18.6**) [17].

Like some other recent tubulin inhibitors, mivobulin (**19.5**) is water soluble and, perhaps more important, crosses the blood–brain barrier. The compound is also quite effective against a selection of human cancers in tissue culture experiments and as xenografts. The first task in the preparations of this agent comprises joining of the two moieties. Thus, displacement of chlorine in the highly substituted pyridine by the amine in resolved amino alcohol (**19.2**). The hydroxyl in the product is next oxidized to the

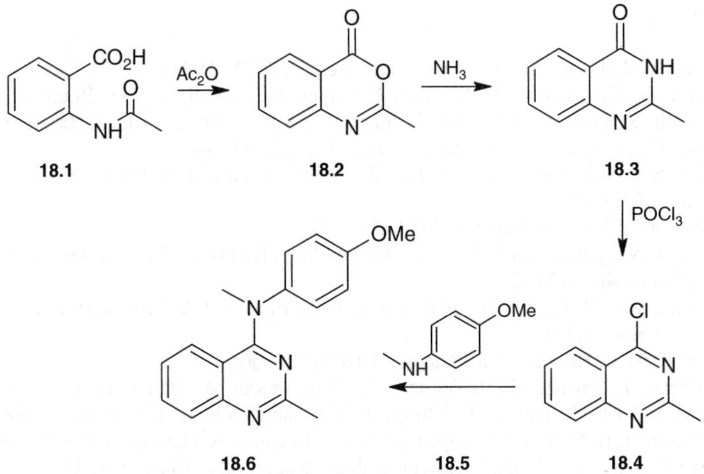

18.1 18.2 18.3 18.6 18.5 18.4

Scheme 5.18 *Verubulin.*

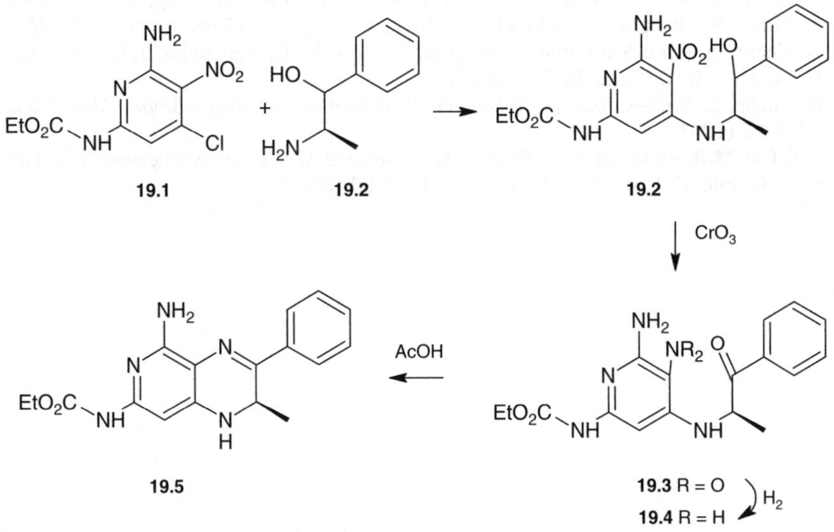

19.1 19.2 19.2

19.5 19.3 R = O
19.4 R = H

Scheme 5.19 *Mivobulin.*

corresponding ketone by means of chromium trioxide in acetic acid; the acidic medium likely protects the primary amine as its salt. Catalytic hydrogenation then reduces the nitro group to a primary amine. Heating the thus obtained intermediate in acetic acid leads to the formation of a Schiff base between the carbonyl group on one fragment and the primary amine on the other. The resulting pyridopiperazine is the tubulin inhibitor **mivobulin (19.5)** [18].

References

[1] R.A. Holton, J.H. Liu, *Bioorg. Med. Chem. Lett.* **3**, 2475 (1993).

[2] R.A. Holton, C. Somoza, H.-B. Kim, F. Liang, R.J. Biediger, P.D. Boatman, M. Shindo, C.C. Smith, S. Kim, H. Nadizadeh, Y. Suzuki, C. Tao, P. Vu, S. Tang, P. Zhang, K.K. Murthi, L. N. Gentile, J. H. Liu., *J. Am. Chem. Soc.* **116**, 1597 (1994).

[3] J.H. Heo, S.J. Park, J.H. Kang, I.S. Lee, J.S. Lee, Y.J. Park, K.S. Kim, J.Y. Lee, *Bull. Korean Chem.* **30**, 25 (2009).

[4] R.A. Holton, P. Vu, U.S. Patent 8,003,812 (2011).

[5] S.X. Cai, B. Nguyen, S. Jia, J. Herich, J. Guastella, S. Reddy, B. Tseng, J. Drewe, S, Kasibhatla, *J. Med. Chem.* **46**, 2474 (2003).

[6] D.J. Chaplin, C.M. Garber III, R.R, Kane, K.G. Pinney, J.A. Prezioso, K. Edvardsen, U.S. Patent ,919,324 (2013).

[7] G.R. Pettit, M'R. Rhodes, U.S. Patent, 7,018,987 (2006).

[8] W. Amberg, T. Barlozzari, H. Bernard, E. Buschmann; A. Haupt; H.G. Hege, B. Janssen, A. Kling, H. Lietz, K. Ritter, M., Ullrich, J. Weyman, T. Zierke, U.S. Patent 7,368,528 (1999).

[9] A. Yamashita, E.B. Norton, J.A. Kaplan, C. Niu, F. Loganzo, R. Hernandez, C.F. Beyer, T. Annable, S. Musto, C. Discafani, A. Zask. S. Ayral-Kalous, *Bioorg. Med. Chem. Lett.* **14**, 5317 (2004).

[10] W. Distelmans, P. VanGinckel, J. Heeres, L.J.E. Van der Vecken, *Drugs Fut.* **16**, 577 (1991).

[11] Y. Iliashi, M.A. Palladino, J. Grodberg, U.S. Patent 7,064,201 (2006).

[12] C.J. Burns, A.F. Wilks, M.F. Harte, H. Sikanyika, E. Fantino, U.S. Patent, 7,981,900 (2011).

[13] G. Lebaut, C. Menciu, B. Kutcher, P. Emig, S. Szeleny, K. Brune, U.S. Patent 6,008,231 (1999).

[14] S. Chen, Z. Xia, M. Nagai, R. Lu, E. Kostik, T. Przewloka, M. Song, D. Chimmanamada, D. James, S. Zhang, J. Jiang, M. Ono, K. Koya, L. Sun, *Med. Chem. Commun.* **2**, 176 (2011).

[15] N. Zhang, S. Ayral-Kaloustian, T. Nguyen, J. Afragola, R. Hernandez, J, Lucas, J. Gibbons, C. Beyer, *J. Med. Chem.* **50**, 319 (2007).

[16] N. Zhang, S. Ayral-Kaloustian, T. Nguyen, R. Hernandez, C. Beyer, *Bioorg. Med. Chem. Lett.* **17**, 3003 (2007).

[17] S.X. Cai, M.B. Anderson, A. Willardson, N.S. Sirisoma, U.S. Patent Application 2008/0,051,398.

[18] C.G. Temple, G.A. Rener, *J. Med. Chem.* **32**, 2089 (1989).

6

Matrix Metalloproteinase Inhibitors

6.1 Introduction

The extracellular matrix comprises the open region between cells in tissues. This region functions first as a physical support for the cells. Among other functions, the matrixin provides the medium for cell-to-cell communication. A group of peptidase enzymes that depend on the presence of zinc or occasionally other metals, termed matrix metalloproteinases or matrixin enzymes, play an important part in various cell functions such as differentiation, angiogenesis, and host defense. One or more steps in carrying out some of those functions can involve the digestion of the matrix. The proliferation of tumors may, for example, require the destruction of the matrix to make room for cancerous cells. This hypothesis is bolstered by evidence that levels of metalloproteinases are elevated in tumors. This has led to the proposition that these proteinases play an important role in the proliferation and dissemination of cancer tumors. A compound that inhibited these enzymes would be expected to impede the progression of cancerous tissues. A sizeable number of small molecules that inhibited the action of matrix metalloproteinases in fact showed good antineoplastic activity against human cancers both in cell culture assays and in grafts in laboratory animals [1]. A large number of these inhibitors have been subjected to at least Phase I clinical trials. None has to date been approved by regulatory agencies for treating cancer.

Most of the examples that follow incorporate centers of asymmetry. In the interest of brevity, the discussions of the schemes of those syntheses repeat the structural diagram found in the sources without any further elaboration.

Antineoplastic Drugs: Organic Synthesis, First Edition. Daniel Lednicer.
© 2015 John Wiley & Sons, Ltd. Published 2015 by John Wiley & Sons, Ltd.

6.2 Hydroxamates

6.2.1 Agents with an Isobutyl Moiety

Four of the following metalloproteinase inhibitors carry an isobutyl substituent at a remove of two carbon atoms from the hydroxamate function. The syntheses for preparing those compounds perforce follow closely related pathways.

The synthesis of batimastat (**1.9**) begins with the preparation of the bromo carboxylic acid that will introduce the isobutyl moiety. Thus, treating valine with nitrous acid (actually sodium nitrite and acid) followed by bromine gives the halogenated acid where bromine has the reversed configuration from the amine in alanine (**1.2**). That intermediate is then converted to *tert*-butyl ester (**1.3**).

The displacement of bromine in that intermediate by the anion from bis-benzyl malonic ester yields alkylation product (**2.1**). The product from that reaction is then used to acylate *N*-methyl phenylalanine (**2.2**). Catalytic hydrogenation over palladium then removes the benzyl protecting groups on the malonate moiety to afford the corresponding dicarboxylic acid; that intermediate, on warming, then ejects one of acids as carbon dioxide. The acid is then converted to some unspecified ester (**2.4**). Reaction intermediate with formaldehyde and base adds a carbinol to the position adjacent to the methylene group next to the carboxyl function. That transient carbinol dehydrates to leave behind an ethylene conjugated to the ester. The treatment of that intermediate with 3-suflfhydrylthiophene leads to conjugate addition of the thiol to the end of the new vinyl group (**2.6**). The reaction of this last intermediate with *O*-benzylhydroxylalmine in the presence of a carbodiimide adds that function by way of an amide. Catalytic hydrogenation then leads to the loss of the benzyl protecting group and formation of batimastat (**2.8**) [2].

The synthesis of the inhibitor ilomastat follows a similar course that comprises assembly of preformed moieties. The first stage of the synthesis of this inhibitor, not shown, would consist of the reaction of *O*-benzylhydroxylamine with isobutylsuccinic acid to yield the first segment of the target molecule. That fragment is used to acylate the indole bearing an *N*-methyl α-aminoamide (**3.2**). This compound likely serves as a mimic of phenylalanine. Catalytic hydrogenation over palladium of the intermediate removes the benzyl protecting group. The metalloproteinase inhibitor **ilomastat** (**3.4**) is thus obtained [3, 4].

A relatively concise scheme affords access to enantiomerically pure marimastat if one ignores the chemistry required to produce starting material (**4.1**). That key compound was prepared from tartaric acid in some seven successive reactions.

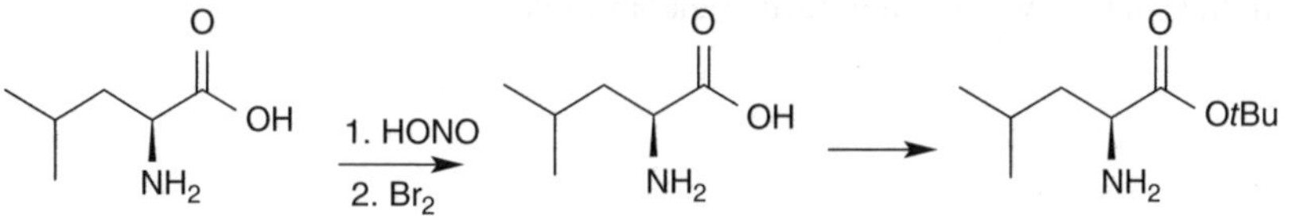

1.1   1. HONO   **1.2**   **1.3**
                 2. Br$_2$

Scheme 6.1   Brominated acid.

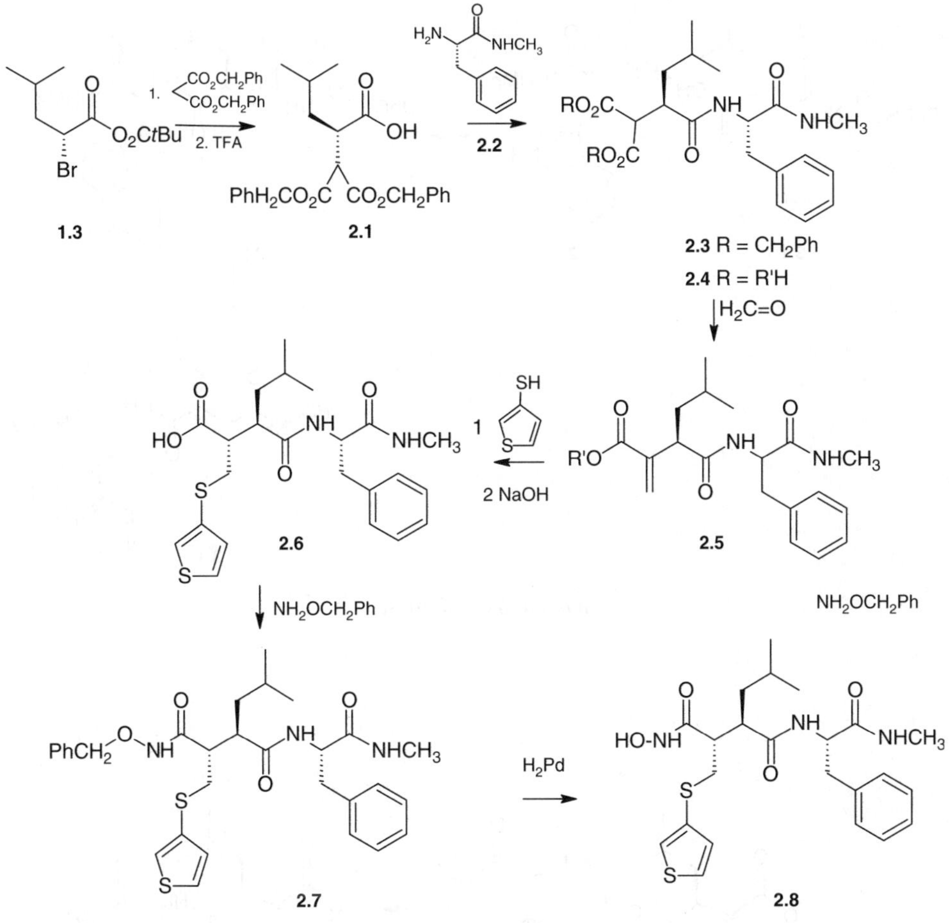

Scheme 6.2 *Batimastat.*

The key first step in the concise synthesis for preparing marimastat (**4.7**) comprises the acylation of the *t*-butyl glycine derivative **4.2** to the acid chloride (**4.1**) that traces back to tartaric acid. The treatment of the product from that reaction with the *O*-benzyl ether of hydroxylamine results in the attack on the ester carbonyl of the ester/acetal ring. Catalytic hydrogenation of the product then reductively removes the benzyl protecting group. **Marimastat (4.7)** is thus obtained [5].

The construction of the succinic acid that comprises one of the moieties of the matrix metalloproteinase inhibitor tosedostat starts by protecting both the carboxylic acid and the hydroxyl group in succinic acid (**5.1**) by reaction with ethylene glycol to afford the cyclic ester/acetal. The remaining carboxyl group is then esterified with pentafluorophenol in order to enhance its reactivity. The other α-amino acid moiety, phenylglycine, is first converted to its cyclopentyl ester by reaction with the corresponding alcohol. (Back in the middle of the twentieth century, it had been noted that a cyclopentyl group offered the maximal steric hindrance against hydrolysis.) The reaction of esterified succinic ester (**5.2**) with amino

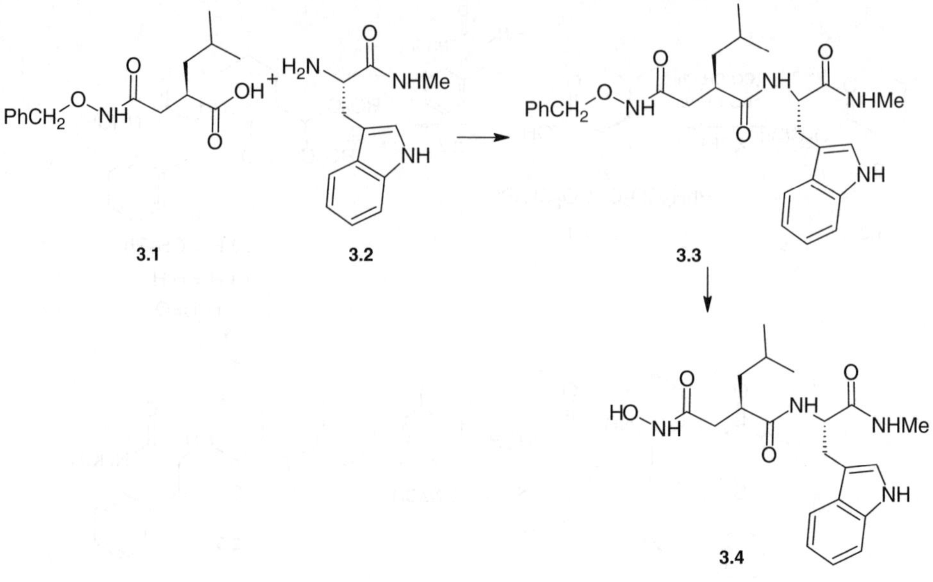

Scheme 6.3 Ilomastat.

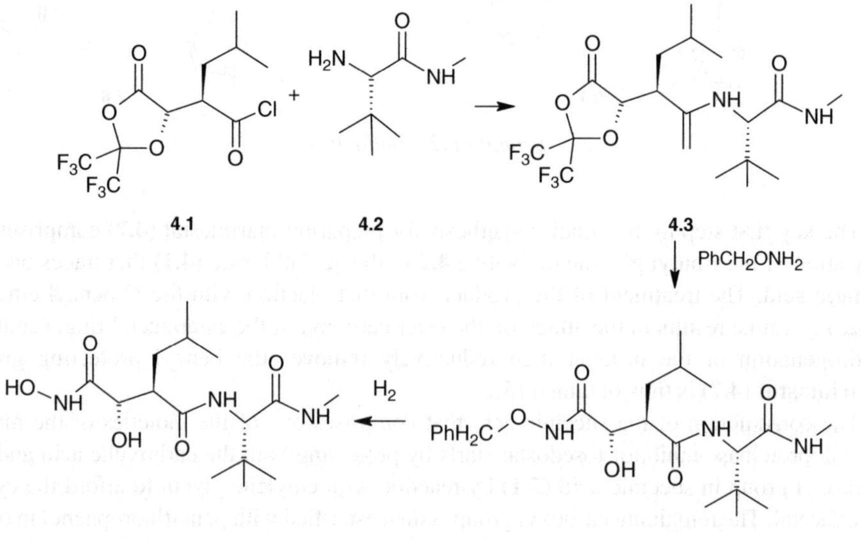

Scheme 6.4 Marimastat.

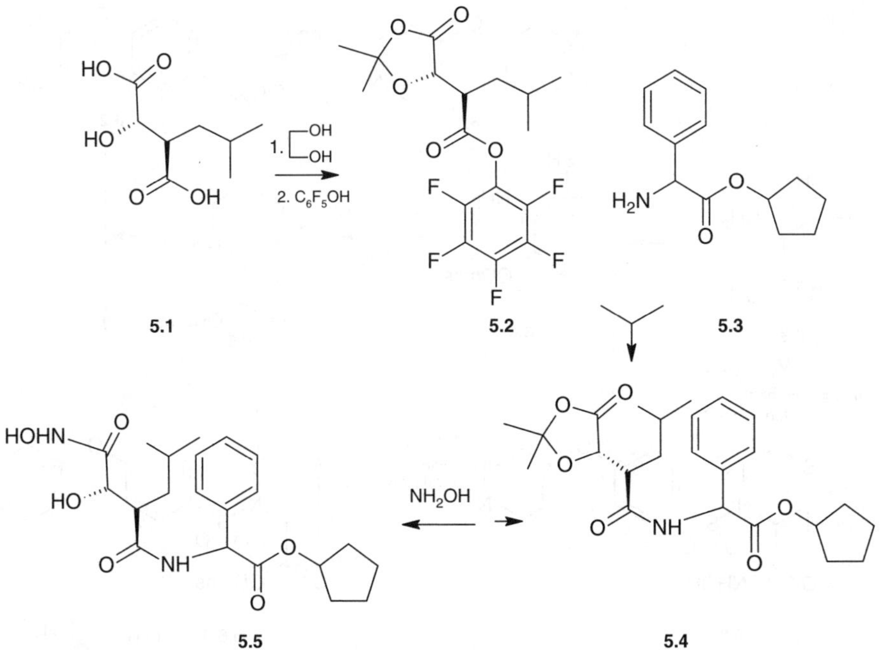

Scheme 6.5 *Tosedostat.*

acid (**5.3**) leads to the displacement of the pentafluorophenyl moiety by the basic amine and formation of an amide; that group now connects the two moieties (**5.4**). The treatment of this last intermediate with hydroxylamine leads to the addition of the reagent to the ring carbonyl function and subsequent formation of an oxamate. The product (**5.5**) goes by the nonpropri-etary name **tosedostat** (**5.5**) [6].

6.2.2 A Thiomorpholine

The relatively uncommon oxamate group present in the prior examples may play an impor-tant role in the antineoplastic activity manifested by those compounds. The same is probably true for the sulfonamide function. It is fitting that both functional groups occur in prinomastat. This compound showed reasonable activity in many of the *in vitro* and tissue culture preclinical experiments. The compound however failed to improve the outcome of patients in a Phase III trial against non-small cell lung cancer. At this point, the sponsor halted the trial.

One arm of the converging synthesis comprises the chloromethylation of the phenyl group in ether (**6.1**) itself produced by the displacement of halogen in 4-chloropyridine by lithium phenoxide. The construction of the thiomorpholine moiety begins by masking of the carboxylic acid in penicillamine (**6.3**) by reaction with cyclohexyldimethylsilyl chloride (Dmhs). The treatment of that intermediate with 1,2-dichloroethylene leads to the displace-ment of the halogen atoms by sulfur and basic nitrogen in the penicillamine so as to form the

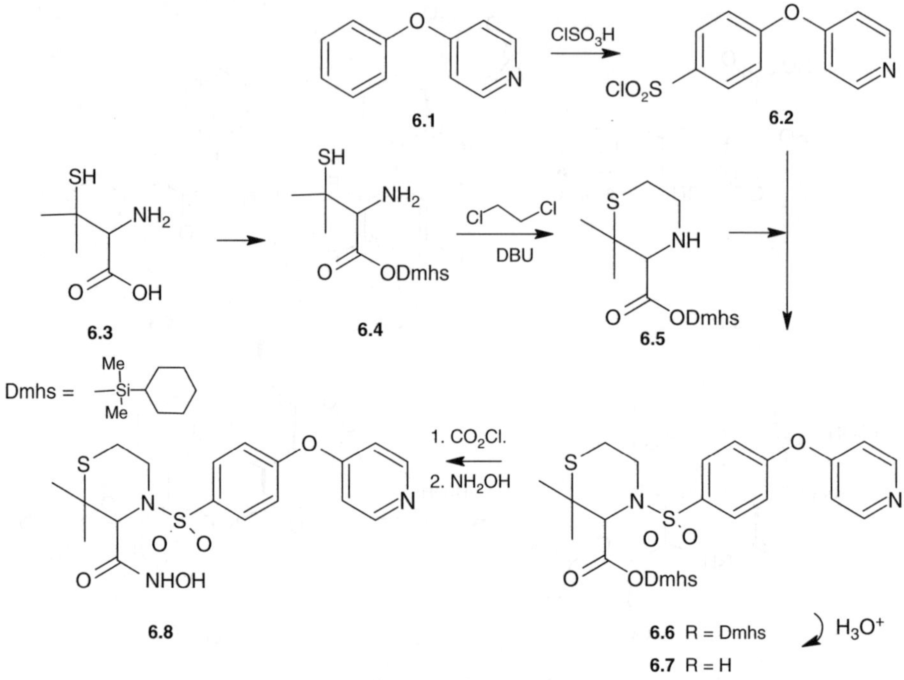

Scheme 6.6   *Prinomastat.*

thiomorpholine as a single enantiomer (**6.5**). Reaction between the chlorosulfonyl (**6.2**) and the basic nitrogen in the thiomorpholine connects the two moieties via a sulfonamide function. Exposure to mild acid then serves to free carboxylic acid (**6.7**). Reaction with oxalyl chloride then converts that carboxylic acid to the corresponding acid chloride. Treatment with hydroxylamine completes the preparation of the matrix metalloproteinase **prinomastat** (**6.8**) [7].

6.2.3   Sulfamates

Though the chemical structures of the molecules containing a sulfamate group differ markedly from the oxamates, these compounds, nonetheless, also inhibit matrix metalloproteinase enzymes. The preparation of the inhibitor pevonedistat (**7.10**) involves a relatively long sequence, much of which is devoted to attaching and subsequently removing a hydroxyl group in the carbon equivalent of a furanose.

The reaction of the starting material (**7.1**) with phosphorus oxychloride, the heterocyclic chemists favored reagent, serves to displace the hydroxyl group in the six-membered ring by chlorine. The newly introduced halogen is then itself displaced by nitrogen in aminoindene (**7.3**). The next step involves attaching the trihydroxycyclopentyl system to the deazapurine moiety (**7.4**). The treatment of that intermediate, protected by a benzyl group as well as a *ter-tiary*butyldimethylsilyl function, with sodium hydride removes the relatively acidic proton on the ring nitrogen. The resulting anion then attacks oxygen the cyclic sulfate-protected glycol; that ring then opens connecting the two fragments in the process. The treatment of this last

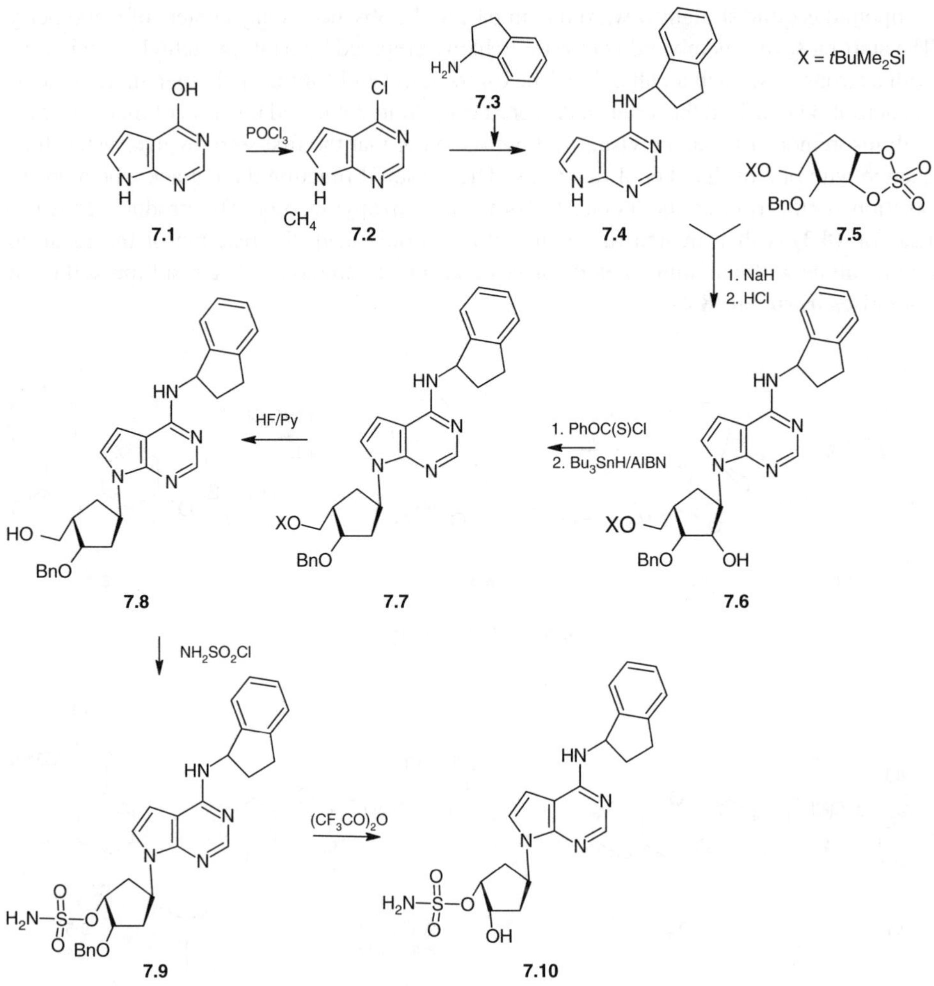

Scheme 6.7 *Pevonedistat.*

product with acid serves to hydrolyze the sulfate that is left behind, affording the free alcohol. The reaction of this last intermediate with phenylchlorothionoformate in the presence of the radical initiator AIBN leads to the corresponding xanthate; treatment with tetrabutyltin then reduces the xanthate, in effect removing the hydroxyl group. The silyl protecting function is next removed by means of hydrogen fluoride in pyridine. The resulting free hydroxyl group is now converted to its sulfamate by reaction with sulfamoyl chloride. The scission of the benzyl ether group with trifluoroacetic anhydride completes the synthesis of **pevonedistat** (**7.10**) [8].

The inhibitor irosustat (**8.4**), in addition to the inhibition of metalloproteinases, was found to have some affinity for the estrogen receptor. The chemical structure of this compound does slightly resemble some of the nonsteroid estrogen antagonists. This led to the focus of early studies on estrogen-positive breast cancer. The preparation of this compound is quite straightforward due in part to the absence of any centers of asymmetry. The starting seven-membered keto ester is likely prepared by treating methyl sebacic ester with a strong base such an alkoxide. The esters at the ends of the chain then undergo aldol condensation to afford the seven-membered ring. Acid-catalyzed reaction of that keto ester with resorcinol (**8.1**) can be visualized by assuming that the first step involves ester inter-change with one of the phenol functions. The carbonyl function then attacks the adjacent position on the ring as its enolate to form the benzopyran ring. The product from that reaction (**8.3**) is then treated first with chlorosulfonic acid; the reaction of the resulting chlorosulfate with ammonia then displaces chlorine by ammonia. The resulting sulfamate comprises **irosustat (8.4)** [9].

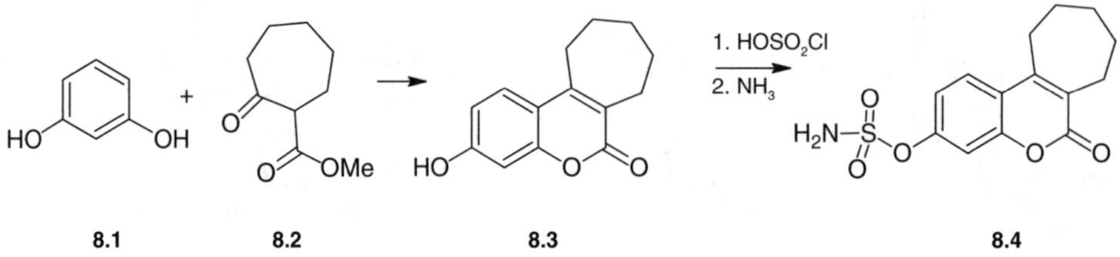

Scheme 6.8 Irosustat.

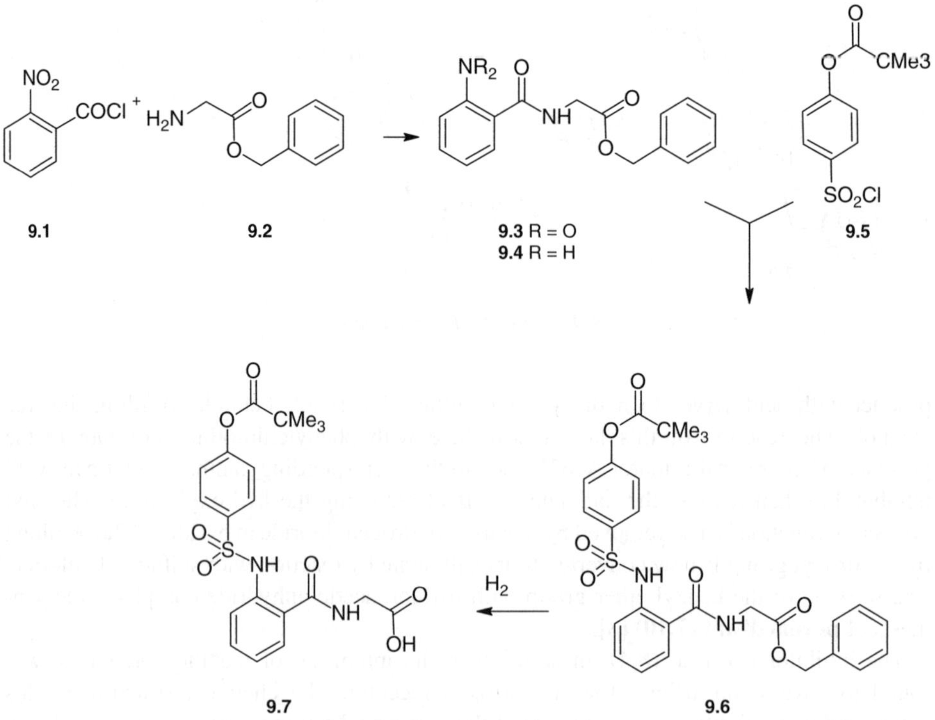

Scheme 6.9 Sivelestat.

The matrix metalloproteinase activity of the compound sivelestat (**9.7**) is expressed principally by its inhibition of neutrophil elastase, a protein-digesting enzyme secreted by serum neutrophils. The drug is marketed outside the United States for just that indication under the trade name Elaspol®. As a result, trials for efficacy in such diverse uses from an adjunct to liver surgery to the treatment of erectile dysfunction have exceeded trials for treating various malignancies. The relatively concise synthesis opens with the acylation of the benzyl ester of glycine with 2-nitrobenzoyl chloride. Controlled catalytic hydrogenation then reduces the nitro group to an amine (**9.4**) while leaving the benzyl ester intact. The reaction of the resulting intermediate with the pivaloyl ester of 4-hydroxysulfonyl chloride then yields the intermediate (**9.6**) that now holds all three requisite moieties. A second hydrogenation serves to remove the benzyl protecting group. **Sivelestat** (**9.7**) is thus obtained [10].

6.2.4 Miscellaneous Compounds

The metalloproteinase inhibitor rebimastat has been characterized as one of the second-generation metalloproteinase inhibitors. Clinical trials are currently under way to assess the antineoplastic efficacy of the compound investigation in the current clinical trials. A glance of the chemical structure of this compound shows that it is composed of four fragments of quite diverse chemical structure. One synthesis of the drug relies on an interesting twist on solid-phase chemistry. In the overwhelming majority of solid-phase syntheses, the molecule to be elaborated is attached to an insoluble bead and then subjected to serial operations or, more specifically, reactions. The procedure is reversed in the case at hand. In this application, reagents for each of the transforms are attached to one of a series of columns; a solution of the synthetic object is then percolated one at a time through reagent-bound columns. Scheme 6.9 however depicts the convergent synthesis in traditional form. Reagents have not been included in the scheme since they could be different from those used in traditional solution-based chemistry.

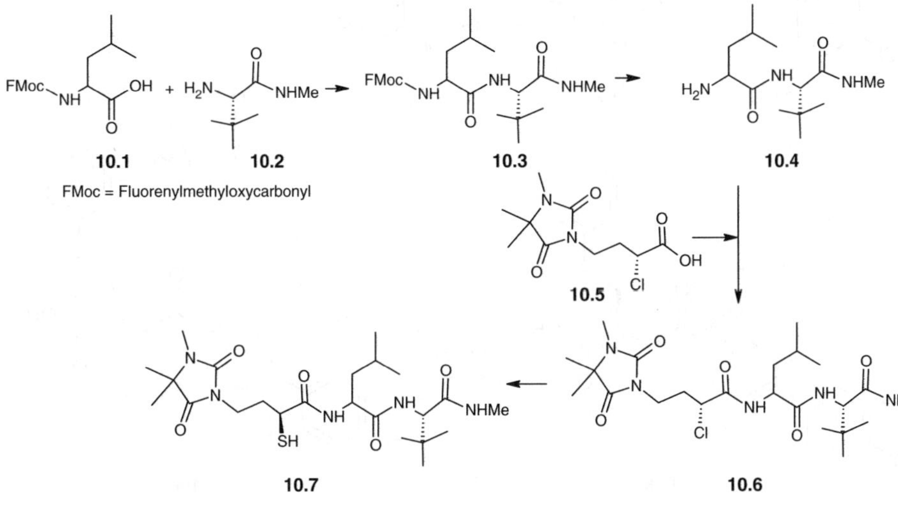

Scheme 6.10 Rebimastat.

The sequence begins with amide formation between fluorenylmethyloxycarbonyl-protected valine and 2-*tertiary*butyl glycine. The protecting group in the resulting amide (**10.3**) is then removed by means of base (**10.3**). The newly revealed amine is next acylated with the substituted 4-imidazodionebutyric acid (**1.4**) to afford the amide **10.5**. The displacement of the halogen in this intermediate by a reagent that will afford a mercaptan proceeds with the inversion of the configuration to afford the matrix metalloprotein inhibitor **rebimastat (10.6)** [11].

Data from tissue culture and laboratory studies on the biphenyl carboxylic acid derivative tanomastat (**11.5**) suggested that the compound would show activity against neoangiogenesis in solid tumors; the data also suggested that drug might act against metastases. Clinical studies proceeded as far as Phase III. That randomized study was discontinued when there was no increase in overall survival. One stereoselective synthesis of this metalloproteinase inhibitor starts with aluminum chloride-mediated acylation of chlorobiphenyl with itaconic anhydride (**11.2**). The treatment of the product with thiophenol in the presence of mild base leads to conjugate addition of the thiophenol to the end of the enone group. That last product is then resolved as its (+) cinchonine salt. This procedure then affords **tanomastat (11.5)**.

Therapeutic agents that incorporate boron in their chemical structure can probably be counted on the proverbial fingers of one hand. One of those rare species consists of the matrix metalloproteinase inhibitor talabostat. In addition to that activity, the drug also inhibits peptidyl peptidases, a class of enzymes implicated in adult onset diabetes, as well as the fibroblast activation protein. The latter has been reported to contribute to tumor growth. In 2006,

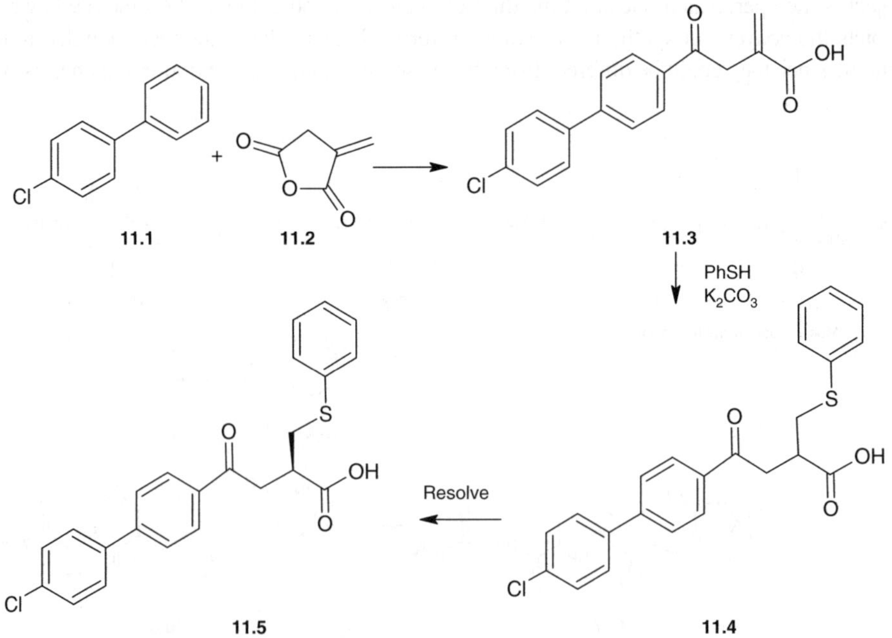

Scheme 6.11 *Tanomastat.*

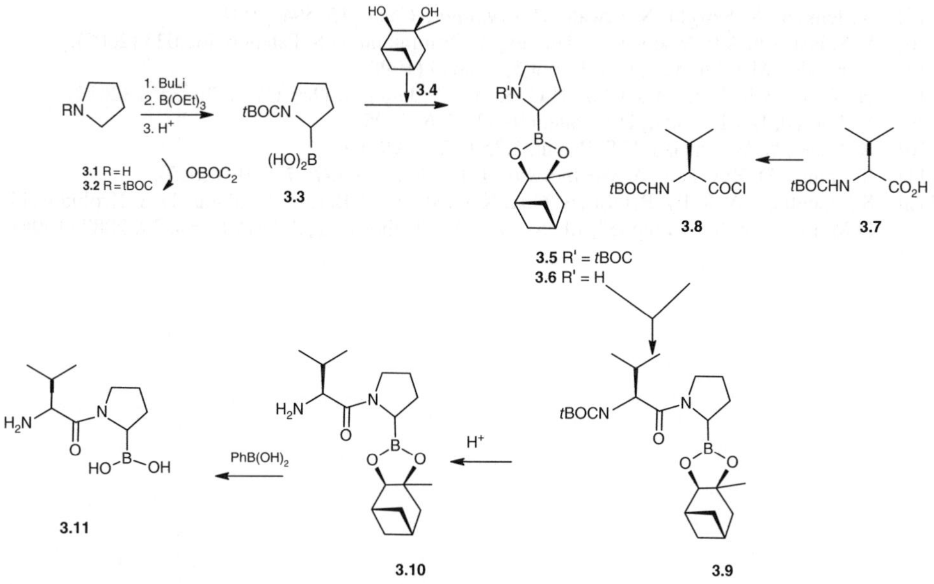

Scheme 6.12   *Talabostat.*

the FDA granted the drug fast track status for treating patients with non-small cell lung cancer. The stereospecific synthesis begins with the protection of the basic nitrogen in pyrrolidine as its *tertiary*butylcarbonyl (BOC) amide. The position adjacent to the amine is sufficiently acidic to be removed by butyl lithium. The treatment of the anion with triethyl borate leads to the displacement of one of the ethoxy groups by the anion, in effect attaching boron to the pyrrolidine. In this case, the overall stereochemistry is controlled by attaching a chiral auxiliary to boron. Acid hydrolysis of the ethoxy groups on boron, in effect esters, frees the two hydroxyls on the borate (**12.2**). When allowed to react with chiral naturally occurring (+)pinanediol, this forms the acetal-like derivative (**12.4**). The BOC group is then removed by acid hydrolysis and the newly revealed amine acylated with the acid chloride of BOC-protected valine (**12.7**); the BOC masking group is then removed by acid hydrolysis. The chiral auxiliary is then removed by exchange with phenylboronic acid (**12.10**) [12]. The final product **talabostat (12.10)** is often handled as a salt to avoid forming a zwitterion.

References

[1]   S. Zucker, J. Cao, *Cancer Biol. Ther.* **8**, 2371 (2009).
[2]   A. Oliva, G. De Cillis, F. Grams, V. Livis, G. Zimmerman, E. Mnta, H.-W. Krell, U.S. Patent 6,355,332 (2002).
[3]   C. Jeanpetit, D. Prigant, P.-A. Settembre, M.-M. Traancart, U.S. Patent 6,344,457 (2002).
[4]   D.E. Levy, F. Lapierre, W. Liang, W. Ye, C.W. Lange, X. Li, D. Grobelny, M. Casabonne, D. Tyrrell, K. Holme, A. Nadzan, R.E. Galardy, *J. Med. Chem.* **41**, 199 (1998).

[5]   K. Jenssen, K. Sewald, N. Sewald, *Bioconjugate Chem.* **15**, 594 (2004).
[6]   L.A. Pearson, A.P. Ayscough, P. Huxley, A. Drummond, U.S. Patent 6,462,023 (2002).
[7]   L. Bender, M.J. Melnick, U.S. Patent 5,753,653 (1998).
[8]   H.W. Lee, S.K. Nam, W.J. Choi, H.O. Kim, L.S. Jeong, *J. Org. Chem.* **76**, 3557 (2011).
[9]   M.J. Reed, B.V.L. Potter, U.S. Patent, 6,921,776 (2005).
[10] K. Imaki, H. Wakatsuka, U.S. Patent, 5,359,121 (1994).
[11] S. France, D. Bernstein, A. Weatherwax, T. Lactka, *Org. Lett.* **7**, 3009 (2005).
[12] S.J. Coutts, T.A. Kelly, R.J. Snow, C.A. Kennedy, R.W. Barton, J. Adams, D.A. Krolikowski, D.M. Freeman, S.J. Campbell, J.F. Skiazek, W. Bachowchin, *J. Med. Chem.* **39**, 2087 (1996).

7

Histone Deacetylase Inhibitors

7.1 Introduction

The problem of packing the macroscopically lengthy chain of DNA into the microscopic cell nucleus is touched upon earlier in this volume. The introduction to Chapter 4, on topoisomerase inhibitors, dealt with the problem of retrieving information from that tightly packed macromolecule. Histones comprise a class of dedicated small proteins that form a coating on supercoiled DNA that squeeze that unit even more tightly in order that it may fit into the cell nucleus. These peptides occur as five or more distinguishable classes. The positive charge carried by histones contributes to the tight binding of this coating protein to negatively charged DNA. The resulting unit, called chromatin, is itself coiled. The activity of genes during cell duplication is regulated by a set of enzymes that chemically modify the histones; two of these comprise histone acetylases and histone transferases. The acetylation of histones removes some of the positive charges on the histone, relaxing the tight binding to DNA. This leads to a greater level of gene transcription. The decrease in deacetylation caused by histone deacetylase (HDAC) inhibitors, on the other hand, will foster tighter binding; this causes a decrease in the level of genome transcription and as a result a decrease in cell proliferation. The antineoplastic activity of these inhibitors is presumably due to the latter effect.

The chemical structure of several of these antineoplastic agents is very close to those of the matrix metalloproteinase inhibitors found in the preceding chapter. Five of those agents, in fact, contain the same hydroxamide acid functional groups that characterize the metalloproteinases.

Antineoplastic Drugs: Organic Synthesis, First Edition. Daniel Lednicer.
© 2015 John Wiley & Sons, Ltd. Published 2015 by John Wiley & Sons, Ltd.

7.2   Hydroxamates

The simplest hydroxamate in terms of chemical structure of the HDAC inhibitors that have been accorded nonproprietary names happens to be the only one of the group that has been approved by the US FDA. This compound, vorinostat (**1.4**), was licensed in 2006 for treating cutaneous T-cell lymphoma. It is available under the trade name Zolinza®.

The preparation of this drug is as simple as its structure. The acylation of aniline with readily available suberic acid monomethyl ester affords the corresponding anilid (**1.3**). The treatment of the latter with hydroxylamine results in replacement of methanol by hydroxyl-amine to yield **vorinostat** (**1.4**) [1].

Belinostat inhibits three classes of HDAC enzymes. As often happens, the drug's anti-neoplastic activity can be attributed to several mechanisms in addition to the inhibition of HDAC. The compound showed good activity in several Phase III clinical trials. In early 2014, the FDA granted belinostat priority review status for its activity against T-cell lymphoma. The trade name for this compound is Beleodaq®. The scheme for preparing that compound begins with the sulfonation of benzaldehyde with fuming sulfuric acid. The product from that reaction is then treated with sodium hydroxide to afford sodium salt (**2.2**). The reaction of that intermediate with the ylide from Emmons reagent extends the aldehyde by two carbon atoms to afford cinnamate (**2.3**). Thionyl chloride then converts the sulfonate to reactive sulfonyl chloride. Reaction with aniline then attaches the extra benzene ring by way of a sulfonamide (**2.5**). The treatment of that intermediate with base hydrolyzes the cinnamate ester. The resulting carboxylic acid is then converted to an acid chloride by reaction with oxalyl chloride. Hydroxylamine reacts with the acid chloride to hydroxamate (**7.2**) and thus **belinostat** [2].

The HDAC inhibitor abexinostat (**3.8**) has shown activity in both *in vitro* and *in vivo* assays against a wide selection of cancers. In addition to inhibiting the deacetylase enzyme, this orally active compound also acts as a radiosensitizer. The synthesis of this compound begins with the Mitsunobu condensation of 4-hydroxymethylbenzoate

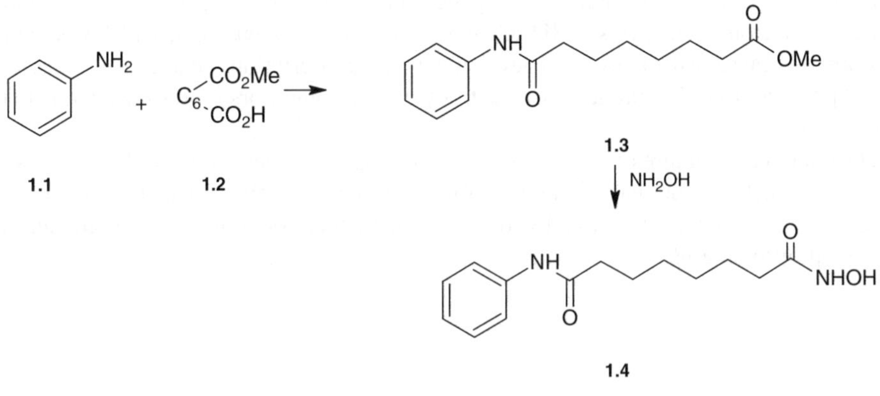

Scheme 7.1   *Vorinostat.*

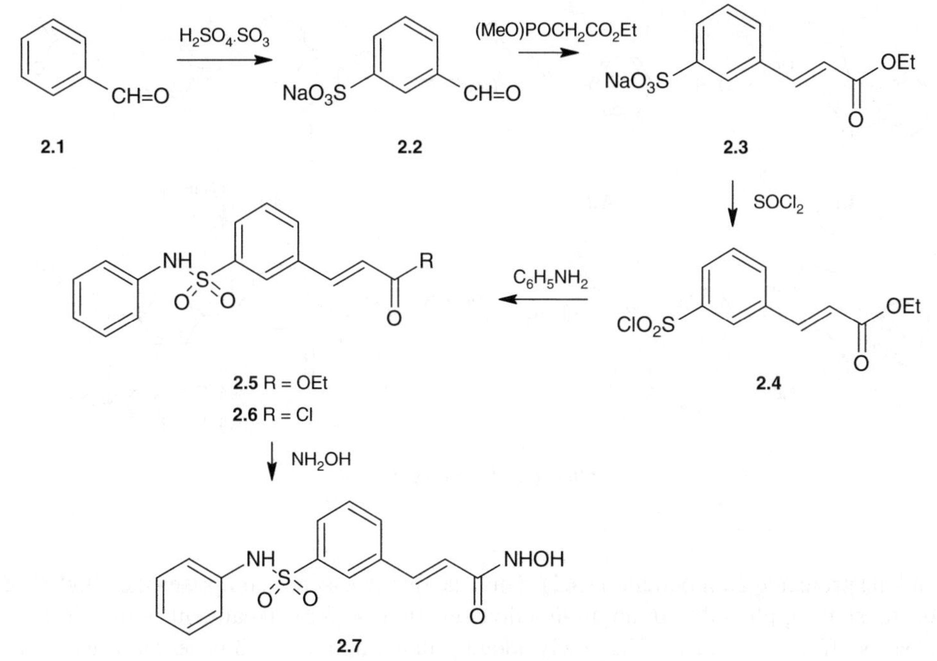

Scheme 7.2 *Belinostat.*

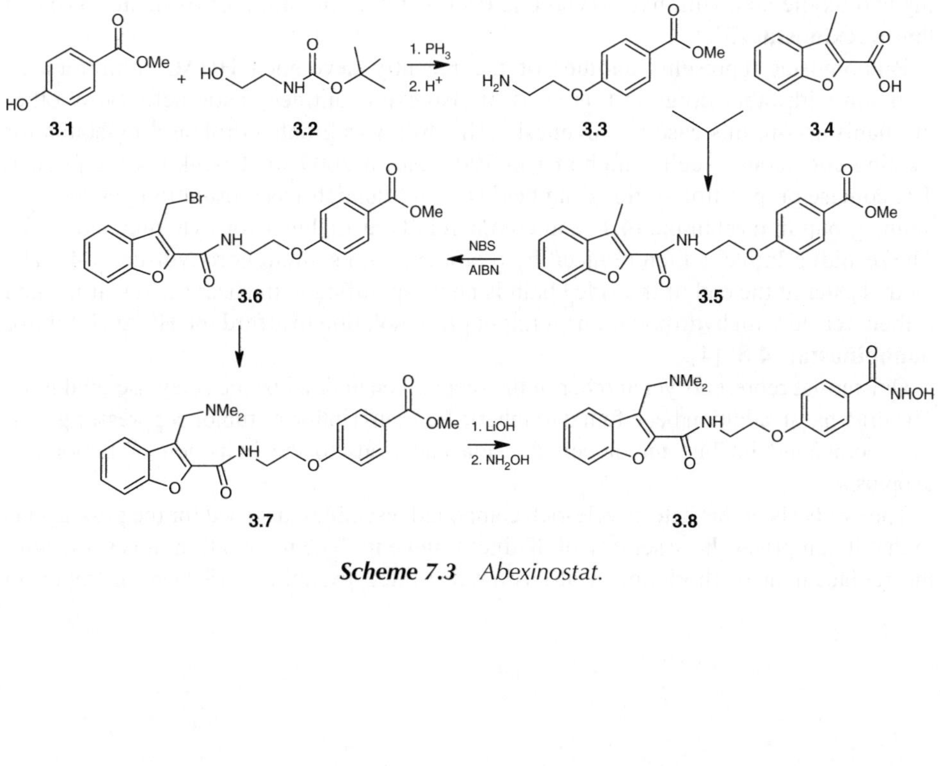

Scheme 7.3 *Abexinostat.*

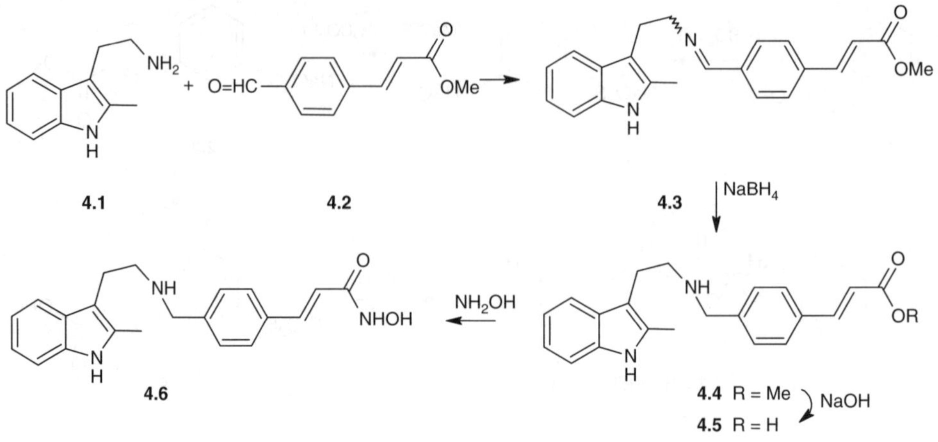

Scheme 7.4 *Panobinostat.*

and the protected aminoethanol (**3.2**). This reaction in essence comprises the alkylation of the starting phenol with an aminoethyl substituent (**3.2**). Treatment with acid then cleaves off the carbamate. The newly added primary amine in **3.3** is next acylated with benzofuran-2-carboxylic acid (**3.4**). The treatment of that intermediate (**3.5**) with bromine and AIBN results in the bromination of the sole methyl group. The reaction of the product with dimethylamine displaces the newly added halogen by basic nitrogen (**3.7**). There remains the task of adding the critical oxime function. The ester in the last intermediate (**3.6**) is then saponified with lithium hydroxide; the reaction of the resulting carboxylic acid with hydroxylamine results in the formation of oxamate (**3.8**) and thus **abexinostat** [3].

Panobinostat represents another of the recently developed HDAC inhibitors. In common with other drugs in this class, it also exerts antineoplastic activity by other mechanisms—in this case angiogenesis. The drug was granted orphan drug status for treating cutaneous T-cell lymphoma in 2007 and in 2009 for Hodgkin's lymphoma. The concise preparation of this drug begins with Mannich-base formation between the amine group in tryptamine (**4.1**) and commercially available 4-formylcinnamate (**4.2**). The resulting imine is then reduced by treatment with sodium borohydride (**4.4**). The methyl ester at the end of the side chain is next saponified with base; the resulting acid is then treated with hydroxylamine without prior isolation to afford the HDAC inhibitor **panobinostat** (**4.8**) [4].

Quisinostat represents yet another of the second-generation histone acetylase inhibitors. The drug has in addition been found to inhibit the transcription of tumor suppressor genes. This compound inhibits tumor cell division and leads to the induction of tumor cell apoptosis.

The synthesis of this heterocycle-rich compound resembles that used for the prior agents in that it comprises the assembly of distinct fragments. The preparation thus starts with the replacement of the halogen atom in substituted pyrimidine (**5.1**) by nitrogen in

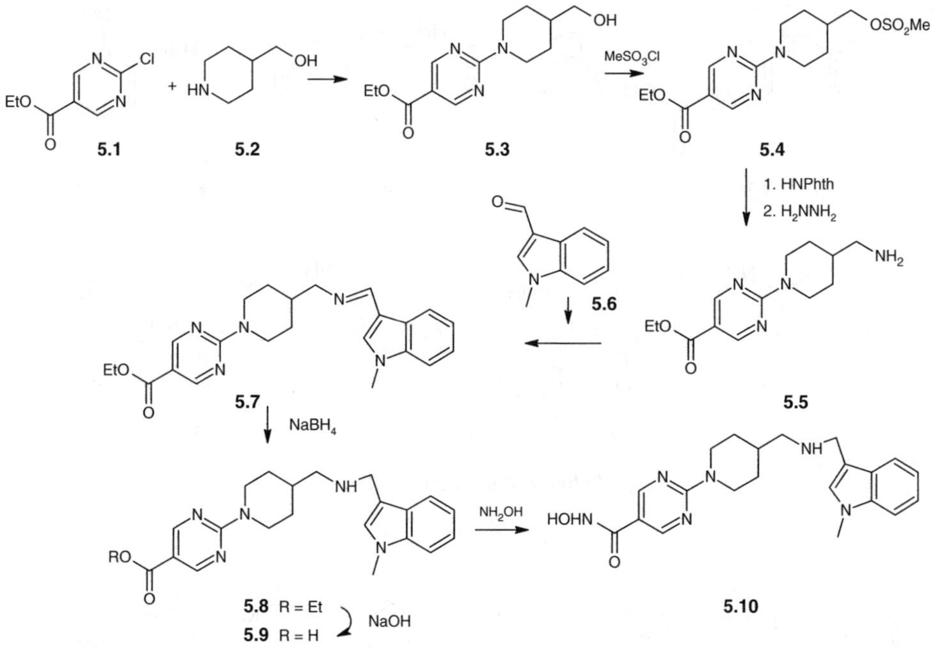

Scheme 7.5   *Quisinostat.*

4-hydroxymethyl piperidine (**5.2**). The carbinol at position 4 in the latter moiety is next prepared for displacement by conversion to a mesylate by reaction with methanesulfonyl chloride. The treatment of that intermediate (**5.4**) with phthalimide displaces the sulfonyl derivative by nitrogen in the reagent. Reaction with hydrazine, a standard reagent for freeing amines in phthalimides, does so in this case, providing amine (**5.5**). The reaction of that intermediate with the formyl group in indole (**5.6**) then connects the two moieties as a Mannich base; sodium borohydride reduces the unsaturation. The ester is next saponified and the acid treated with hydroxylamine. The hydroxamate **quisinostat** (**5.10**) is thus obtained [5].

7.3   Phenylenediamines

In addition to the quintet of HDAC inhibitors whose structures share a hydroxamic function, several HDAC inhibitors lack that function and compounds feature an *ortho*-diamino benzene moiety in their structure.

The inhibitor tacedinaline (**6.5**) has shown antineoplastic activity against a wide group of cancers in both *in vitro* and in cell culture screens. Clinical trials were under way in 2014 aimed at testing the efficacy of this drug against non-small cell lung cancer. One concise scheme for preparing the compound begins by converting the carboxylic acid to the chloride

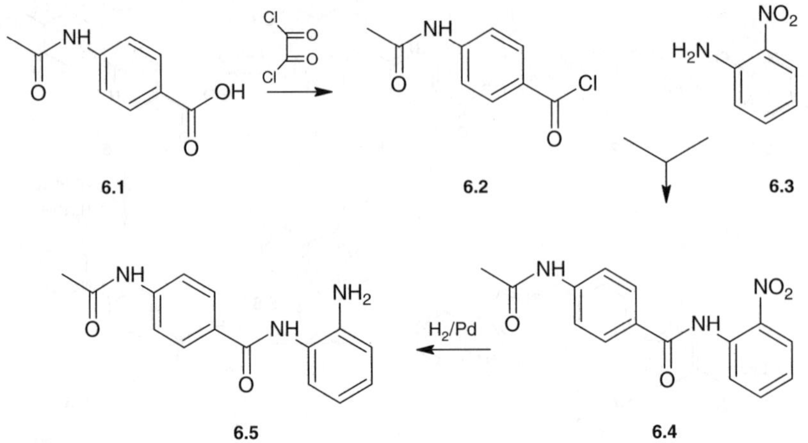

6.1 6.2 6.3 6.5 6.4

Scheme 7.6 *Tacedinaline.*

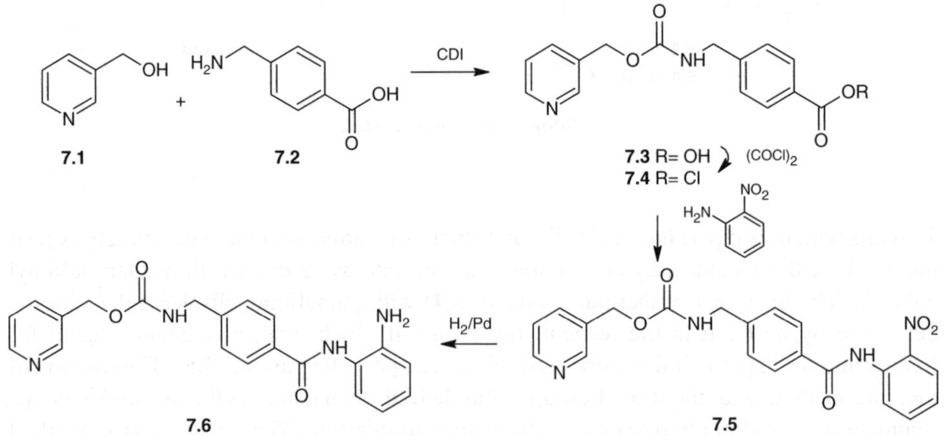

7.1 7.2 7.3 R= OH 7.4 R= Cl 7.6 7.5

Scheme 7.7 *Entinostat.*

by reaction with oxalyl chloride. That reactive intermediate is then reacted with 2-nitroaniline to form the corresponding amide (**6.4**). Catalytic hydrogenation over palladium reduces the nitro group, affording **tacedinaline** (**6.5**) [6].

In 2013, the FDA designated entinostat (**9.6**) as breakthrough therapy for the treatment of estrogen-positive breast cancer when administered in conjunction with the aromatase inhibitor exemestane (*see* Chapter 3). This designation is intended to expedite the development and review of drugs for serious or life-threatening conditions.

The synthesis of this compound in some ways resembles that used for tacedinaline. The initial step comprises connecting 3-hydroxymethylpyridine (**7.1**) and 4-aminomethyl benzaldehyde (**7.2**) by way of a carbamate. Combining the two moieties with carbonyl-diimidazole (CDI) affords the linked intermediate (**7.3**). Treating the product of that reaction with oxalyl chloride then converts the carboxyl group to the more reactive

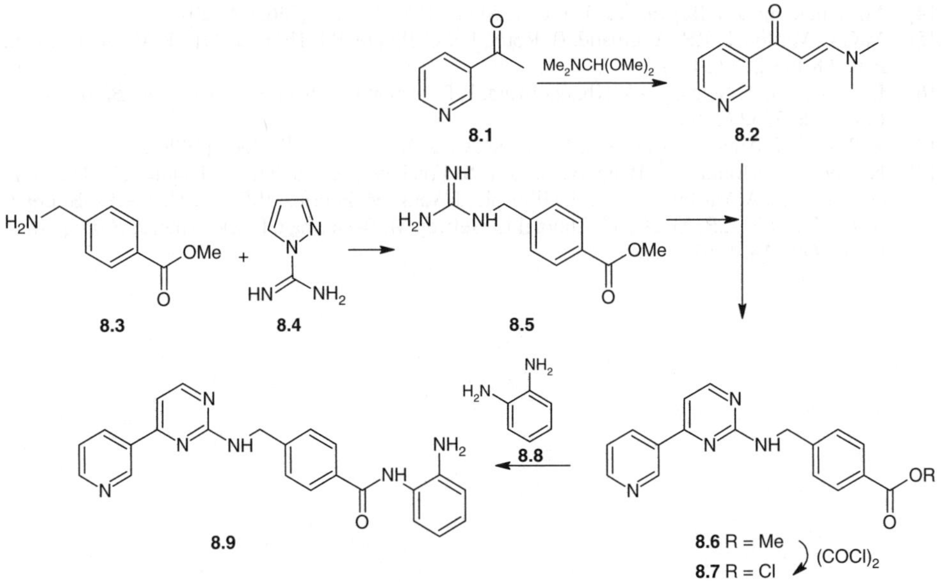

Scheme 7.8 *Mocetinostat.*

acid chloride (**7.4**). That intermediate is then treated with 2-nitroaniline to afford the corresponding benzamide (**7.5**). Catalytic hydrogenation reduces the nitro group to an amine. The HDAC inhibitor **entinostat** (**7.5**) is thus obtained [7].

The heterocycle-rich HDAC inhibitor mocetinostat (**8.9**) has shown good antineoplastic activity in tissue culture assays and implants of human cancers in laboratory animals. Clinical trials on this inhibitor were initiated in 2009. One short branch of the convergent synthesis comprises the reaction of 3-acetylpyridine with DMF acetal to afford enamide (**8.2**). The preparation of the other moiety involves the transfer of a guanidine function to aminomethylbenzoic ester (**8.3**) from pyrazole (**8.4**). The reaction of the guanidine group in the product (**8.5**) with enamide (**8.2**) results in the formation of a pyrimidine that now links the two fragments (**8.5**). The transform can be visualized by assuming that one of the guanidine nitrogen atoms adds to the end of the enamide function with loss of nitrogen; the condensation of the enamide carbonyl with the other guanidine nitrogen completes the formation of the pyrimidine (the order of the two steps may be reversed). The ester is next saponified and the resulting acid converted to its acid chloride by treatment with oxalyl chloride. The reaction of that last intermediate with 1,2-diaminobenzene concludes the synthesis on **mocetinostat** (**8.9**) [8].

References

[1] J.C. Stowell, R.I. Huott, L. Van Voastt, *J. Med. Chem.* **38**, 1411(1995), *this reference reflects the first publication rather than synthesis method.*
[2] I. Kalvinsh, E. Loza, V. Gallite, U.S. Patent 7,557,140 (2009).
[3] E.J. Verner, M. Sendzik, C. Baskaran J.J. Buggy, J. Robinson, U.S. Patent 7,420,089 (2008).

[4] M. Acemoglu, J.S. Bajwa, D.J. Parker, J. Slade, U.S. Patent, 8,536,346 (2013).
[5] M.G.C. Verdonck, P.R. Angibaud, B. Roux, I.N.C. Pilatte, P.T. Holte, J. Arts, K.V. Emelen, U.S. Patent 8,193,205 (2012).
[6] L.K. Gediya, A. Belosaya, A. Khandelwala, P. Purushottamachara, V.C.O. Njarab, *Bioorg. Med. Chem.* **16**, 3352 (2008).
[7] T. Suzuki, T. Ando, K. Tsuchiya, N. Fukazawa, *J. Med. Chem.* **42**, 3001 (1999).
[8] N. Zhou, O. Moradei, S. Raeppel, S. Leit, S. Frechette, F. Gaudette, I. Paquin, N. Bernstein, G. Bouchain, A. Vaisburg, Z. Jin, J. Gillespie, J Wang, M. Fournel, P.T. Yan, M. Trachy-Bourget, A. Lu, J. Rahil, A.R. MacLeod, Zuomei Li, Jeffrey M. Besterman, D. Delorme, *J. Bioorg. Med. Chem. Lett.* **18**, 1067 (2008).

8

Enzyme Inhibitor, Part I, Tyrosine Kinases

8.1 Introduction

The presence or absence of a phosphate group at a specific site on selected enzymes and related proteins acts as an instruction to turn on or cease some operation. Kinases comprise a large group of enzymes that attach phosphate groups (PO_4) by a covalent bond to the hydroxyl group of a specific amino acid on a polypeptide. Three of the essential amino acids, tyrosine, threonine, and serine, each of which has a hydroxyl group, comprise the recipients of kinase-mediated phosphates. The phosphate group itself comes from a high-energy source of that ion, such as adenosine triphosphate (ATP). The proteins receiving the phosphate groups are involved in many basic cellular processes such as regulation of the cell cycle. Protein kinases accordingly play a crucial role in cellular proliferation, differentiation, and various regulatory processes. Kinases are named after the amino acid in the peptide that is to be phosphorylated (the enzymes that later remove phosphate groups are termed phosphorylases). The kinases that originate from malignant tumors are the cause many of the ills caused by cancer that include unregulated cell proliferation, leukocyte activation, cell migration, and not least, pain. Kinin inhibitors have been the target for many research programs over the past several decades aimed at developing new antineoplastic agents. Such inhibitors, it was felt, might offer a more specific target than those of the earlier antineoplastic and, as such, might evoke fewer unwanted side effects. Tyrosine kinases comprise a sizeable group of proteins that phosphorylate the hydroxyl group on tyrosine. Chapter 9 describes a number of kinase that add phosphates to serine and threonine. The factors and their acronyms are listed in Table 8.1. The section that follows lists the "small-molecule" compounds that have been assigned a nonproprietary name. Compounds are arranged by the class of tyrosine kinase receptors with which the drug interacts. Tyrosine kinase inhibitors (TKI), it will be noted, have been an unusually

Antineoplastic Drugs: Organic Synthesis, First Edition. Daniel Lednicer.
© 2015 John Wiley & Sons, Ltd. Published 2015 by John Wiley & Sons, Ltd.

Table 8.1   *Tyrosine kinase factors*

ALK	Anaplastic lymphoma kinase
EGF	Epidermal growth factor
ErbB	Epidermal growth factor
FGF	Fibroblast growth factor
IGF-1	Insulin growth factor
MEK	Mitogen-activated protein kinase
MET	Hepatocyte growth factor
PDGF	Platelet-derived growth factor
SRC	Nonreceptor protein kinase
VEGF	Vascular endothelial growth factor

rich class of compounds that have been approved by the FDA for treating malignant tumors; just over twenty TKIs have been cleared for use in the clinic. Note however that those approvals currently tend to be very detailed and restrictive. The drug might be approved for treating patients with non-small cell lung cancer whose disease no longer responded to cisplatin. Since this is not a prescribing manual, this book stays with the broader definition. It need be noted in passing that these drugs are usually administered in solid dosage form since they are orally active. This is in contrast to many older antineoplastic agents that are often administered as intravenous infusion.

8.2   Epidermal Growth Factor Inhibitors

The tyrosine kinases that target epidermal growth factor (EGF) play a major role in the proliferation of tumor cells as well as formation of blood vessels in those tumors. This subclass of TKI has proven to be a fertile source for FDA-approved antineoplastic agents. The three drugs that have been licensed by the agency for treating cancer patients comprise 50% of the epidermal growth factor receptor (EGFR) inhibitors that have entered clinical trials.

Erlotinib (**1.7**) was the second kinase inhibitor to be approved by the FDA. This drug reacts reversibly with the EGFR that occur frequently in cancer cells. Clinical trials showed positive effects against a selection of solid malignant tumors. The FDA first approved the drug for use against pancreatic cancer when used in combination with gemcitabine. Use against other cancers followed. The agent is said to be well tolerated: the most frequent side effect being a rash on the neck and face. This agent is sold under the trade name Tarceva®.

Virtually all compounds that inhibit tyrosine kinase receptors feature a quinazoline moiety; building that ring system thus comprises an important part of any scheme for preparing that ring systems. The synthesis of erlotinib starts with a step to install one of the nitrogen atoms in the future fused ring. Thus, nitration of benzoic ester (**1.1**) with the traditional mixture of nitric and sulfuric acids provides the 2-nitro derivative (**1.2**). Catalytic hydrogenation then reduces the newly introduced nitro group to a primary amine. Treating that product with ammonium formate builds the requisite new ring. This transform can be rationalized by assuming that the reaction starts with the addition of the formate to the amine in (**1.3**). Ammonia from the ammonium salt then displaces the ethoxy group from

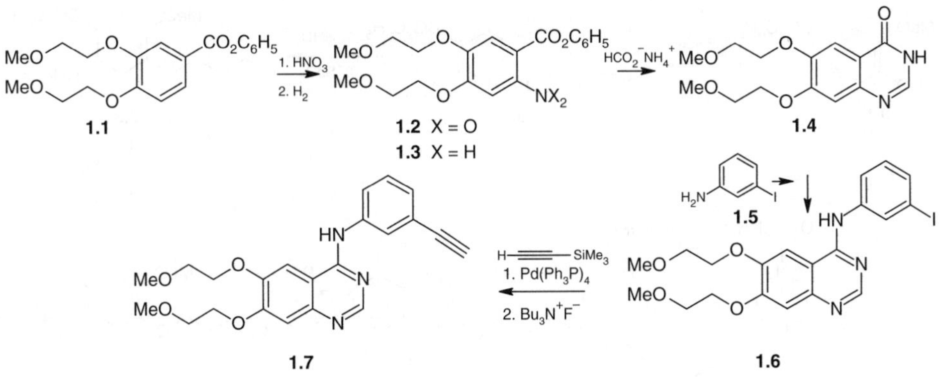

Scheme 8.1 *Erlotinib.*

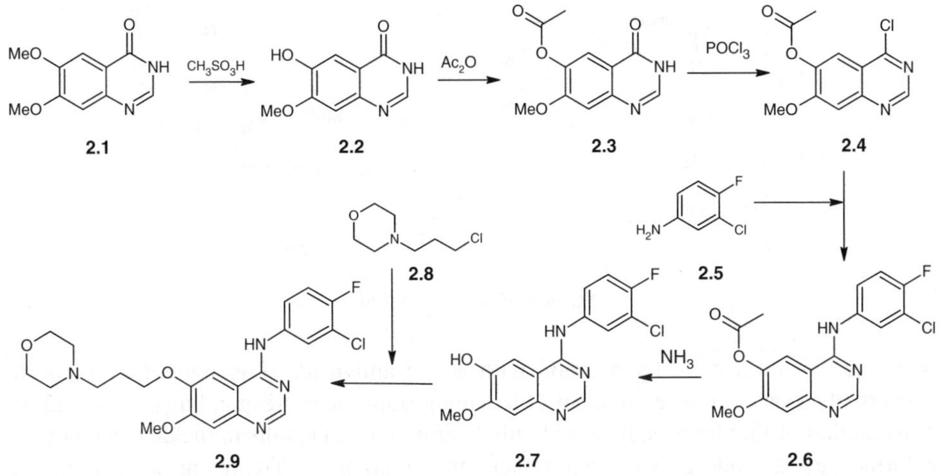

Scheme 8.2 *Gefitinib.*

the ester. Addition–elimination between the amides closes the ring to afford quinolone (**1.4**). The reaction of 3-iodioaniline with the last intermediate precedes to add that moiety to the heterocyclic ring, speculatively again by an addition–elimination scheme. This is followed by palladium-/triphenylphosphine-catalyzed displacement of iodine in a Suzuki exchange. The removal of silane at the end of the acetylene side chain with a fluoride ion completes the synthesis of **erlotinib** (**1.7**) [1].

Gefitinib was the first TKI to receive FDA approval; the agency approved the drug for use in treating cancer patients. More specifically, the drug was licensed for use in treating non-small cell lung cancer. By the same token, the drug acts by binding to EGFR. Like most other TKI, gefitinib is orally active. The drug is available under the trade name Iressa®.

The synthesis of gefitinib begins with a reaction that differentiates the two oxygens on the fused benzene ring. The treatment of that compound with methylsulfonic acid results in demethylation of the most electron-rich ether (**2.2**). The phenol in that product

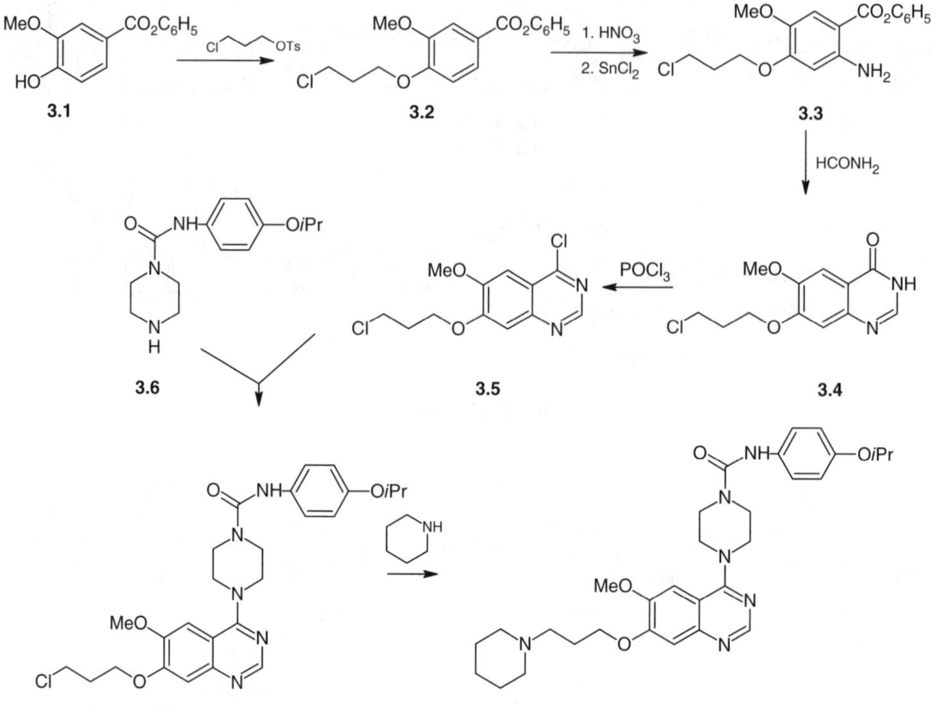

Scheme 8.3 *Tandutinib.*

is masked as its acetate by reaction with acetic anhydride. Phosphorus oxychloride converts the carbonyl group to a better leaving group: the corresponding chloride (**2.4**). The reaction of that intermediate with dihalo aniline (**2.5**) results in the displacement of chlorine and the addition of that moiety to quinazoline. Treatment with ammonia hydrolyzes the masking group to afford free phenol (**2.7**). The reaction of that intermediate with base followed by *N*-(3-chloropropyl)morpholine (**2.8**) affords the inhibitor **gefitinib** (**2.9**) [2].

The structurally more complex TKI tandutinib also binds to EGFR. The drug also inhibits autophosphorylation of selected tyrosine kinases. The first few steps in the synthesis resemble those used to prepare erlotinib with the exception that one of the oxygens occurs as a free phenol (**3.1**). The reaction that builds the fused pyrimidine ring uses formamide rather than ammonium formate as the source of the required carbon atom. Reaction with phosphorus oxychloride converts the carbonyl group, shown as its keto tautomer, to a better leaving group (**3.5**) than the side chain halogen. The reaction of that last intermediate with piperazine urea (**3.6**) leads to the displacement of chlorine by piperazine nitrogen (**3.6**). The next step consists of another displacement, this time of the side chain chlorine by piperidine. The TKI **tandutinib** (**3.8**) is thus obtained [2].

Lapatinib is the next TKI to receive FDA approval for treating cancers. In this case, the drug was licensed in 2007 for treating breast cancer patients who were already taking gemcitabine. In 2001, the agency further increased the indications by approving the use of

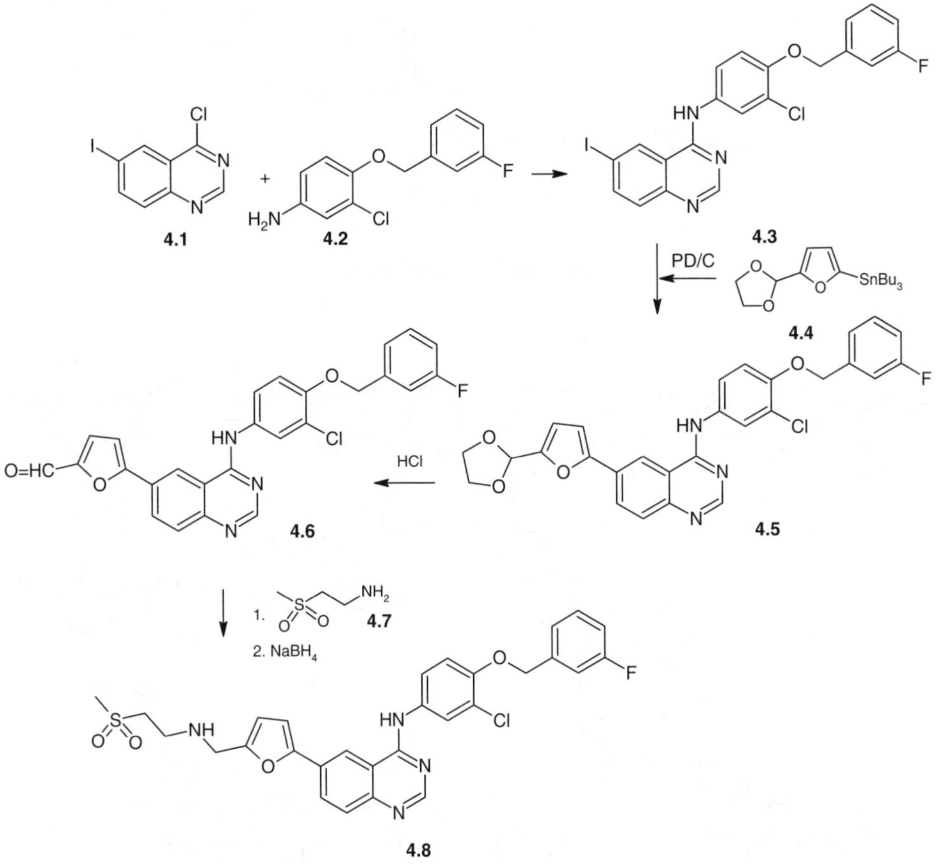

Scheme 8.4 *Lapatinib.*

the drug in combination with the aromatase inhibitor letrozole (*see* Chapter 3). The drug is sold under the trade name Tykerb®.

Although the structure of the TKI lapatinib looks forbidding at first sight, the synthesis actually consists of a series of displacements. The key reaction comprises the displacement of chlorine in quinazoline (**4.1**) by the amine in aniline (**4.2**). The selectivity for that chlorine over iodine can be rationalized by the fact that chlorine comprises an enol chloride. This intermediate is the subjected to cross-linking the reaction of the iodide on the quinazoline with tributyltinfuran (**4.4**). The acetal on the latent aldehyde on the furan is then opened by acid-catalyzed hydrolysis. The condensation of the newly revealed aldehyde with aminosulfone (**4.7**) joins the two moieties via a Mannich base: that link is then reduced to an amine by reaction with sodium borohydride. This product comprises the inhibitor **lapatinib** (**4.8**) [4].

The biologic activity of the TKI varlitinib is quite similar to that of the preceding agents. The drug, for example, binds reversibly to EGFR. In addition the compound inhibits the vascular endothelial growth factor (VEGF). In the absence of a source for the scheme used to prepare varlitinib the structure is displayed sans synthesis.

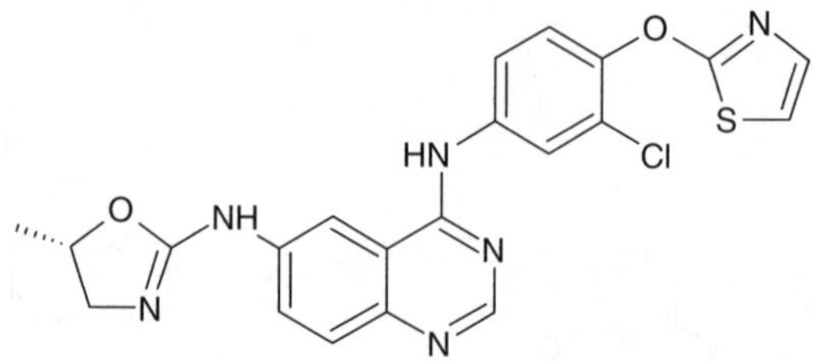

Scheme 8.5   *Varlitanib.*

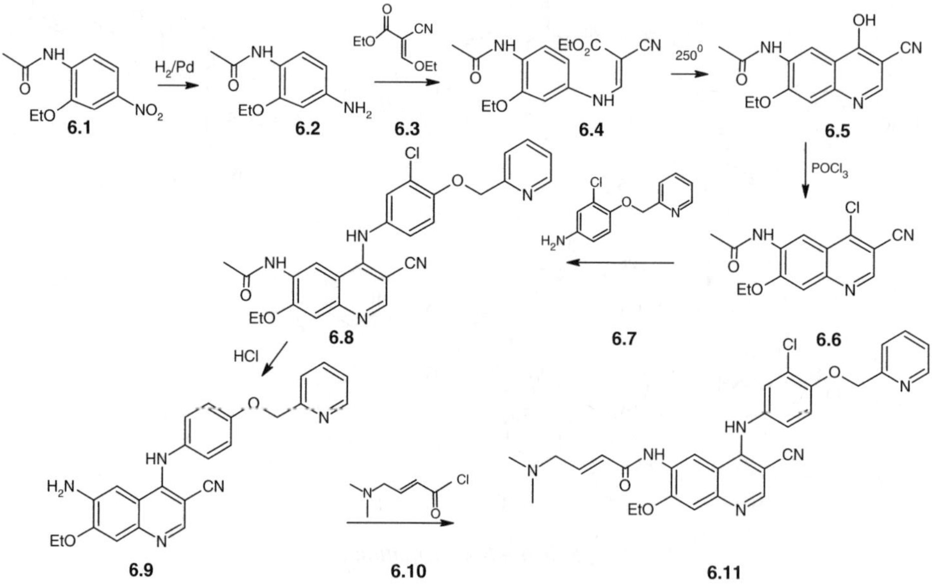

Scheme 8.6   *Neratinib.*

Neratinib is another TKI that binds to EGFR. This binding is however irreversible as is the case for the predecessor. The drug is fully active when taken by mouth; clinical trials thus involved that route of administration. Data from Phase II trials on breast cancer showed encouraging activity.

The synthesis of this kinase inhibitor begins with the construction of the quinoline core. Catalytic hydrogenation reduces the nitro group in the starting material to an amine (**6.2**). The reaction of that product with enol ether (**6.3**) leads to the adduct (**6.4**); this reaction likely involves an addition–elimination sequence. Heating the adduct to 250° then causes the ring to close and form the desired quinazoline. In order to ease addition of a major substituent, the hydroxyl group at C4 is converted to a chloride by means of phosphorus oxychloride. The newly introduced halogen is then displaced by the amine in pyridiloxyl ether (**6.7**). Treatment with acid then causes the acetyl group to hydrolyze affording amine (**6.9**). The acylation of that

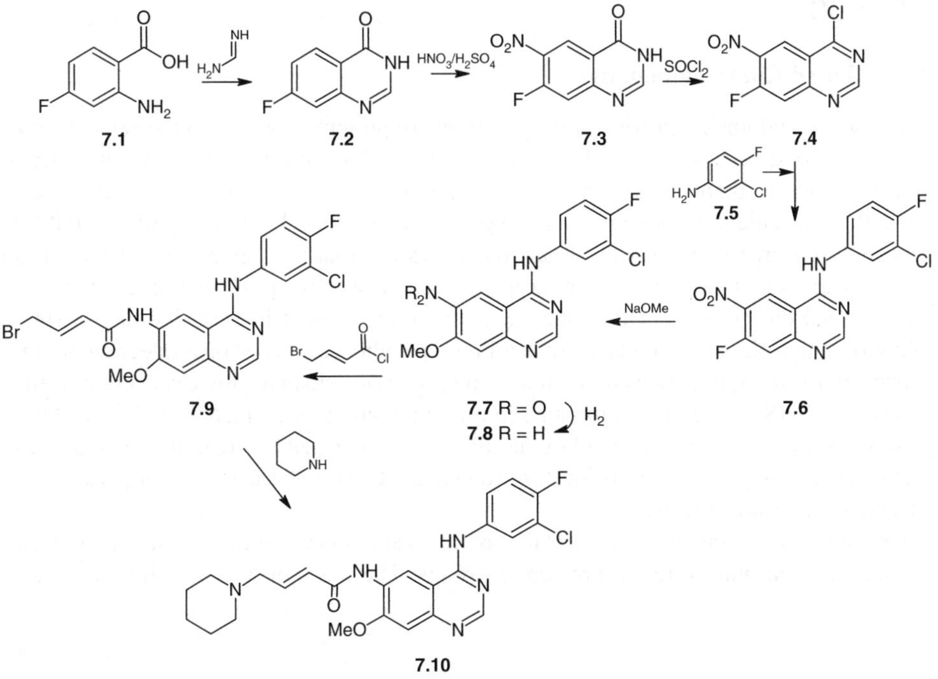

Scheme 8.7   *Dacomitinib.*

function with 4-dimethylaminobut-2-enyric chloride affords the corresponding amide. **Neratinib (6.11)** is thus obtained [6].

In vitro assays showed that the TKI dacomitinib specifically and irreversibly binds to and inhibits human EGFR subtypes. The drug also exhibited encouraging results in various *in vivo* studies in laboratory models of various cancers. Results from clinical trials such as those targeted against non-small cell lung cancer failed to show results that were better than those achieved with existing treatment.

The preparation of the TKI dacomitinib follows the general route used to prepare neratinib. Thus, the synthesis starts with the formation of the quinazoline nucleus. The condensation of the substituted anthranilic acid (**7.1**) with amidine leads to the requisite quinazoline (**7.2**). The treatment of that intermediate with the traditional mixture of nitro and sulfuric acids affords the nitro intermediate (**7.3**) which will be retained as such until the penultimate step in the synthesis. The carbonyl function in the heterocyclic ring is next converted to a better leaving group by reaction with thionyl chloride instead of the customary phosphorus oxychloride. The newly introduced halogen is next displaced by the amine in the substituted aniline (**7.5**) to afford the intermediate (**7.7**). The treatment of that product with sodium methoxide then serves to displace the fluorine substituent and thus introduces a methyl ether at that position. Catalytic hydrogenation then serves to reduce the nitro group to an amine (**7.8**). The only basic amine in this intermediate is next acylated with 4-bromobut-2-enebutyril chloride. The displacement of the pendant bromine by piperidine affords **dacomitinib (7.10)** [7].

8.3 VEGF

8.3.1 Fused Ring Compounds

The vascular endothelial factor comprises a group of proteins that control the cells that produce new blood vessels in case of need such as the lack of oxygen created by an injury or a disease such as diabetes. The rapid proliferation of malignant tumors can quickly outgrow their vasculature, thus creating oxygen-poor interiors. Inhibiting VEGF would be expected to decrease the growth of new blood vessels, that is, angiogenesis. This will, in theory, starve the tumor's supply of oxygen. In contrast to the prior sections, compounds are arranged on the basis of their chemical structures. It is of interest that only a single inhibitor in this class is based on a quinazoline. The others are based on a selection of other benzo heterocycles; one such compound is composed of a chain of single ring components.

Vandetanib (8.9), the only compound in this small group of antineoplastic agents based on quinazoline, has been approved by the FDA; it is indicated for treating medullary thyroid cancers. The package insert for this drug named Caprelsa® carries a strong warning of adverse cardiac side effects.

One synthesis of this drug starts with a preformed quinazoline nucleus that carries two phenols masked with different protecting groups. The reaction of that quinazoline (**8.1**)

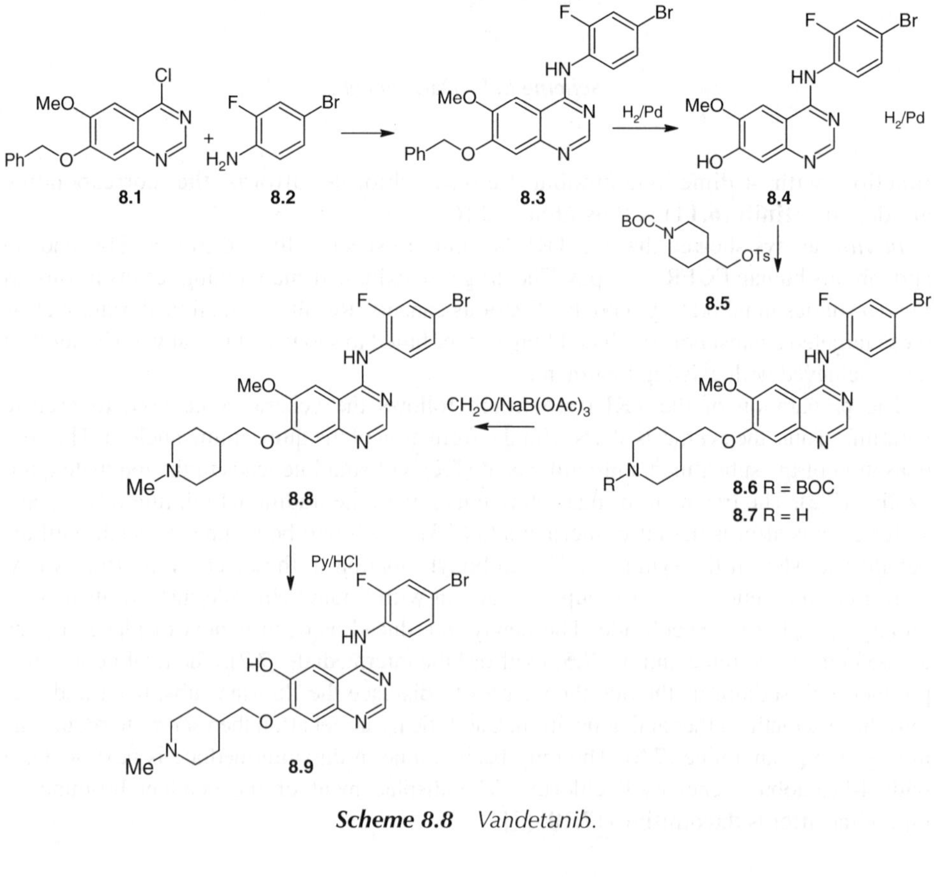

Scheme 8.8 *Vandetanib.*

with the substituted aniline (**8.2**) results in the displacement of quinazoline chlorine to form the linked intermediate (**8.3**). Catalytic hydrogenation removes the benzyl protecting group present on one of the phenols to afford free phenol (**8.4**). The hydroxyl group is then alkylated with piperidine 4-methyltoluenesulfonate to yield piperidyl ether (**8.6**). The tertiary butylcarbonyl group on piperidyl ether is then removed by exposure to acid. The newly revealed free secondary amine is then alkylated by reaction with formaldehyde and sodium triacetoxy borohydride (formaldehyde first forms a Mannich base which is then reduced by borohydride). Scission of methyl ether completes the synthesis of **vandetanib** (**8.9**) [8].

The TKI lenvatinib is another agent that acts on the VEGF receptor. The drug has been granted orphan drug status by the FDA in the United States as well as in Japan by the Japanese Ministry of Health and Welfare. A Phase III clinical trial to treat thyroid cancer in a specifically selected group of patients showed improvement in progression-free survival [9].

The concise preparation of this inhibitor begins with the preparation of the carbamate from the substituted *para*-aminophenol (**9.1**). Thus, the reaction of the latter with chlorocarbamate (**9.2**) yields phenoxy carbamate (**9.3**). The treatment of that intermediate with cyclopropylamine results in the displacement of the phenoxide by cyclopropylamine to form urea (**9.4**). The reaction of that product from that reaction with chloroquinolone (**9.5**) in the presence of potassium *tert*-butoxide results in the displacement of chlorine in the

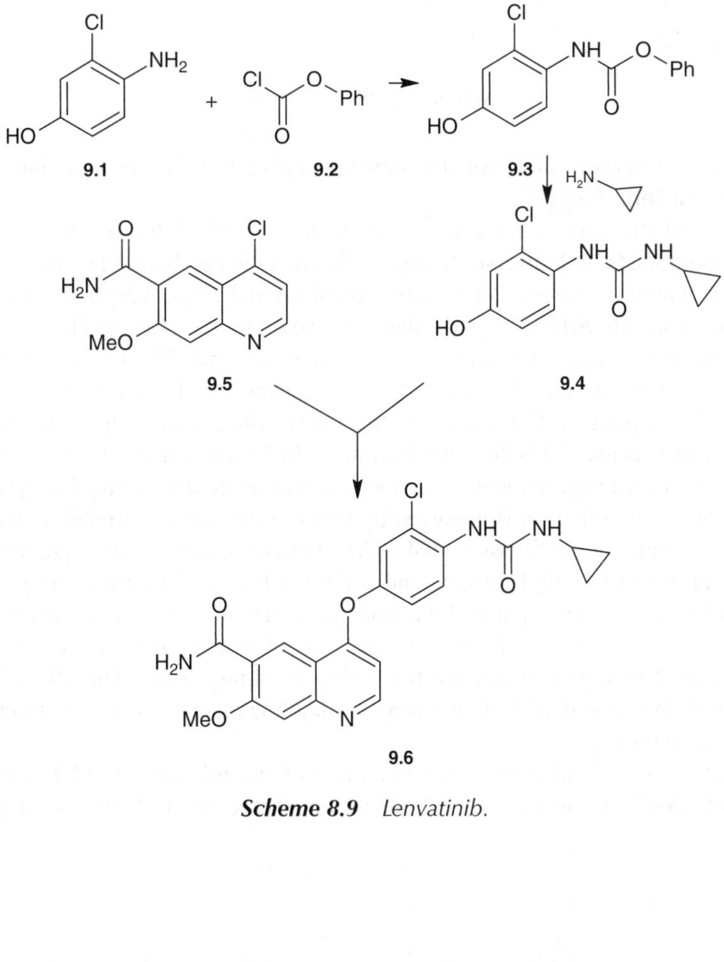

Scheme 8.9 *Lenvatinib.*

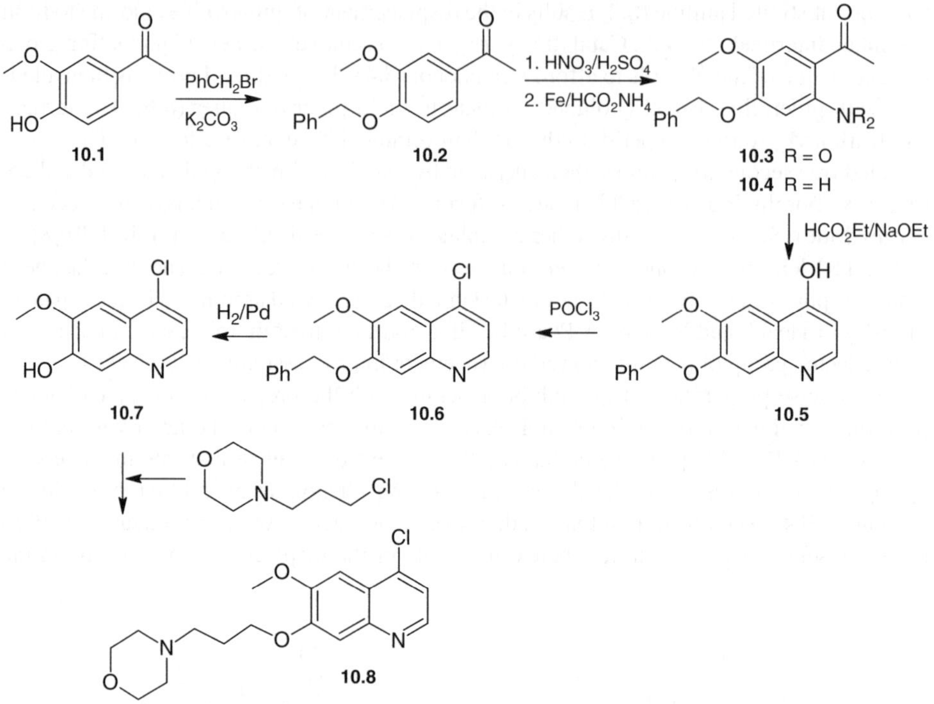

Scheme 8.10   Foretinib I

quinolone by the oxygen anion from the substituted phenol (**9.4**). The now-linked moieties comprise **lenvatinib** (**9.6**) [10].

The chemical structure of many good TKI suggests that high-throughput screening of a library of compounds with diverse structures obtained via combinatorial chemistry played a major role in their discovery. The scientists involved in the discovery and development of the inhibitor foretinib (**8.11**) have published an account to that effect [11].

The preparation of this agent involves a convergent scheme. The synthesis of the quinoline moiety begins with the alkylation of the free phenol in **10.1** with benzyl bromide in order to mask that oxygen. The reaction of that intermediate with the traditional mixture of nitric and sulfuric acids yields the nitro derivative (**10.3**). The common means for reducing that function, catalytic hydrogenation, is ruled out by the presence of the benzyl protecting group. Instead, the reduction is achieved by treating the nitro intermediate with iron in the presence of ammonium formate (**10.4**). The additional carbon atom required for closing the ring is provided by ethyl formate and sodium ethoxide. The hydroxyl group in the resulting quinolone is then replaced by a more active leaving group, chlorine, by means of phosphorus oxychloride. There remains the task of adding the side chain. Catalytic hydrogenation then serves to remove the benzyl masking group. The alkylation of the newly formed phenol with *N*-3-chloropropylmorpholine completes the construction of the quinoline-based moiety.

The reaction of the anion from the treatment of benzyl alcohol (**11.1**) with sodium hydride and 3,4-difluoronitrobenzene leads to the displacement of fluorine at position 4.

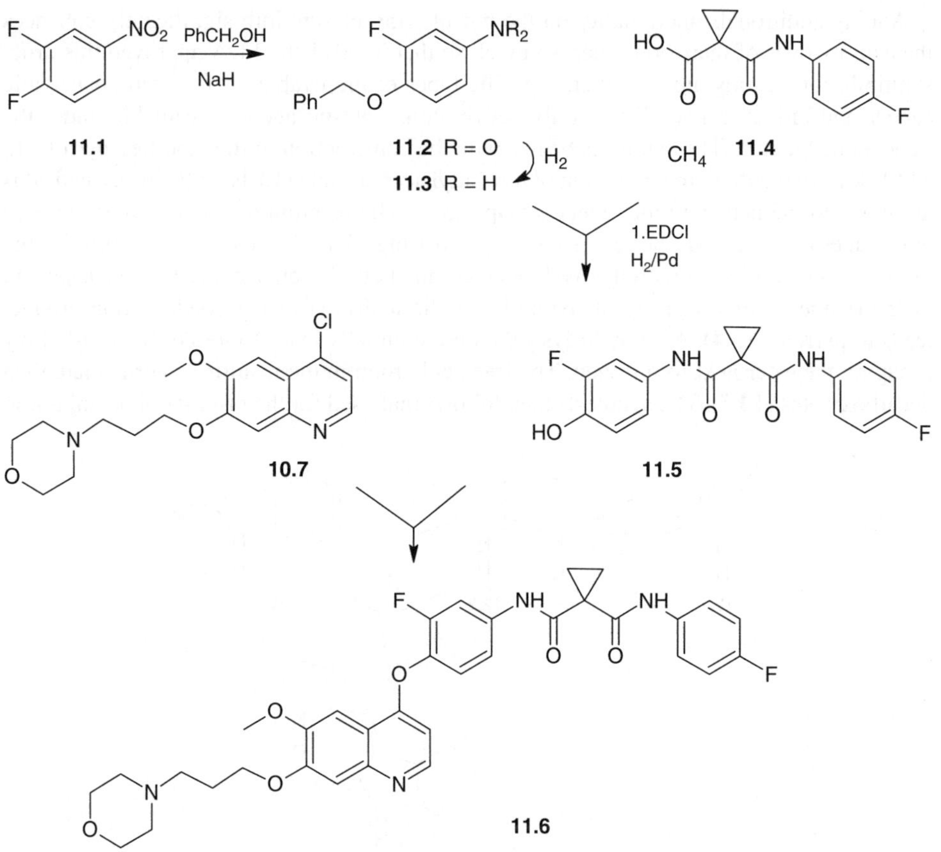

Scheme 8.11 *Foretinib II.*

The nitro group is next reduced to the corresponding amine by catalytic hydrogenation (**11.3**). The condensation of the newly formed substituted aniline with the carboxylic acid (**11.4**) mediated by a carbodiimide connects the fragments by way of an amide (**11.5**). The benzyl protecting group is next cleaved by catalytic hydrogenation to afford the corresponding phenol (**11.5**). Yet another displacement reaction on the anion from the treatment of phenol (**11.5**) with the quinoline from the other arm of the synthesis affords the long strung product **8.11**. That product comprises the inhibitor **foretinib** [12].

The structure of indolone-based antineoplastic agent semaxanib obviously departs radically from the previous TKI. The compound shows quite selective binding to VEGF receptors. That activity translates to antiangiogenic activity in animal models. Trials on the drug have been halted due to poor results as well as early promising data on the closely related inhibitor sunitinib.

The scheme for preparing this uncharacteristically simple TKI begins with a modern version of the Vilsmeier reaction. Thus, the treatment of 3,5-dimethylpyrrole (**12.1**) with DMF and phosphorus oxychloride serves to introduce a formyl group at C2. Base-catalyzed condensation of this product with indolone (**12.3**) affords **semaxanib** (**12.4**) [13].

Adding additional functionality on the pyrrole ring of sunitinib significantly enhances the activity over of that forerunner, so much so that in 2011 the FDA approved this drug, sunitinib, for treating among others, specific types of inoperable or metastatic pancreatic cancer and kidney cancer. This orally active antineoplastic agent is available under the trade name Sutent®. The synthesis begins with the construction of the modified pyrrole. In a little known reaction, the treatment of *tert*-butyl acetoacetate (**13.1**) with nitrous acid adds an oxime to the activated methylene group (**13.2**). The reaction of this intermediate with ethyl acetoacetate can be envisaged by positing that the carbonyl group in the intermediate condenses with the methylene group of ethyl acetoacetate. The displacement of the oxime hydroxyl group of oxime by enolic hydroxyl of the condensation product leads to pyrrole (**13.4**). Acid hydrolysis then preferentially leads to loss of the *tert*-butoxy group on the ester next to nitrogen. The free acid group in the resulting intermediate then decarboxylates (**13.5**). The synthesis then follows that used for the prior indolone inhibitor.

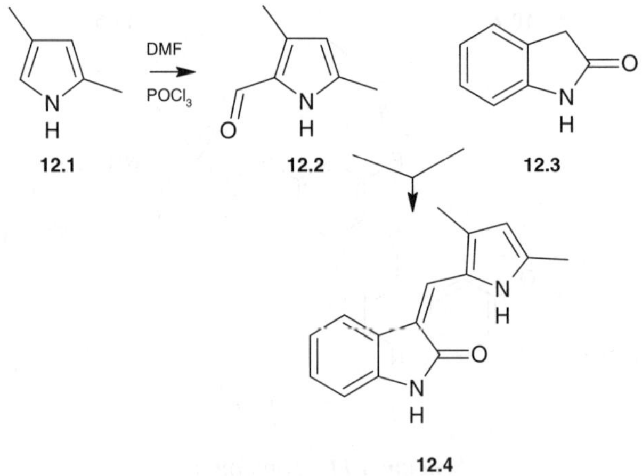

Scheme 8.12   *Semaxanib.*

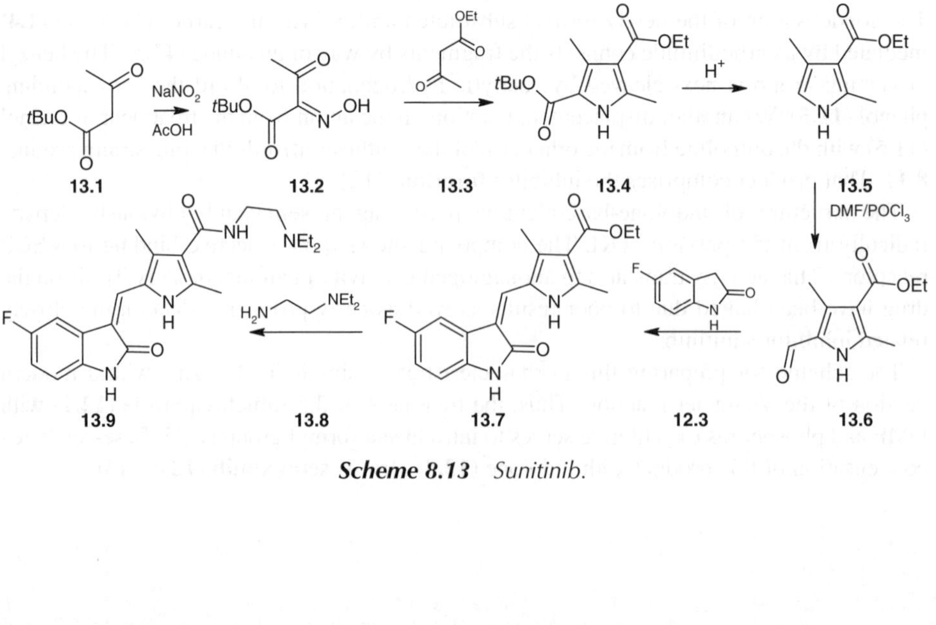

Scheme 8.13   *Sunitinib.*

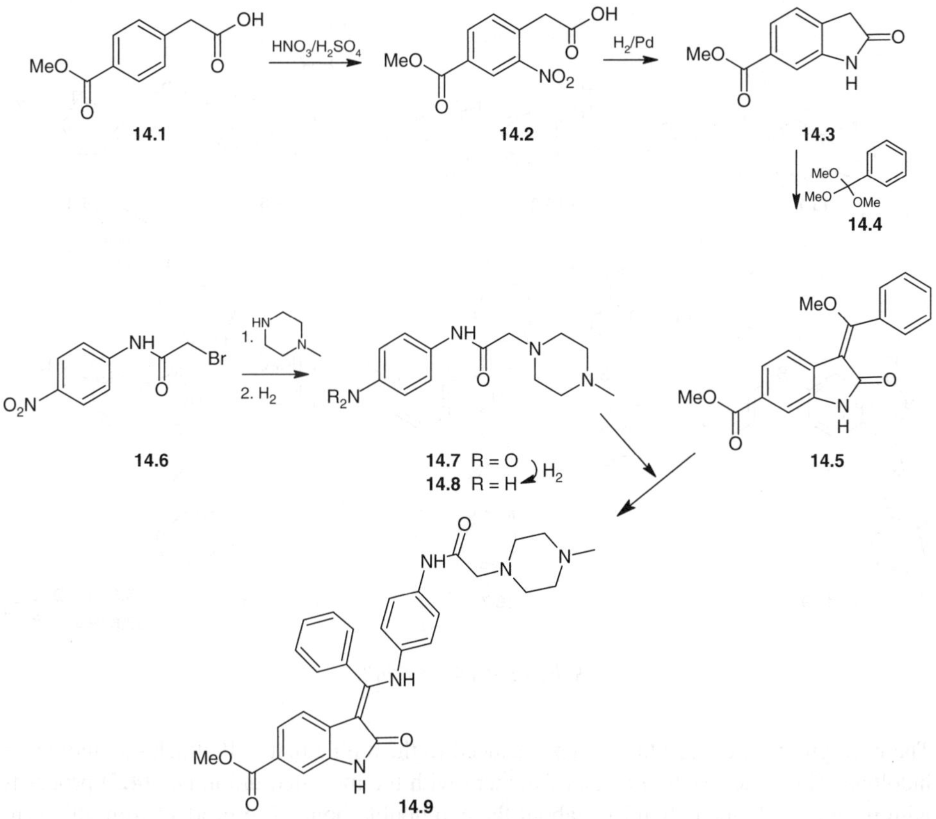

Scheme 8.14 *Nintedanib.*

The required formyl group is added by reaction with DMF and phosphorus oxychloride. This is then condensed with indolone (**12.3**) to form the requisite carbon skeleton. The ester–amine interchange of this last intermediate with *N,N*-diethyl ethylenediamine adds the basic side by way of an amide. The tyrosine inhibitor **sunitinib** (**13.9**) is thus obtained [14].

The TKI nintedanib also binds selectively to VEGF receptors. The resulting inhibitions reduces angiogenesis in tumors. Clinical studies are underway for determining the activity in a wide selection of cancers. VEGF receptors are also involved in idiopathic pulmonary fibrosis, a debilitating noncancerous disease for which there is no approved treatment. In 2011, the FDA granted the drug orphan status for treating that illness. The trade name for this agent is Vargatef®.

The convergent scheme for preparing this compound starts with the construction of the indolone nucleus. The treatment of homopterephthalic ester (**14.1**) with a mixture of nitric and sulfuric acids introduces a nitro group at the position adjacent to the methylene group (**14.2**). Catalytic hydrogenation then reduces that function to an amine. This then reacts to displace the ester affording indolone (**14.3**) in one fell swoop. Base-catalyzed condensation of the indolone with orthoformate (**14.4**), derived from benzoic acid, adds that group to the only methylene group as an enolate. The other branch of the synthesis involves first the displacement of bromine in the methylene group on acetamide (**14.6**) by *N*-methylpiperazine.

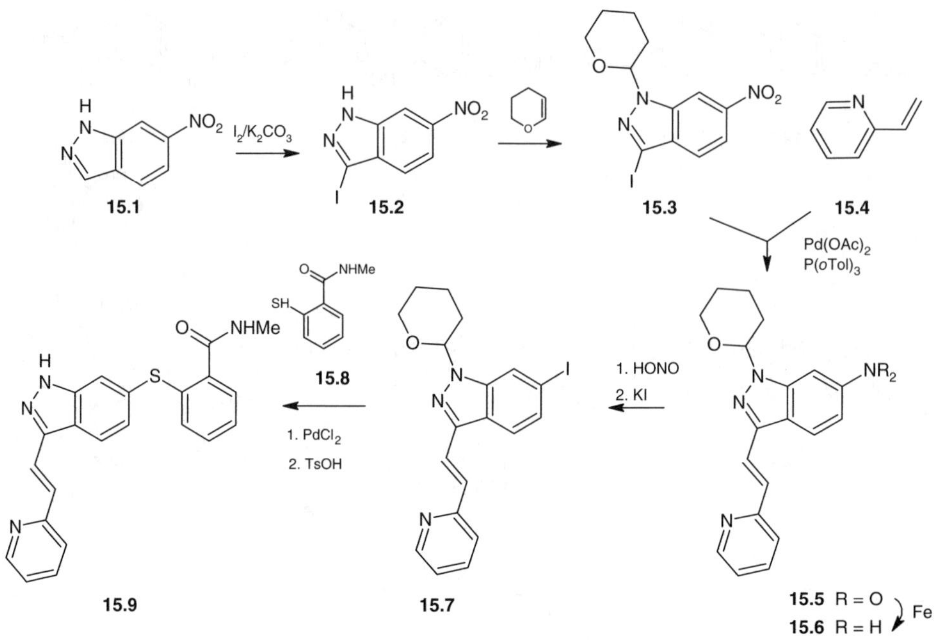

Scheme 8.15   *Axitinib.*

The nitro group in the product is next reduced to the amine that will attach this moiety to indolone. Thus, the reaction of that fragment with the modified indolone (**14.5**) proceeds with inversion of stereochemistry about the exo double bond. The product from this synthesis (**14.9**) comprises **nintedanib** [15].

The TKI axitinib binds relatively selectively to VGFR receptors. This leads to the inhibition of angiogenesis and consequent regression of tumors due to the resulting anoxia. In common with the majority of other kinase inhibitors, this drug is orally active; it is consequently administered as tablets. In 2012, the FDA approved axitinib for the treatment of renal cell carcinoma for patients who had undergone a prior course of chemotherapy that had failed. The drug is available under the trade name Inlyta®.

An early scheme for preparing the VRGF receptor inhibitor axitinib relies on organometallic chemistry or, more specifically, the Hecht reaction for two steps. The synthesis begins by the addition of iodine to 1,2-diazaindene (**15.1**) by reaction with iodine under weakly basic conditions. The proton on the heterocyclic ring is next masked by reaction with dihydropyran (**15.3**). In the first application of the Hecht procedure, the 1,2-diazaindene moiety (**15.3**) is coupled with azastyrene (**15.4**) mediated by palladium–phosphine complex. The insertion of iodine for the second coupling starts with the reduction of the nitro group to the corresponding amine with iron, the presence of the styrenoid bond precluding hydrogenation. The new amine is next replaced by iodine: that group is thus diazotized with nitrous acid in the presence of iodide ion. The mercaptan-substituted benzamide (**15.8**) is then coupled to diazaindene by means of aromatic nucleophylic displacement (**15.7**). The treatment of the coupled product with aqueous toluenesulfonic acid removes the pyran masking group. The VEGF receptor inhibitor **axitinib** (**15.9**) is thus obtained [16].

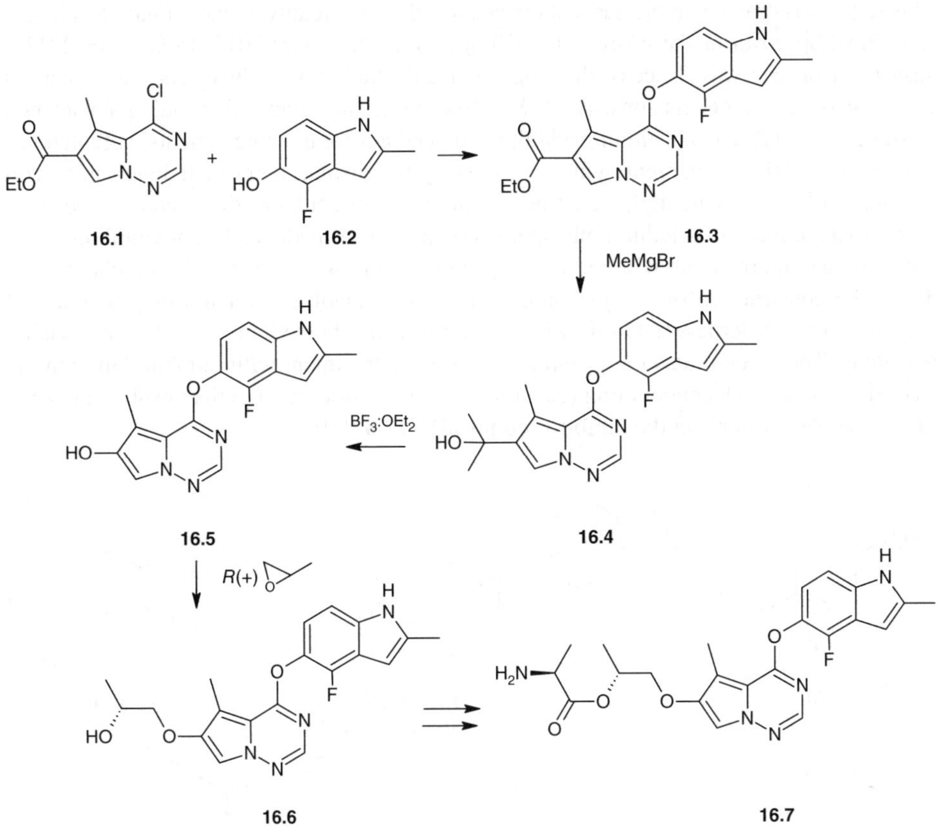

Scheme 8.16   *Brivanib.*

The VEGF receptor inhibitor brivanib alaninate showed good activity against malignant cells in the customary in vitro assays as well as in animal models. On the basis of results from initial trials, the European Commission gave the drug orphan status for the treatment of hepatocellular carcinoma. A sizeable Phase III trial of brivanib against hepatocellular cancer, however, failed to show activity against that cancer.

The displacement of chlorine from pyrrolotriazine by the oxygen anion from hydroxyindole (**16.2**) leads to the intermediate that incorporates the major part of the target molecule. The reaction with methylmagnesium bromide then converts the ester to a tertiary carbinol (**16.4**). The treatment of that compound with boron tribromide etherate results in an oxidative rearrangement whereby the entire carbinol group departs and is replaced by oxygen (**16.5**). The treatment of that product with a single enantiomer of propylene oxide leads to ether (**16.6**) and thus **brivanib** [17]. For reason of pharmacokinetics clinical trials were actually performed with a more water-soluble derivative. The terminal side chain hydroxyl group was thus coupled in several steps with alanine to yield **brivanib alaninate** (**16.7**) [18].

Toward the end of 2012, the FDA granted the VEGF receptor inhibitor drug, pinatinib accelerated approval for treating patients suffering from several narrowly defined leukemias.

This status was granted on the basis of the positive data of a relatively large Phase III clinical trial. Final approval of the 45 mg tablets followed in December 2012. In October 2013 a suspension of sales was placed on the drug, bearing the trade name Iclusig®, because it caused an excess of vascular occlusions. The FDA allowed sales to resume after adding a black box warning on the label and instituting additional procedures for tracking patients' vasculatures.

One arm of the convergent synthesis starts by the displacement of halogen in benzyl bromide (**17.1**) by *N*-methylpiperazine. Catalytic hydrogenation then reduces the nitro group to an amine. A palladium–phosphine complex next mediates Hecht coupling of the amine in that intermediate with iodobenzoyl chloride (**17.4**), to afford the side-chain-to-be **17.5**. The construction of the pyrrolodiazine moiety involves first the displacement of bromine in the heterocycle (**17.6**) with the anion from the sodium salt of trimethylsilyl acetylene. The silicon masking group is removed by treatment with tetrabutylammonium fluoride. A second Hecht coupling reaction attaches the side chain to the unsubstituted end of the acetylene affording the inhibitor **ponatinib** (**17.8**) [19].

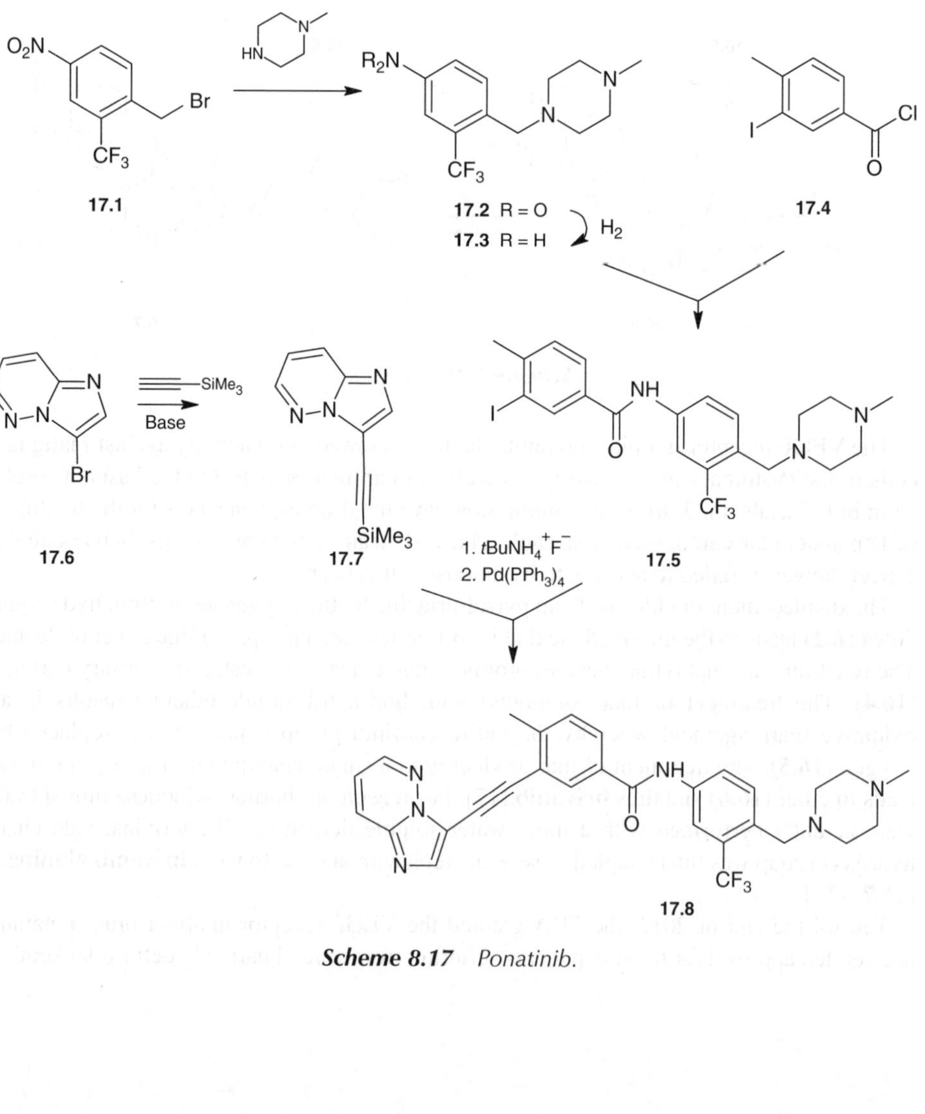

Scheme 8.17   *Ponatinib.*

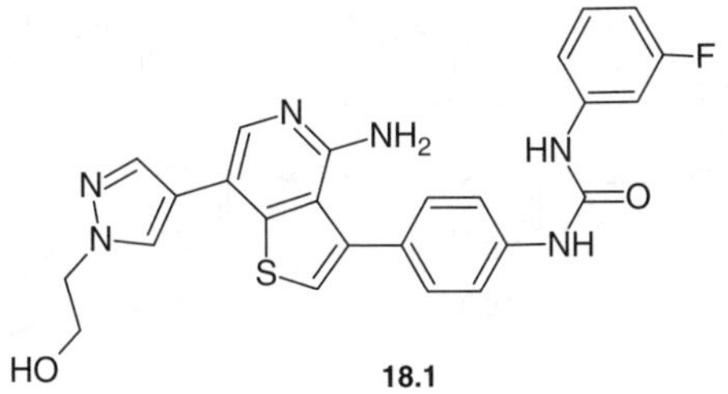

Scheme 8.18 *Ilorasertib.*

Ilorasertib (**18.1**) is a relatively new TKI that interacts with both VEGF and platelet-derived growth factor (PDGF) receptors. The compound has been found effective in models for both solid and hematological cancers. It has been designated by the FDA as an orphan drug for the treatment of ovarian cancer. There exist a series of patents (*see* [20] for preparing thiopheno-pyridine antineoplastic agents), however none seems to relate specifically to ilorasertib (**18.1**).

8.3.2 Linear Arrays

Golvatinib comprises yet another tyrosine receptor inhibitor that binds to the VEGF receptor. This drug also interacts with the subclass hepatocyte growth factor receptors (MET). The agent shows good activity in various *in vitro* assays as well as in animal models and implanted human tumors. This drug interestingly differs from the other VEGF inhibitors in that the chemical structure consists of a string of monocyclic rings. One scheme for preparing golvatinib starts with forming aryl ether (**19.3**); the displacement of chlorine in the substituted pyridine by the anion from the treatment of phenol (**19.2**) with base affords ether (**19.3**). The reaction of that intermediate with phosgene in the presence of base then replaces the amine with the corresponding isocyanate (**19.4**). The addition of *N*-methylpiperazine-piperidine (**19.5**) to the isocyanate attaches the amine by way of a urea. The sequence continues with the reduction of the nitro group in **19.6** that tagged along, so far unchanged, to an amine. This functional group is next acylated with the mono-acid chloride of cyclopropyl malonate. The protecting group on the other carboxyl group is then removed by catalytic hydrogenation (**19.9**); the newly revealed carboxylic acid is then concerted to the corresponding acid chloride. The reaction of that intermediate (**9.10**) with 4-fluoroaniline leads to the corresponding amide **golvatinib** (**19.11**) [21].

The kinase inhibitor sorafenib, (20.7) in common with all agents in this section, targets VEGF receptors. The compound ranks among the earliest TKI with sufficient *in vitro* and preclinical activity in laboratory animals to be taken to the clinic. Available under the trade name Nexavar®, the drug has been approved by the FDA for treating kidney cancer (2005), liver cancer (2007), and most recently thyroid cancer, all with special qualifications.

The concise preparation of this kinase inhibitor starts with displacement of fluorine the nitrobenzene (**20.2**) by the anion from treatment of the pyridine (**20.2**) with sodium hydride. The nitro group is next reduced to an amine by hydrogenation. The condensation with aryl isocyanate (**20.5**) links the moieties by means of a urea (**20.6**). The carbethoxy function is then saponified by means of base. Treatment with methylamine converts the resulting acid to a methyl amide. The kinase inhibitor **sorafenib** (**20.7**) is thus obtained [21*].

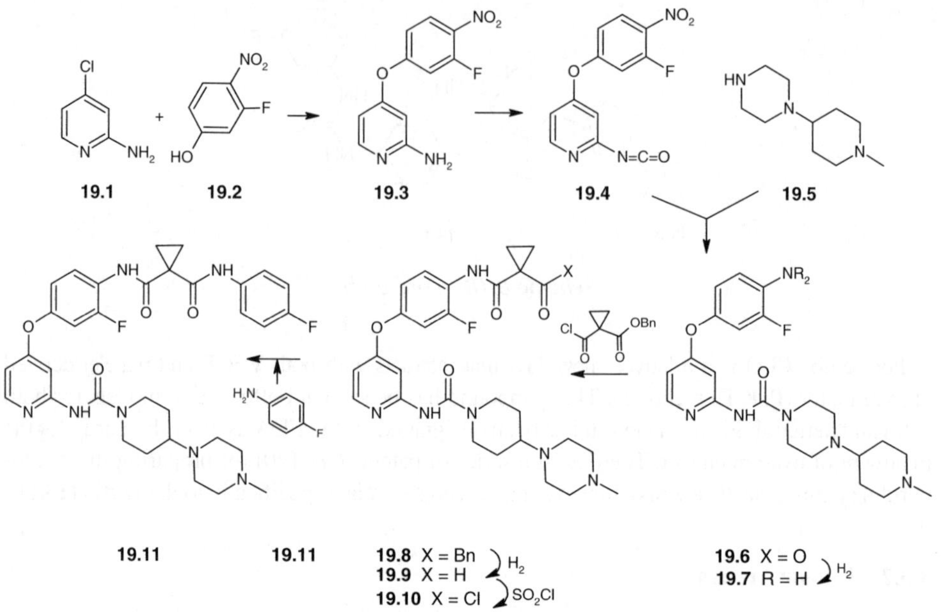

Scheme 8.19 *Golvatinib.*

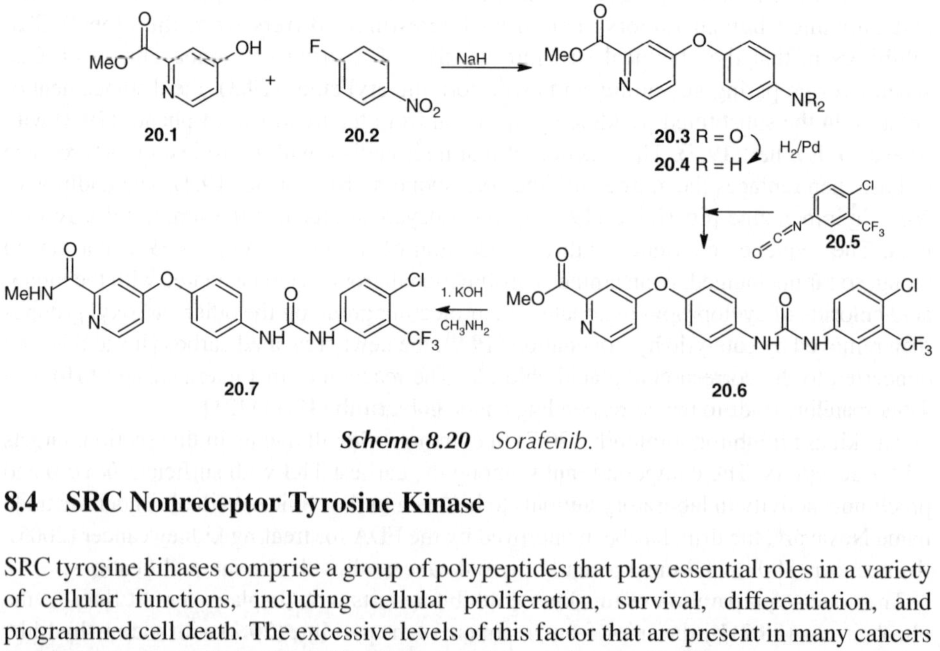

Scheme 8.20 *Sorafenib.*

8.4 SRC Nonreceptor Tyrosine Kinase

SRC tyrosine kinases comprise a group of polypeptides that play essential roles in a variety of cellular functions, including cellular proliferation, survival, differentiation, and programmed cell death. The excessive levels of this factor that are present in many cancers lead to the uncontrolled cell proliferation that characterizes malignant tumors.

The drug dasatinib was found to be active *in vitro* against a series of kinases. It was especially active *in vitro* against several leukemias. In 2009, the FDA granted this drug, trade named Spycell®, approval for treating leukemias.

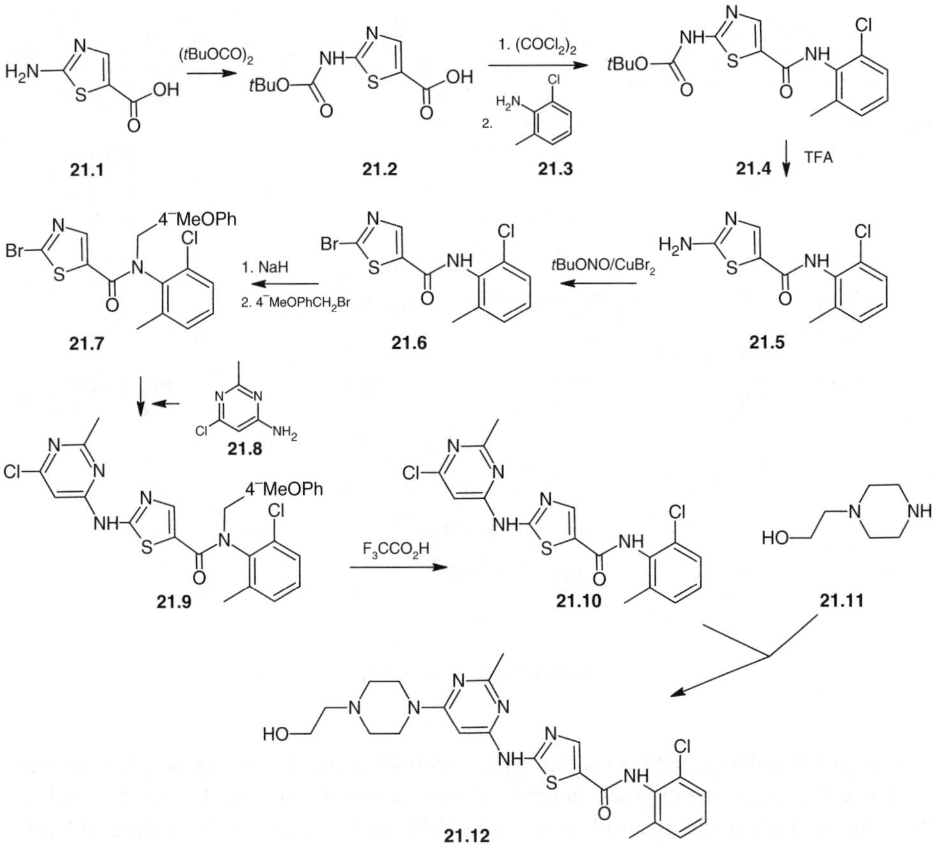

21.1 **21.2** **21.3** **21.4** **21.7** **21.6** **21.5** **21.8** **21.9** **21.10** **21.11** **21.12**

Scheme 8.21 *Dasatinib.*

One somewhat involved preparation of dasatinib begins with the reaction of aminothiazole carboxylic acid (**21.1**) with BOC anhydride to afford the derivative with a masked primary amine. The carboxylic acid is next converted to its acid chloride by means of oxalyl chloride; this activated intermediate is then allowed to react with the substituted aniline (**21.3**) to afford anilide (**20.4**). The BOC protecting group was next removed by treatment with trifluoroacetic acid. The resulting amine was then replaced by bromine by the use of a variant of the traditional Sandmeyer method: the treatment of the amine with *t*ertbutyl nitrite in the presence of cuprous bromide accomplishes the desired conversion. The amide nitrogen is then protected by serial reaction with sodium hydride followed by 4-methoxybenzyl bromide (**21.7**). The next ring system in the chain, 4-aminopyrimidine (**21.8**), is attached by straightforward alkylation. The treatment of the product from the reaction with trifluoroacetic acid cleaves off the benzyl protecting group. The replacement of halogen on the pyrimidine ring by free amine on hydroxyethyl piperidine (**21.11**) then affords the kinase inhibitor **dasatinib** (**21.12**) [22].

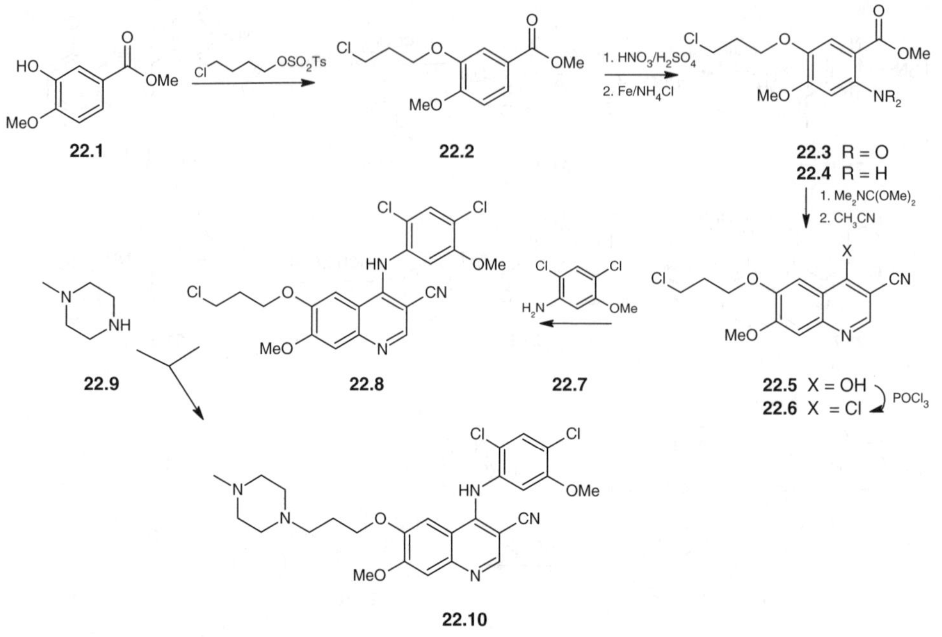

Scheme 8.22 *Bosutinib.*

The quinoline-based antineoplastic agent bosutinib is another compound that interacts with SRC receptors; the drug also inhibits Abl oncogenes. Clinical trials demonstrated that the compound was active against leukemias. In 2012, the FDA approved bosutinib tablets for treating patients with a specific leukemia. The drug is available under the trade name Bosulif®.

In an interesting move, phenol (**22.1**) is first alkylated with a chloropropyl fragment; this serves as a protecting group for the phenol through the various steps involved in the synthesis; this fragment then serves as a point for attaching a piperidine moiety in the last step. The resulting intermediate (**22.2**) is then nitrated by reaction with a mixture of nitric and sulfuric acids. The newly installed function is next reduced to an amine by reaction with iron in the presence of ammonium chloride (**22.4**). Serial reaction with dimethyl formamide acetal and then acetonitrile then generates the fused pyridine ring (**22.5**). The reaction can be rationalized by assuming that the DMF derivative reacts with the amine to add a carbon atom as an amidine; the anion from acetonitrile then adds to amidine. The negative charge on the adduct closes the ring by displacing methoxide on the ester. The hydroxyl group on the quinoline is next converted to a better leaving group, chlorine, by treatment with phosphorus oxychloride. The reaction of the product from that transform (**22.6**) with the substituted aniline (**22.7**) takes place preferentially on the ring halogen rather than the side chain aliphatic chlorine (probably because the former can be viewed as an enol chloride) to afford the adduct (**21.8**). The substitution of the side chain chlorine with free amine in *N*-methylpiperazine (**21.9**) completes the preparation of **bosutinib** (**22.10**) [23].

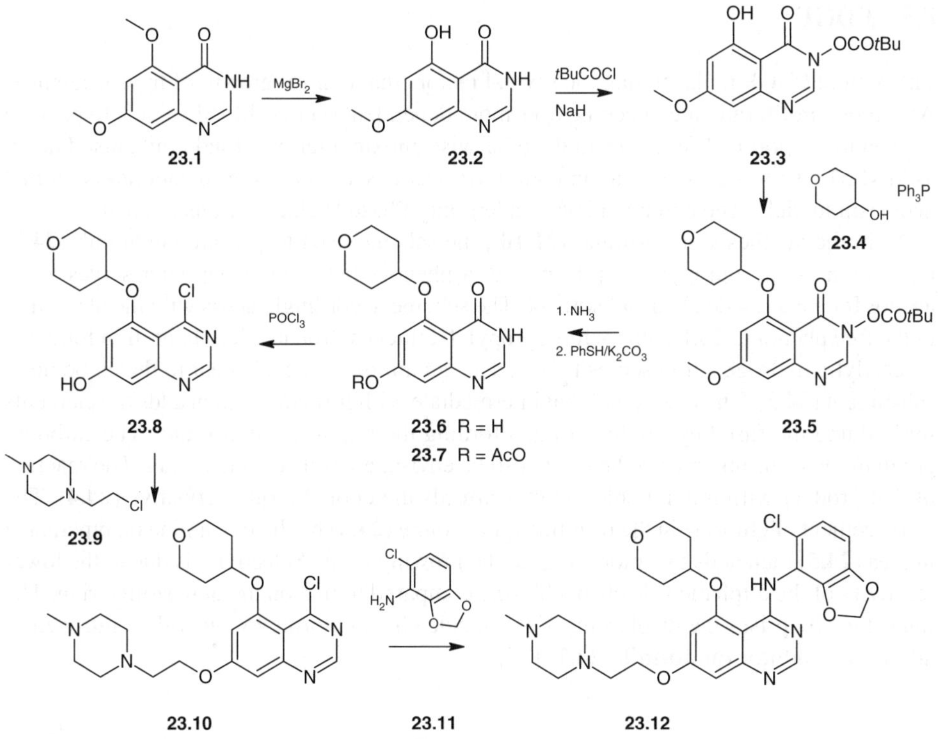

Scheme 8.23 *Saracatinib.*

Like its predecessor, the TKI saracatinib also acts on both SRC and Abl kinases. This orally active compound shows good antineoplastic activity in cell culture of a series of human cancers. The drug also shows activity in a series of xenografts of human cancer tissues.

The somewhat lengthy scheme for preparing saracatinib starts with the selective demethylation of one of the pair of methyl ethers in quinazoline (**23.1**) by means of magnesium bromide. As might be expected, the reaction favors the ether adjacent to the carbonyl group. The intermediate is then treated with BOC chloride in order to mask the amine on the fused pyrimidine. The resulting protected intermediate is next reacted with 4-hydroxypyran under modified Mitsunobu conditions in the presence of triphenylphosphine to afford the corresponding pyranyl ether (**2.5**). Treatment with ammonia then removes the protecting group on pyrimidine nitrogen (**23.6**). The remaining methyl ether is then cleaved by reaction with alkaline phenyl mercaptan; acetic anhydride next converts the newly formed phenol to its acetate. Phosphorus oxychloride then converts the enol form of the carbonyl group in the fused pyrimidine to the corresponding chloride. Treatment with base results in restoration of free phenol (**23.8**). The oxygen anion from the reaction of this last intermediate with base, followed by *N*-chloroethyl- *N'*-methylpiperazine (**23.9**), attaches the basic side chain (**23.10**). The displacement of the ring chlorine by the primary amine in piperonal (**23.11**) adds the last moiety. The protein kinase inhibitor **saracatinib** (**23.12**) is thus obtained [24].

8.5 PDGF

The protein PDGF ranks among the several factors that control growth of the vasculature. As tumors proliferate, they become ever more dependent on new blood vessels to provide oxygen to support cell functions in the otherwise anoxic interiors. Tandutinib, also known as MNL518, shows good activity in both *in vitro* assays and in studies in laboratory animal neoplasm models. The compound was undergoing Phase II clinical trials in 2014.

As in the synthesis of bosutinib (**21.10**), the scheme used to prepare tandutinib (**24.7**) relies on a 3-chloropropyl group to protect a phenol; that same group later serves as an anchor for a basic side chain substituent. The scheme accordingly starts with the alkylation of the free phenol in **24.1** with 3-chloropropyl-1-toluenesulfonate. Serial nitration followed by catalytic hydrogenation serves to introduce an amine group adjacent to the carbethoxy substituent (**24.3**). The reaction of that intermediate with formamide then adds the elements for building the fused pyrimidone ring, affording the required quinazolone. The carbonyl group in the resulting intermediate can also be envisaged as its enol tautomer. The reaction of that product with thionyl chloride thus affords the enol chloride derivative (**24.4**). The displacement of chlorine in the resulting quinazoline (**24.4**) by the free amine on piperazine in urea (**24.5**) then adds two more rings to the growing array. Selectivity is due to the lower reactivity of the aliphatic side chain chlorine compared to that on the quinazoline ring. The final step comprises the displacement of the side chain chlorine by piperidine nitrogen to afford the inhibitor **tandutinib** (**24.7**) [25].

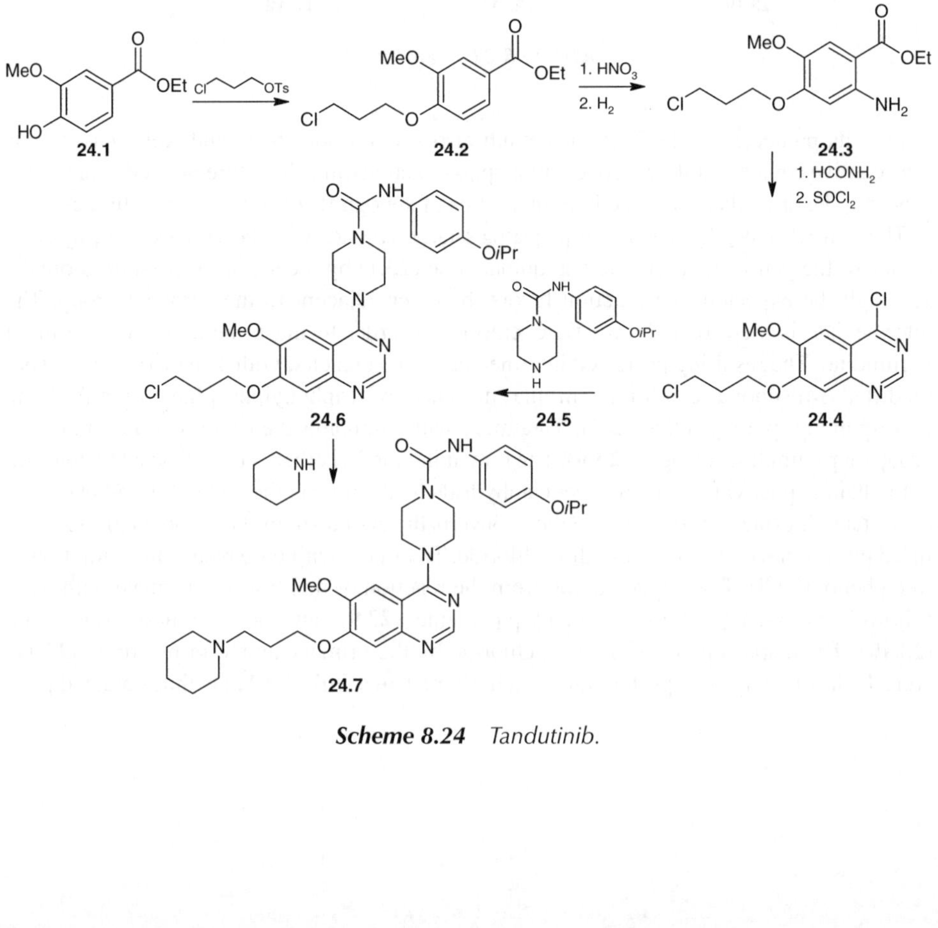

Scheme 8.24   *Tandutinib.*

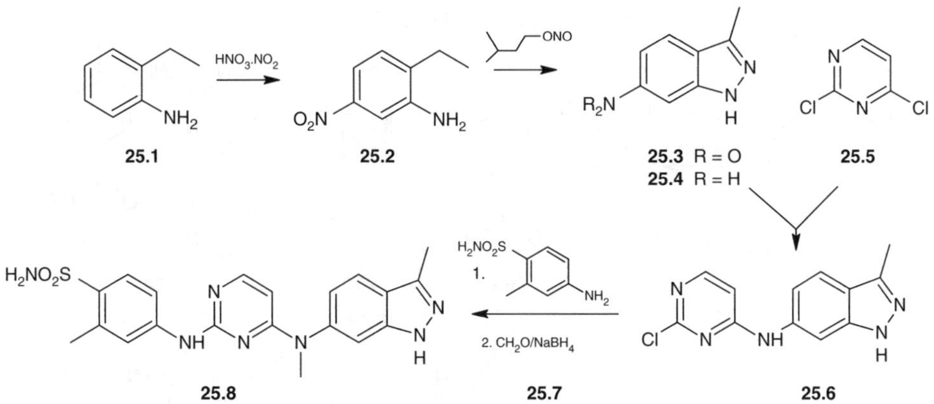

Scheme 8.25   *Pazopanib.*

Pazopanib is a multitarget inhibitor of PDGF receptors: all three variant VEGF receptors as well as several other tyrosine kinases. On the basis of the drug's efficacy in clinical trials, it was approved by the FDA in 2009 for the treatment of patients with renal cancer as well as soft cell carcinomas. The drug is available under the trade name Votrient®.

The concise scheme for preparing pazopanib starts by nitration of 2-ethylaniline with fuming nitric and sulfuric acid. The resulting nitro aniline is then treated with isoamyl nitrite to afford indazole (**25.3**). The initial product in this transform likely comprises the corresponding diazonium salt; this then proceeds to attack the adjacent methylene carbon of the ethyl group to close the fused ring. Note that this requires abstracting a proton from the benzylic methylene on the adjacent ethyl group. The nitro group in the product **25.3** is then reduced to the amine by means of stannous chloride in strong acid (**25.4**). The reaction of that product with 2,4-dichloropyrimidine results in the displacement of chlorine at position 4, linking the two moieties. The treatment of that product with 1-methyl-4-aminobenzenesulfonamide (**25.7**) leads to the replacement of the remaining chlorine on pyrimidine, linking the new fragment. The resulting secondary amine is then methylated by any of a set of procedures, for example, formaldehyde and sodium borohydride. This last transform completes the synthesis of **pazopanib** (**25.8**) [26].

Nilotinib is yet another small molecule that inhibits PDGF receptors as well as those for TGF and several other factors. Data from clinical trials indicated that the drug showed good activity against leukemias. It was approved by the FDA in 2007 for treating patients with chronic myelogenous leukemia. The drug is available under the trade name Tasigna®.

One synthesis for the compound starts with the reaction of carboethoxyaniline (**26.1**) with cyanamid to afford guanidine (**26.2**). The treatment of that intermediate with the enamine from 3-(formylaceto)pyridine transforms the guanidine to the pyrimidine in **26.4**. The displacement of the ethyl ester anion from the amine in (**26.5**) and base links that moiety to the string of five rings. The resulting product is **nilotinib** (**26.6**) [27, 28].

Linifanib, also called ABT-869, comprises another multitargeted TKI. The compound most prominently inhibits PDGF and VEGF. The inhibitor has shown good activity against human cancers in both *in vitro* and in xenografts of the same. In a Phase II clinical trial reported in 2013, the drug exhibited activity against hepatocellular cancer.

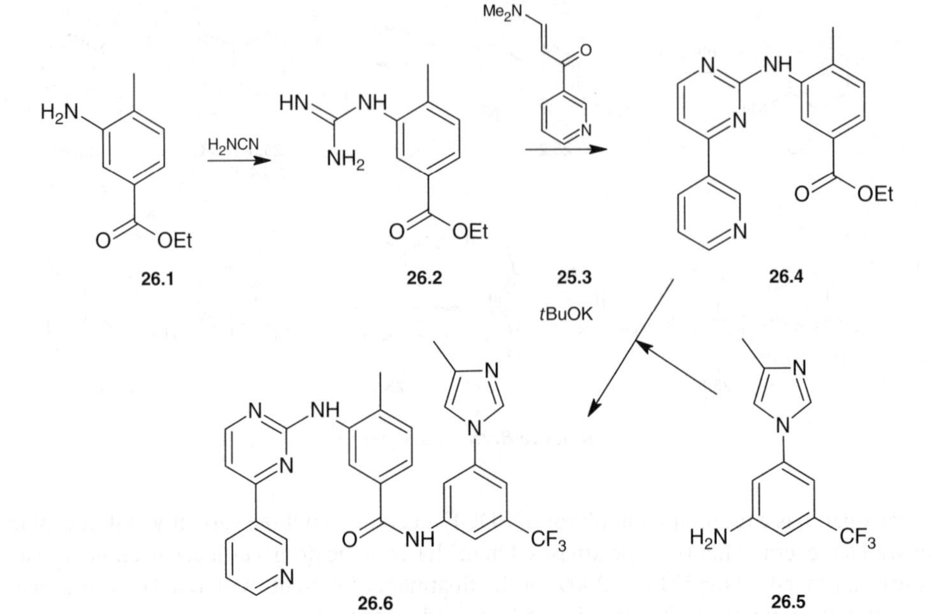

Scheme 8.26 *Nilotinib.*

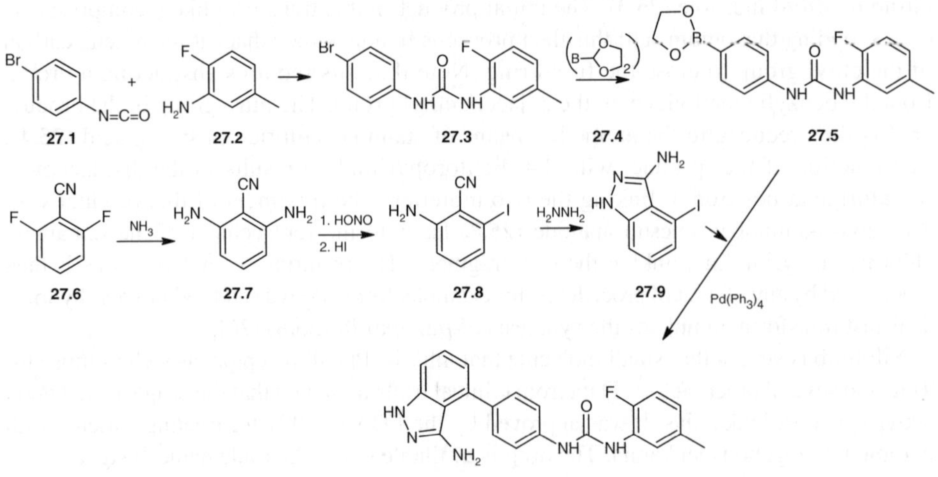

Scheme 8.27 *Linifanib.*

The absence of a single accessible publication on the synthesis of linifanib that is depicted in Scheme 8.27 was stitched together from two sources. Some of the details represent guesses. The resulting scheme starts with the reaction of isocyanate (**27.1**) with fluoroaniline (**27.2**) to afford urea (**27.3**). Bromine in one ring is then replaced by a boron derivative such as the dimer (**27.4**) [29]. In a converging scheme, ammonolysis of

2,5-difluorobenzonitrile (**27.6**) affords the corresponding 1,3-bisaminobenzonitrile (**27.7**). The reaction of that product with nitrous acid apparently leads to the conversion of a single amine to a diazonium salt. The reaction of that intermediate with a source of iodine leads to the iodo derivative (**27.8**). The Suzuki cross coupling of the two moieties then yields the inhibitor **linifanib** (**27.10**) [30].

8.6 EGF

The EGF is a relatively small polypeptide whose activation results in cellular proliferation, differentiation, and survival. It is found in excess in cancerous cells resulting in uncontrolled cell proliferation.

The EGF inhibitor ganetespib showed promising activity in non-small cell lung cancer. On the basis of that data, the FDA granted the drug fast review status in 2013.

The preparation of this drug is surprisingly concise. The reaction of the substituted benzoic acid (**28.1**) with oxalyl chloride affords the corresponding acid chloride (**28.2**). That intermediate is then used to acylate aminoindole (**28.3**). The treatment of that intermediate with Lawesson's reagent (P_4S_{10}) converts the amide carbonyl group to a more reactive thiocarbonyl function (**28.5**). The reaction of this product with hydrazine replaces the sulfur by nitrogen putting in place the requisite three nitrogen atoms (**28.6**). Carbonyldiimidazole then adds the required extra carbon atom. The reagent reacts with the terminal nitrogen and that on the indole in the process closing the ring to a triazolone (**28.7**). Treating that product with pyridine and hydrochloric acid cleaves the phenol ethers to afford the inhibitor **ganetespib** (**28.8**) [31].

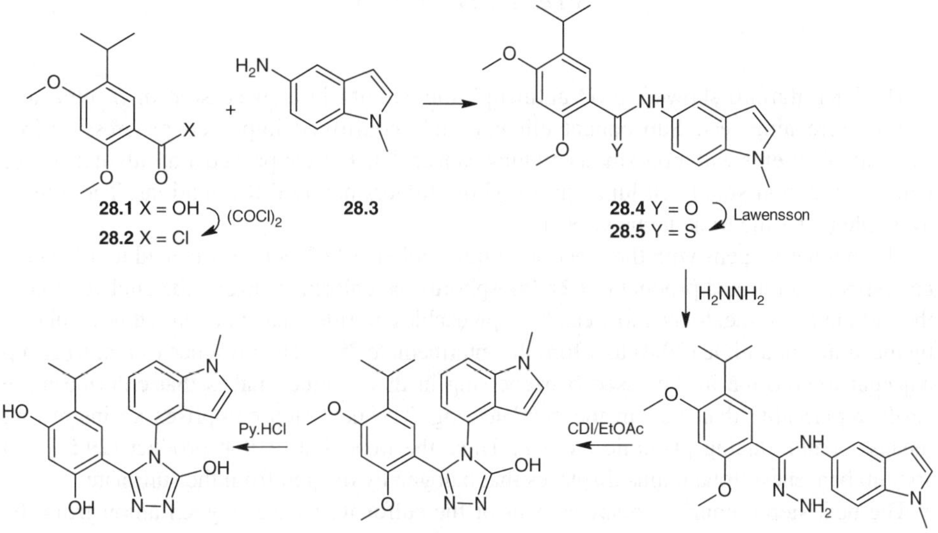

Scheme 8.28   Ganetespib.

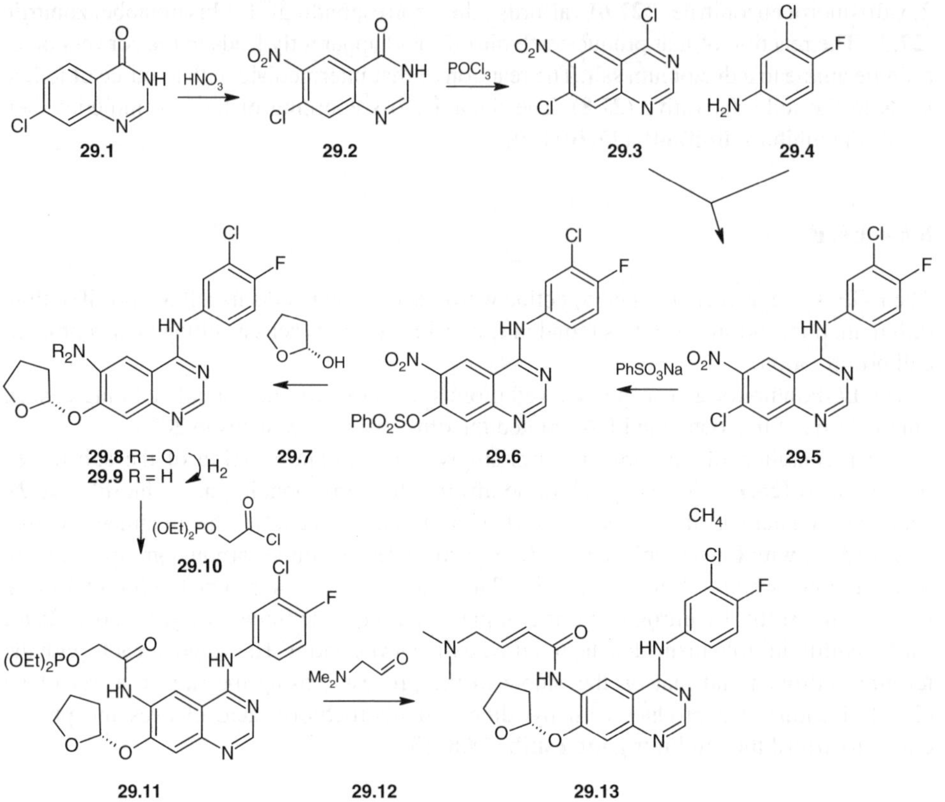

Scheme 8.29   Afatinib.

The TKI afatinib showed good antineoplastic activity in a succession of *in vitro* and *in vivo* screening test. Subsequent clinical trials confirmed improved progression-free survival of patients with non-small cell lung cancer. The FDA approved afatinib in 2013 for patients with non-small cell lung cancer whose tumors had EGFR mutations. The drug is available under the trade name Giotrif®.

The synthesis opens with the reaction of quinazolone (**29.1**) with nitric acid to afford the corresponding nitrated product (**29.2**). Phosphorus oxychloride converts the enol version of the carbonyl tautomer to the more easily displaceable chloride. That halogen is thus displaced by the amine in aniline (**29.4**) to afford the intermediate **29.5**. The presence of a nitro group adjacent to chlorine in the fused benzene ring in that product makes that chlorine more easily replaceable than that in the pendant ring. The next step comprises an interesting means for introducing phenolic oxygen. Thus, the treatment of that product (**29.5**) with sodium benzenesulfonate thus displaces that halogen by oxygen from the sulfonate.

The next step comprises displacement of the sulfonate by the oxygen anion from the reaction of chiral 2-hydroxydihydrofuran (**29.12**). The nitro group that has been tracked thus far through a number of transforms is finally reduced to an amine by hydrogen over Raney nickel. The addition of the required additional side chain starts with the acylation of the newly formed amine with 2-diethylphosphonoacetic acid chloride (**29.10**). The reaction of that intermediate with strong base forms an ylide. The condensation of that species with 2-dimethylaminoacetaldehyde (**28.12**) finally affords **afatinib** (**29.13**) [32].

8.7 Other TKI

A sizeable number of TKI differ from those above as to their principal target; those targets are too varied to group the compounds. They are thus grouped by chemical structures.

8.7.1 Linear Arrays

Mubritinib (**30.8**) is one of the earlier TKI that was being developed for the treatment of solid tumors. This compound was picked out as active by various screening assays as worth pursuing. Mubritinib was then entered in Phase I clinical trials.

The convergent synthesis essentially comprises a series of alkylations and acylations. The first step thus comprises displacement of methanesulfonate in (**30.1**) by the basic amine in 1,2,3-triazine (**30.2**). The benzyl protecting group is then cleaved off by catalytic hydrogenation. The preparation of the other moiety involves the classical procedure for forming oxazoles. Thus, the reaction of 4-trifluoromethylcinnamide (**30.5**) with 1,3-dichloroacetone (**30.6**) leads to chloromethyloxazole (**30.7**). The phenol in (**30.4**) is next converted to the anion by treatment with base. The reaction of this charged species with chloromethyloxazole (**30.7**) displaces chlorine by alkoxide to afford ether (**30.8**). The TKI **mubritinib** (**30.8**) is thus obtained [33].

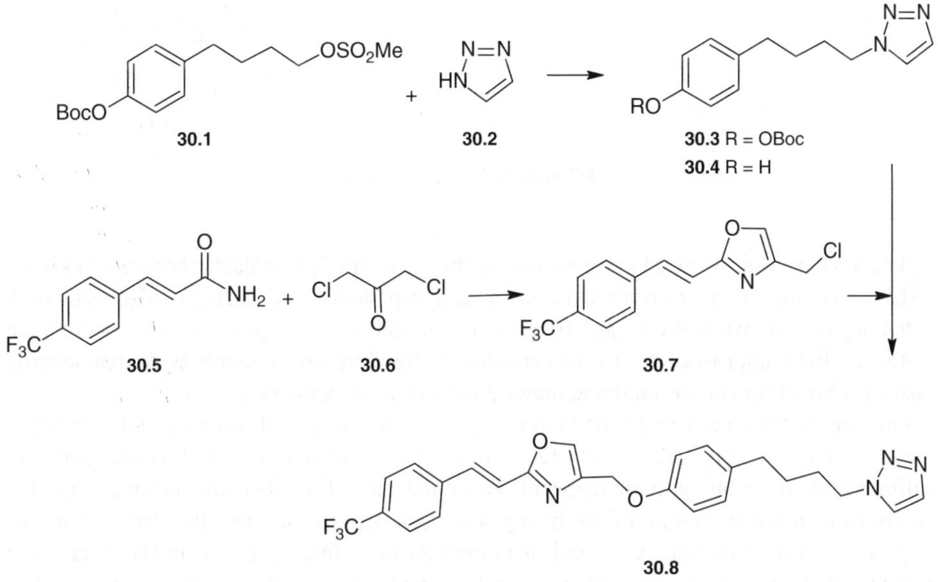

Scheme 8.30 Mubritinib.

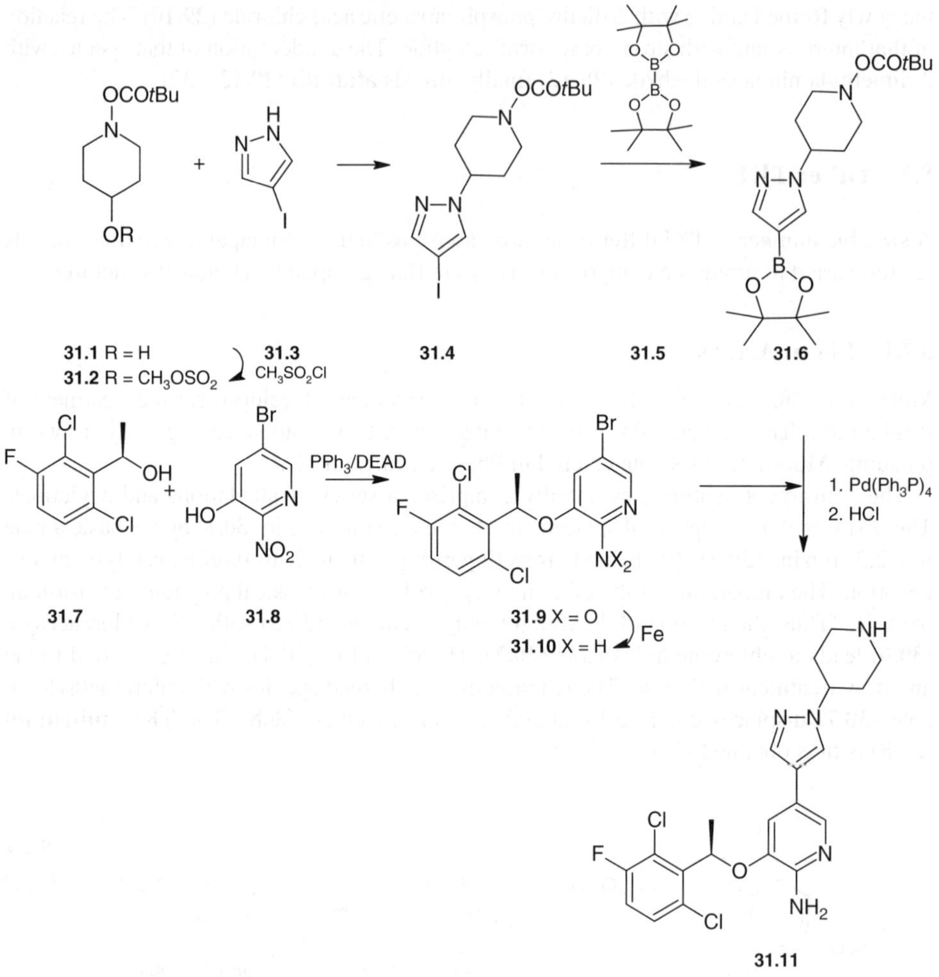

Scheme 8.31   Crizotinib.

The TKI crizotinib (**31.11**) targets and inhibits principally anaplastic lymphoma kinase (ALK) receptors. Clinical trials demonstrated that patients with ALK-positive non-small cell lung cancer exhibited longer progression of free survival periods than controls. In 2013, the FDA approved the use of capsules of this drugs for patients with that specific disease. Crizotinib is now available under the trade name Xalkori®.

One arm of the convergent scheme for preparing the compound involves first converting the hydroxyl group at position 4 in BOC-protected hydroxypiperidine (**31.1**) to a methanesulfonate by reaction with methanesulfonyl chloride (**31.2**). The displacement of that newly formed group by one of the nitrogens in 3-iodopyrazole links the first of the four rings in the final product. A second displacement of iodine by boron in the boranating reagent (**31.5**) yields one of the two moieties that will be linked down the line. The construction of the second half of the final product comprises connecting the alcohol group

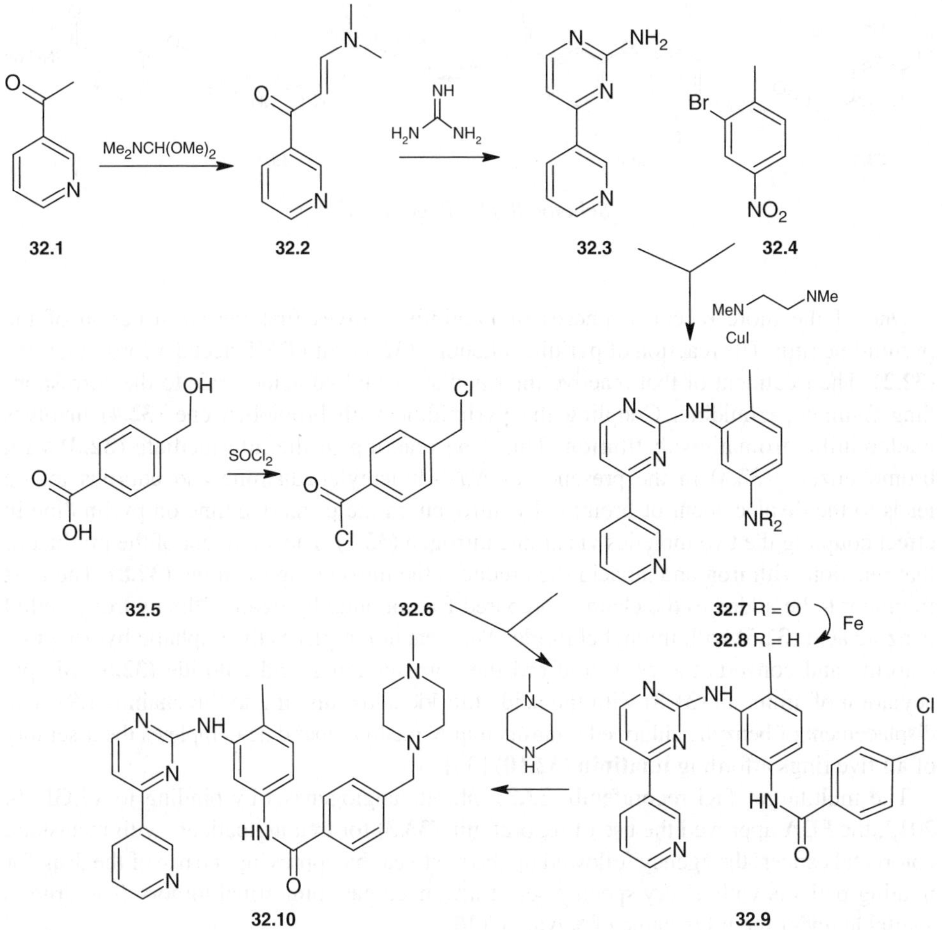

32.1 32.2 32.3 32.4 32.5 32.6 32.9 32.10

Scheme 8.32 *Imatinib.*

in pyridine (**31.8**) with the same function in chirally pure **31.7** via the Mitsunobu reaction (triphenylphosphine and diethyl diazodicarboxylate (DEAD)) to afford the second moiety linked via an ether. The nitro group in the product is then reduced by iron and acetic acid (**31.10**). The two halves (**31.6, 31.10**) are joined via palladium-catalyzed cross coupling reaction. The removal of the BOC protecting group with acid completes the synthesis of **crizotinib** (**31.11**) [34].

Imatinib is one of the earliest TKI. The compound binds to the kinase involved in binding to ATP and activation of growth receptors. The drug might be considered the poster child for TKI under its trade name Gleevec®. A significant part of the cover story of the May 28, 2001, issue of Time magazine is devoted to Gleevec®. The drug was approved by the FDA as far back as 2003 for the treatment of patients with chronic myelogenous leukemia. In common with most of later TKI, the agent was administered in a solid dosage form, in this case tablets.

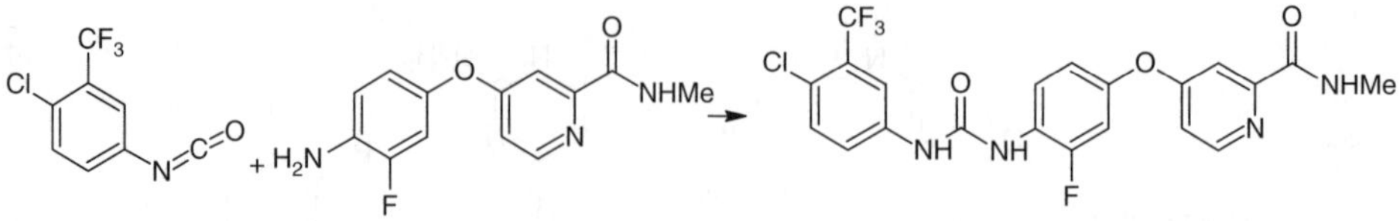

Scheme 8.33   Regorafenib.

One of the more recent syntheses of imatinib involves first the construction of the pyrimidine ring. The reaction of pyridine phenone (**32.1**) with DMF acetal affords enamine (**32.2**). The treatment of that reactive intermediate with hydrazine leads to the correspon-ding 2-aminopyrimidine. Coupling the pyrimidine with bromobenzene (**32.4**) involves nucleophilic aromatic substitution. Thus, the reaction of the intermediate (**32.3**) with bromobenzene (**32.4**) in the presence of *N,N'*-dimethylenediamine and cuprous iodide leads to the displacement of bromine by nitrogen on the primary amine on pyrimidine in effect coupling the two moieties via amine nitrogen (**32.7**). The treatment of the product of that reaction with iron and an acid then reduces the nitro group to amine (**32.8**). The next fragment to be added to the chain is prepared for coupling by treating the hydroxymethyl benzoic acid (**32.5**) with thionyl chloride. This reaction replaces the aliphatic hydroxyl by chlorine and converts the acid to afford the corresponding acid chloride (**32.6**). Simple acylation of amine in (**31.8**) with the acid chloride adds this ring to the chain (**32.9**). The displacement of benzylic chloride by nitrogen in *N*-methyl piperidine completes the assembly of all five rings affording **imatinib** (**32.10**) [35].

The multitarget TKI regorafenib (**33.3**) inhibits angiogenesis by binding to VEGF. In 2012, the FDA approved the use of regorafenib (**33.3**) for treating patients with metastatic colorectal cancer; the agency followed up the next year by approving the use of the drug for treating patients with a very specific set of advanced gastrointestinal tumors. The drug is available under the trade name of Stivarga®[36].

The very simple published preparation describes the condensation of pyridyl aniline (**32.2**) with isocyanate (**33.1**) to yield **regorafenib** (**33.3**) [36].

8.7.2   Compounds with Two Fused Rings

The TKI cabozantinib (**34.9**) is classed as one of the multitarget kinase inhibitors. The drug was approved in late 2012 for patients with metastatic thyroid cancer. The compound has in addition shown promise in clinical trials against several additional cancers. The drug is available under the trade name Cometriq®.

The relatively straightforward synthesis of cabozantinib in essence consists, as in the case of many other kinase inhibitors, of several acylations and alkylations. The scheme begins with the formation of mono acid chloride (**34.2**) from the symmetrical malonic acid (**34.1**) via serial reaction of the compound with triethyl amine and then thionyl chloride. The treatment of that intermediate with 4-fluoroaniline leads to amide (**34.4**). The second ring is attached by the reaction of 4-aminophenol (**34.5**) with the free carboxylic acid from the prior transform. Instead of converting the remaining acid to an acid chloride, the amide

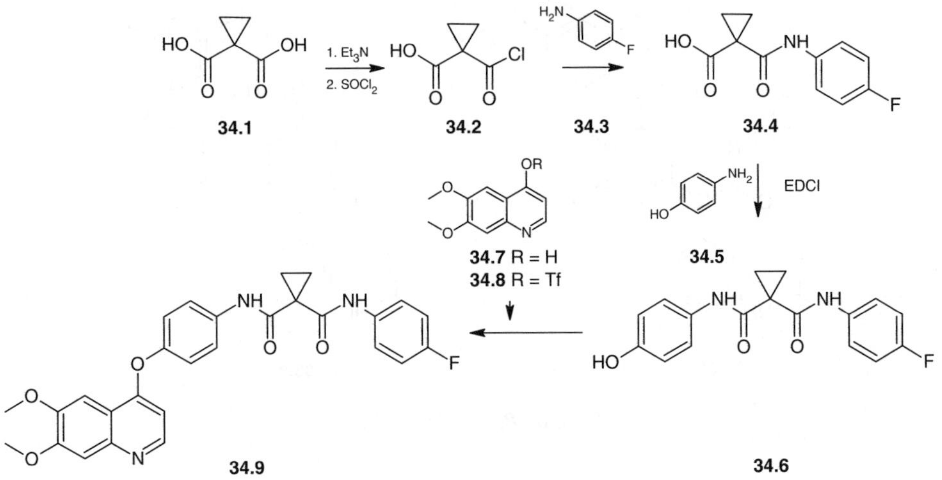

Scheme 8.34   *Cabozantinib.*

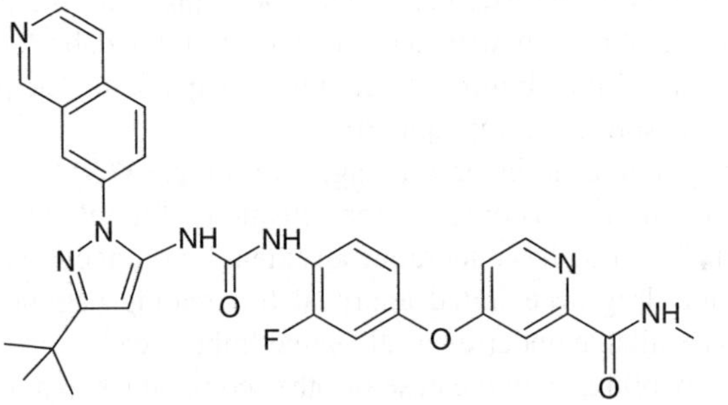

Scheme 8.35   *Rebastinib.*

is formed by reacting the free acid with amine in the presence of ethyl carbodiimide. The quinoline fragment is prepared for inclusion by activating the hydroxyl group as a trifluoromethanesulfonate (triflate) by treatment with the corresponding acid chloride. The reaction of phenol (**34.6**) with triflate (**34.8**) in the presence of 2,6-lutidine leads to the displacement of triflate by the oxygen anion from the phenol linking the two moieties via an ether linkage. The product from that last reaction comprises the inhibitor **cabozantinib** (**34.9**) [37].

The potent orally active antineoplastic agent **rebastinib** (**35.1**) binds to and inhibits to a large selection of factors that fall under the heading of tyrosine kinases. Those targets include kinases that regulate angiogenesis. The compound has shown activity in various *in vitro* assays. The drug increased the lifespan of mice that bore implanted leukemia cells. A Phase I clinical trial had been performed as of early 2014. The chemical structure of the compound lacking a synthetic scheme is displayed due to the lack of sources for the latter.

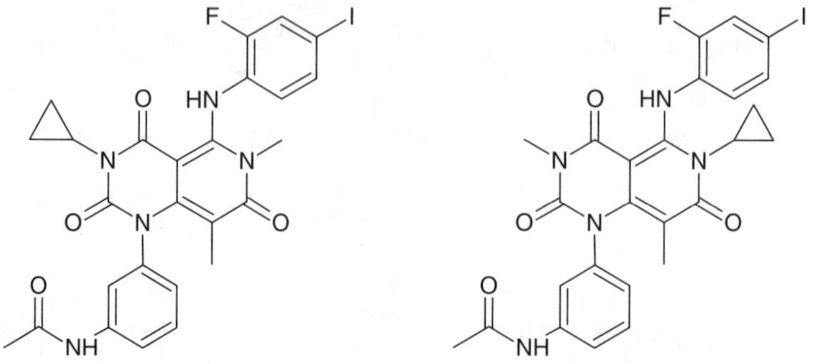

36.1   **36.2**

Scheme 8.36   *Trametinib.*

The tyrosine kinases MEK 1 and MEK 2 (mitogen-activated protein kinase) often occur in excess amounts in tumors. The kinase inhibitor **trametinib** was approved by the FDA for treating patients with inoperable or metastatic melanoma in early 2013. The drug is available under the trade name Mekinist®. The following year, the agency approved the use of this drug in combination with the enzyme inhibitor dabrafenib (*see* Chapter 9). The chemical structure of this drug is depicted in Scheme 8.36. The patent [38] describes the preparation of the isomeric compound 36.2.

The orally active potential antineoplastic agent ibrutinib (**37.9**) binds to an entity known as Bruton's tyrosine kinase. This prevents the activation of the B cells that serve as part of the immune system. The drug has undergone an extensive set of clinical trials. In late 2013, the FDA granted the drug accelerated approval for treating patients with specific lymphoma. The drug is available under the trade name Imbruvica®.

Palladium cross coupling, as in the case of other compounds, plays a crucial role in the synthesis of ibrutinib (**37.9**). The reaction of the commercially available pyrazolopyrimidine (**37.1**) with iodosuccinimide inserts the halogen on the carbon atom in the pyrazole ring (**37.2**). Palladium-catalyzed cross coupling of that iodide with the boronic acid derivative **37.3** attaches the diphenyl ether moiety. In a sequence intended to cover the amine in hydroxypiperidine (**37.5**), nitrogen is acylated with *tert*-butoxycarbonyl anhydride to afford the BOC-protected derivative **37.6**. This derivative is then reacted with the fused pyrazole in the presence of triphenylphosphine and diethyl azodicarboxylate (DEAD) to afford the alkylated derivative **37.7**. Mild acid hydrolysis cleaves off the BOC protecting group. The acylation of the newly revealed amine with acryloyl chloride completes the preparation of **ibrutinib** (**37.9**) [39].

The orally active TKI **linsitinib** (**38**) binds and inhibits insulin growth factors (IGF), a protein that is particularly prominent in a variety of human cancers. The compound is active *in vitro* against a collection of human cancers as well as implants of such cancers in laboratory animals. The synthetic route is difficult to reconstruct from the patent and is thus not shown [40].

The structurally distantly related TKI crenolanib (**39.8**) shows significant activity against tumor cells in a selection of *in vitro* assays. In 2011, this orally active drug entered Phase 2 clinical trials.

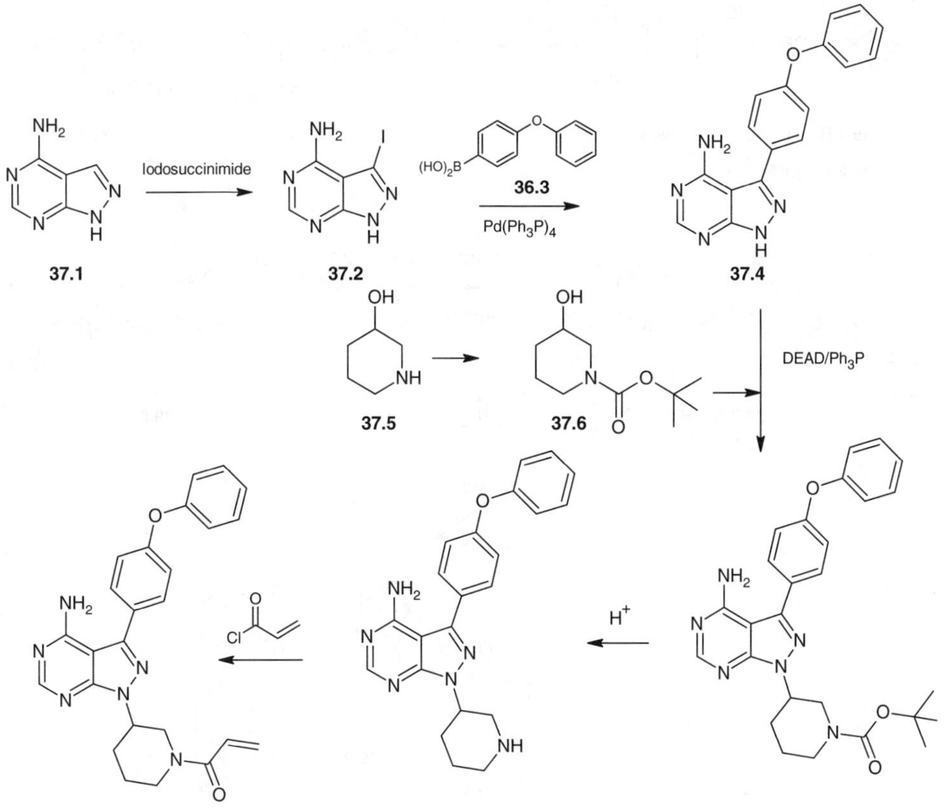

Scheme 8.37 *Ibrutinib.*

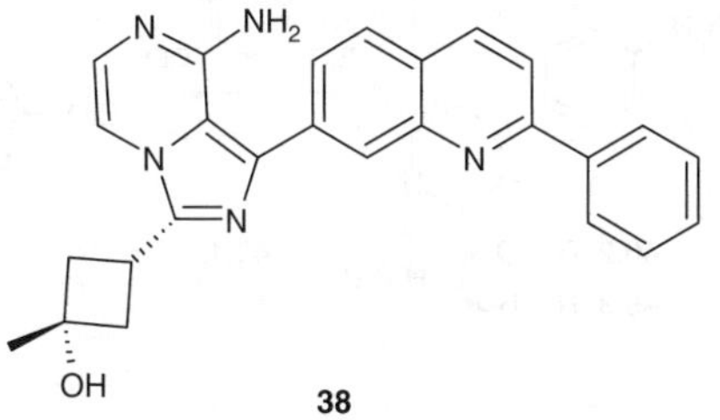

Scheme 8.38 *Linsitinib.*

The synthesis opens with the formation of methanesulfonate of hydroxymethyl oxetane (**39.1**). The displacement of that function by aminophenol (**39.3**) in the presence of cesium carbonate affords the corresponding ether (**39.4**). In a second displacement reaction, chlorine in the quinoline (**39.8**) is replaced by amine in the intermediate (**39.4**). The next few steps convert the adjacent nitro and amino group to an imidazole. The reduction of the

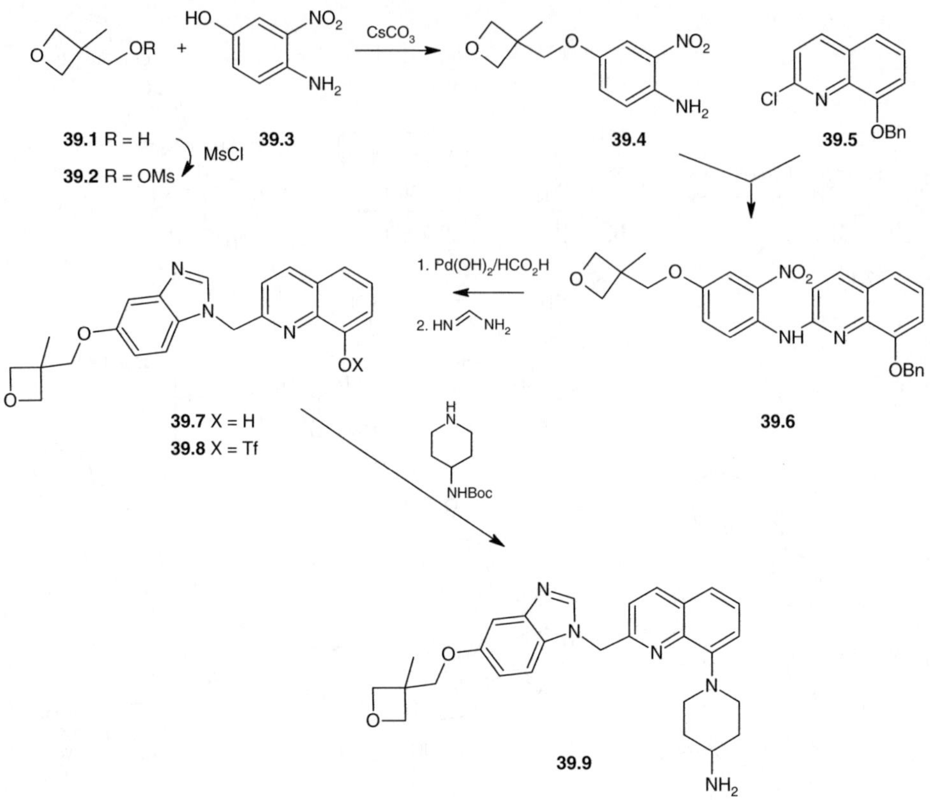

Scheme 8.39   Crenolanib.

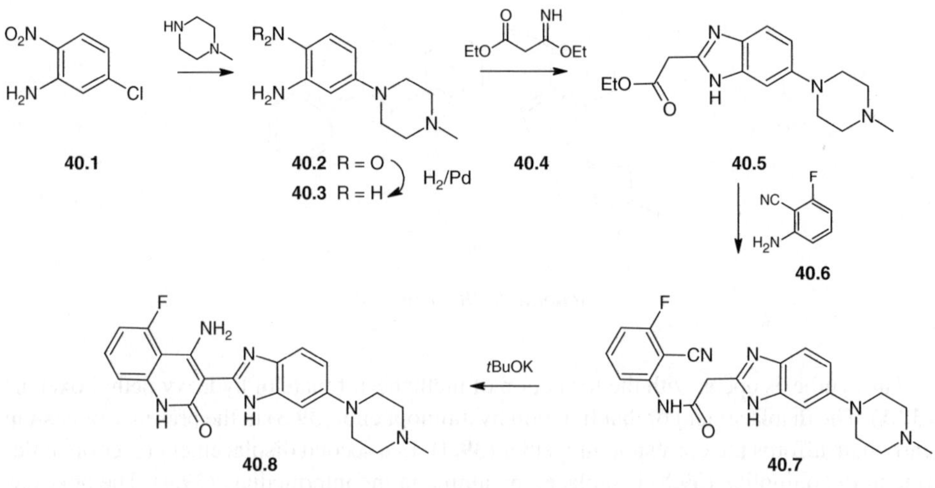

Scheme 8.40   Dovitinib.

nitro group to an amine with palladium and formic acid at the same time cleaves of the benzyl protecting group and reduces the nitro group to the corresponding amine. The reaction of the reduction product with amidine then forms the imidazole fused to the benzene ring. The newly revealed phenol is next converted to a better leaving group by reaction with trifluoromethanesulfonic anhydride to afford the substituted benzimidazole (**39.8**). In yet another replacement, trifluoromethanesulsufonate is replaced by *tert*butoxy-carbonyl-protected 4-aminopiperidine. The removal of the protecting group then affords **crenolanib** (**39.9**) [41].

The TKI dovitinib inhibits the fibroblast growth factor (FGF). The compound shows good activity in various *in vitro* antineoplastic assays. By 2013, it had been tested in several Phase I and Phase II clinical trials. In 2013, it was awarded orphan drug status for treating patients with adenoid cystic carcinoma.

The relatively concise synthesis of dovitinib starts by displacement of the aromatic chlorine in (**40.1**) by the secondary amine in *N*-methylpiperidine. The nitro group that has facilitated the displacement is next reduced to an amine with hydrogen over palladium. The condensation of the thus formed diamine with the half imine of diethyl malonate affords benzimidazole (**40.5**). The other ester group comprises the anchor for the side chain (**40.5**). The ester–amine interchange between that ester and substituted aniline (**40.6**) affords the amide (**38.7**); this intermediate now features substituents poised to form a quinolone. Thus, reaction with strong base removes a proton from the only methylene group in the molecule. The resulting anion then adds to the nitrile on the anilide to close the ring, forming a quinolone. The kinase inhibitor **dovitinib** (**40.8**) is thus obtained [42].

Sufficient preclinical and clinical data was presumably collected on the TKI tivozanib (**41.9**) for the drug to enter a Phase III trial against kidney cancer. The FDA Advisory Committee, however, voted against approving the drug since they considered the results of the trial equivocal. The agency turned down the application for approval for much the same reason.

The concise, brief scheme for preparing this compound starts by the reaction of 3,4-dimethoxyaminoacetophenone with methyl formate. This adds the extra carbon atom to form the fused pyridine, in sum, affording quinolone (**41.2**). The treatment of that intermediate with phosphorus oxychloride converts what had been a carbonyl to an enol chloride; that halogen now constitutes a good leaving group (**41.3**). Coupling the next moiety via an ether group requires that phenol oxygen in hydroxyaniline (**41.4**) react in preference to the aniline nitrogen. Reaction with potassium *tert*butoxide, as expected, preferentially abstracts a proton from the phenol. The resulting phenoxide displaces the ring chloride to afford the corresponding ether (**41.5**). In a convergent arm, the reaction of amino isoxazole (**40.6**) with phenyl chlorocarbonate (**41.7**) leads to carbonate amide (**40.8**). The treatment of the free amine in quinolone (**41.5**) with isoxazole (**41.8**) leads to the displacement of the phenoxide substituent in the carbamate by the amine to yield the corresponding urea. The TKI **tivozanib** (**41.9**) is thus obtained [43].

Yet another compound that features a roughly similar motif is based on a quinazoline nucleus.

The tyrosine kinase inhibitor canertinib (**42.8**) shows good activity in various *in vitro* antineoplastic assays. The compound is irreversibly bound to some kinases, arguably because of the presence of the acrylate moiety. The drug has proceeded as far as Phase II clinical trials.

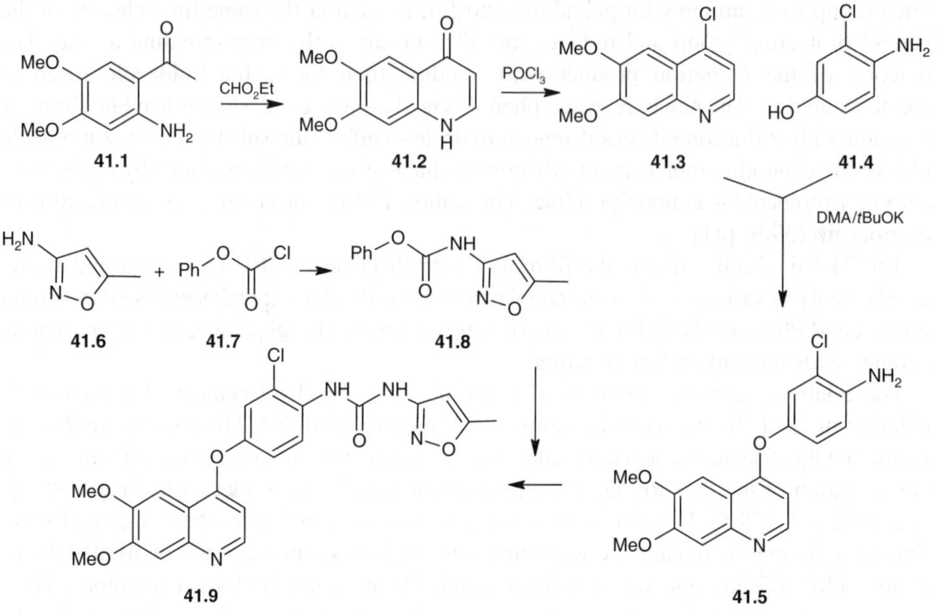

Scheme 8.41 *Tivozanib.*

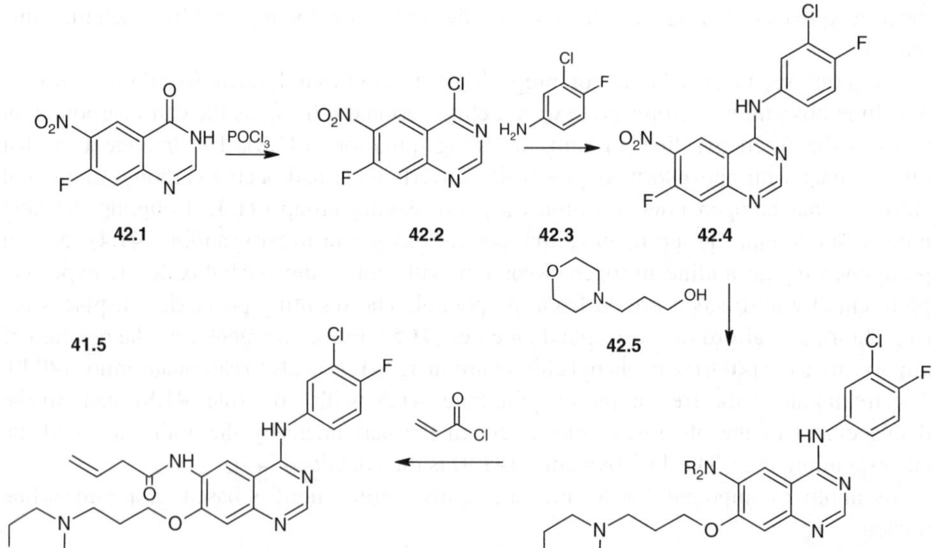

Scheme 8.42 *Canertinib.*

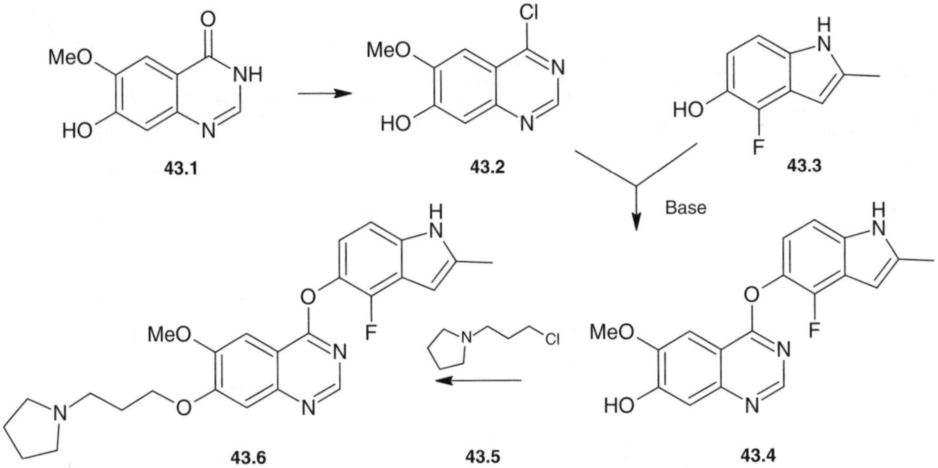

Scheme 8.43   *Cediranib.*

The straightforward synthesis starts by converting the carbonyl group in **42**.1 to enol chloride (**42.2**) by means of phosphorus oxychloride. That halogen atom is then displaced by nitrogen in the substituted aniline (**42.3**). The first side chain is added by nucleophilic aromatic displacement of the relatively labile fluorine in quinazoline (**42.4**) by the alkoxide from aminoalcohol (**42.5**). Catalytic hydrogenation then serves to reduce the nitro group to an amine (**42.7**). The acylation of that amine with acryloyl chloride leads to the corresponding amide. **Canertinib** (**42.8**) is thus obtained [44].

The inhibitor cediranib (**43.6**) in common with other quinazolines showed promising preclinical activity both in *in vitro* and in animal models. Some of the data indicated significant antiangiogenic activity. Data from clinical trials was however equivocal. This drug interestingly bears the trade name Recentin®, though it has not yet been approved by the FDA.

The synthesis of this drug generally follows a similar scheme used for other quinazoline inhibitors. The first step comprises converting the carbonyl group in the starting material (**43.1**) to a chloride. The next major moiety is incorporated by displacing halogen with the anion from the hydroxyl group in the substituted indole (**43.3**). The reaction of the hydroxyl group in the resulting intermediate with base forms the corresponding alkoxide; the treatment of this with chloropropylpiperidine (**43.5**) attaches that fragment via an ether linkage. The kinase inhibitor **cediranib** (**43.8**) is thus obtained [45].

The TKI tivantinib shows good activity in screens for detecting antineoplastic agents as well as *in vivo* assays where human cancer tumors are implanted in laboratory animals. This orally administered drug has shown promising activity in patients with liver cancer.

The synthesis for this apparently complex compound is surprisingly straightforward. The alkylation of tetrahydroquinoline (**42.1**) on nitrogen with ethyl 4-bromoacetoacetate affords the intermediate **44.2**. Treatment with magnesium chloride, a weak Lewis acid, brings about cyclization of the carbonyl via the corresponding enolate at position 2 into the aromatic ring forming a fused indole. The ester is then saponified and the resulting acid caused to decarboxylate. The reaction of that intermediate with the half-acid chloride of

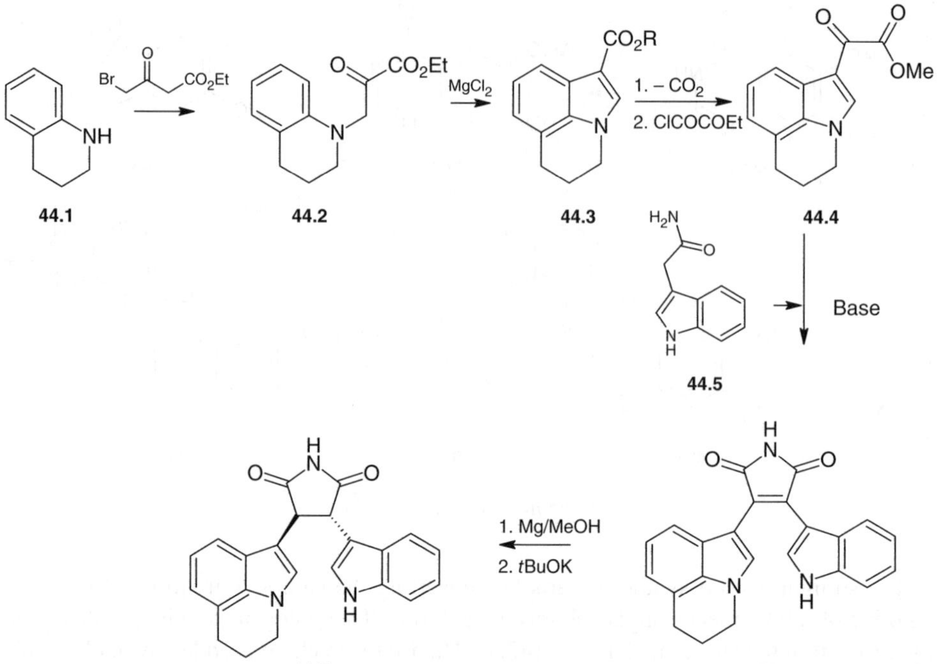

Scheme 8.44   *Tivantinib.*

methyl oxalate puts in place the carbon atoms that will form the base of the future pyrrole ring. The reaction of the resulting oxalate with indole 3-ethylcarboxamide (**44.6**) mediated by the base generates an anion on the methylene group of the reagent. A formal means to rationalize the overall transform is to assume that it first involves an amide ester inter-change that links the two moieties. Ordinary base-catalyzed aldol condensation will then close the ring. The dissolving metal reduction, in this case magnesium, leads to the unde-sired Z configuration for the two indole fragments. Treatment with the strong base, potassium *tert*butoxide, converts the stereochemistry to the E isomer. **Tivantinib (44.7)** is thus obtained [46].

8.8   Janus Kinase Inhibitors

The class of tyrosine kinases dubbed Janus differs significantly from those discussed to this point in that they do not interact with receptors. Cytokines are a group of small pro-teins that are also involved in cell signaling and are especially important in the operation of the immune system; they do not however have any kinase activity of their own. The Janus kinases fulfill that function for the cytokines by adding phosphate groups to selected tyrosines.

In vitro assays confirmed that the momelotinib is a selective inhibitor of Janus kinases 1 and 2. In a mouse model of myelofibrosis, the compound normalizes blood parameters.

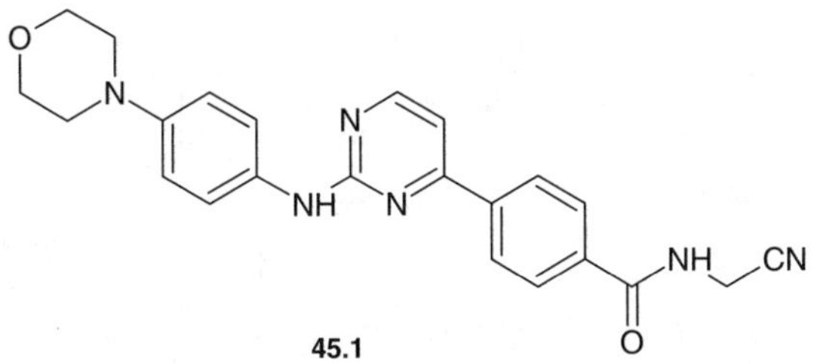

45.1

Scheme 8.45 *Momelotinib.*

Phase I and 2 clinical testing against the chronic blood cancer myelofibrosis showed the drug to be well tolerated; it also reduced some of the disease symptoms.

This is yet another case where the lack of sources for the synthesis leads to the display of only the structure of the Janus inhibitor **momelotinib** (**45.1**).

The Janus kinase inhibitor ruxolitinib normalizes blood parameters in the mouse model of myelofibrosis. This observation was repeated in clinical trials of patients suffering from that disease. In 2011, the FDA approved the drug for the treatment of intermediate- and high-risk myelofibrosis. The compound is available under the trade name Jakafi®.

One branch in the somewhat complex convergent scheme for preparing optically pure ruxolitinib starts with the protection of bromopyrazole as its ethylvinyl ether (**46.2**). The bromine at position 3 is displaced by boron in pinacol borate to afford the borated pyrazole (**46.3**). In a converging branch, the free amine in purine (**46.4**) is treated with DMF acetal to give the corresponding homologated imine. That is the converted in several steps to a protected diaryloxymethylene group (**46.6**). The Suzuki coupling reagent (**46.3**) with the halogenated purine (**44.6**) served to add the pyrazole to the purine. There remains the task of adding the side chain as a single enantiomer. The carbon skeleton of the side chain is constructed by Wittig reaction of cyclopentane carboxaldehyde with the ylide from triphenylphosphine and chloroacetaldehyde.

The key reaction comprises Michael addition of pyrazole nitrogen to the conjugated aldehyde. The reaction is carried out in the presence of the chiral base (**46.10**) derived from R proline. The chirality is transferred to the adduct (**46.11**) which was obtained in close to 90% enantiomeric excess. The aldehyde was converted to the corresponding nitrile by reaction with hydroxylamine and iodine. The Janus kinase inhibitor **ruxolitinib** (**46.13**) [47].

In vitro assays confirm the inhibitory activity against Janus kinase of gandotinib. The compound also inhibits those kinases in human cancer implanted on laboratory animals. The drug has been subjected to a Phase I as well as a Phase II clinical trial in patients with myeloproliferative neoplasms.

The scheme for preparing gandotinib features discrete stages for preparing components of the structure and then assembling those. The reaction of aminopyridazine (**45.1**) with DMF acetal results in the formation of imidine (**47.2**) activated for adding the required fused imidazole. 2-Haloketones are frequently used as a partner in building heterocycles. Haloketone in this case is prepared by treating the substituted acetophenone (**47.3**) with thionyl chloride. The reaction of pyridazine (**47.2**) with chloroacetophenone thus affords the substituted imidazopyridine (**47.5**). In a complex reaction, that intermediate is then treated with N-methylmorpholine oxide in the presence of a vanadium oxide-acetoacetate

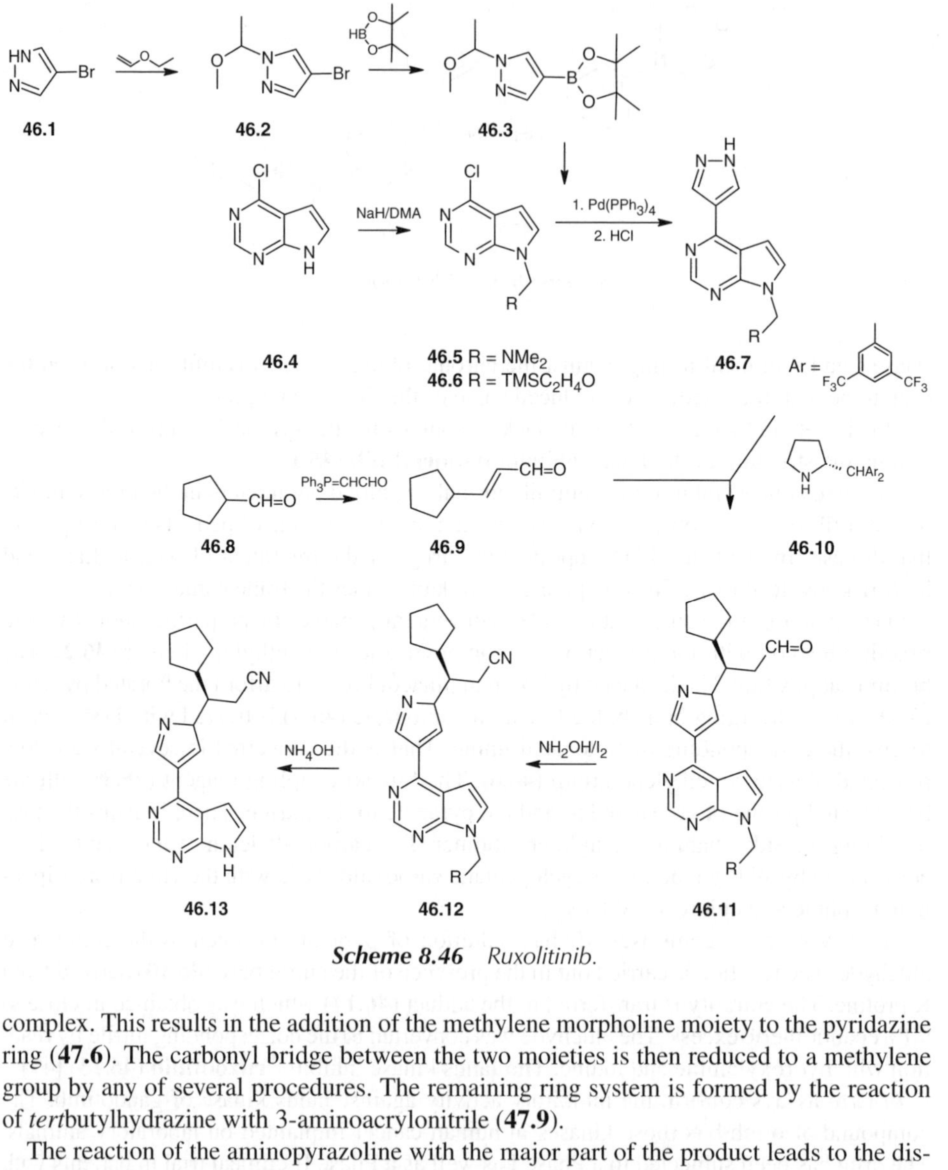

Scheme 8.46   *Ruxolitinib.*

complex. This results in the addition of the methylene morpholine moiety to the pyridazine ring (**47.6**). The carbonyl bridge between the two moieties is then reduced to a methylene group by any of several procedures. The remaining ring system is formed by the reaction of *tert*butylhydrazine with 3-aminoacrylonitrile (**47.9**).

The reaction of the aminopyrazoline with the major part of the product leads to the displacement of chlorine by the amino group in former, thus affording the *tert*-butyl-protected version of the final product. Treatment with trifluoroacetic acid cleaves off that group, affording the Janus kinase inhibitor **gandotinib** (**47.11**) [48].

The chemical structure of Janus kinase inhibitor pacritinib comprises a most unusual eighteen-membered ring. The compound exhibits potent inhibition of Janus kinases *in vitro*. It also shows good activity against human tumors implanted in laboratory animals. A combined Phase I and Phase II trial in patients with myelofibrosis showed encouraging results. A Phase III clinical trial started in late 2013.

One branch of the convergent preparation of this compound begins with the alkylation of the phenol in nitrosalicylaldehyde (**48.1**) with 1-bromo-3-chloropropane; this chain

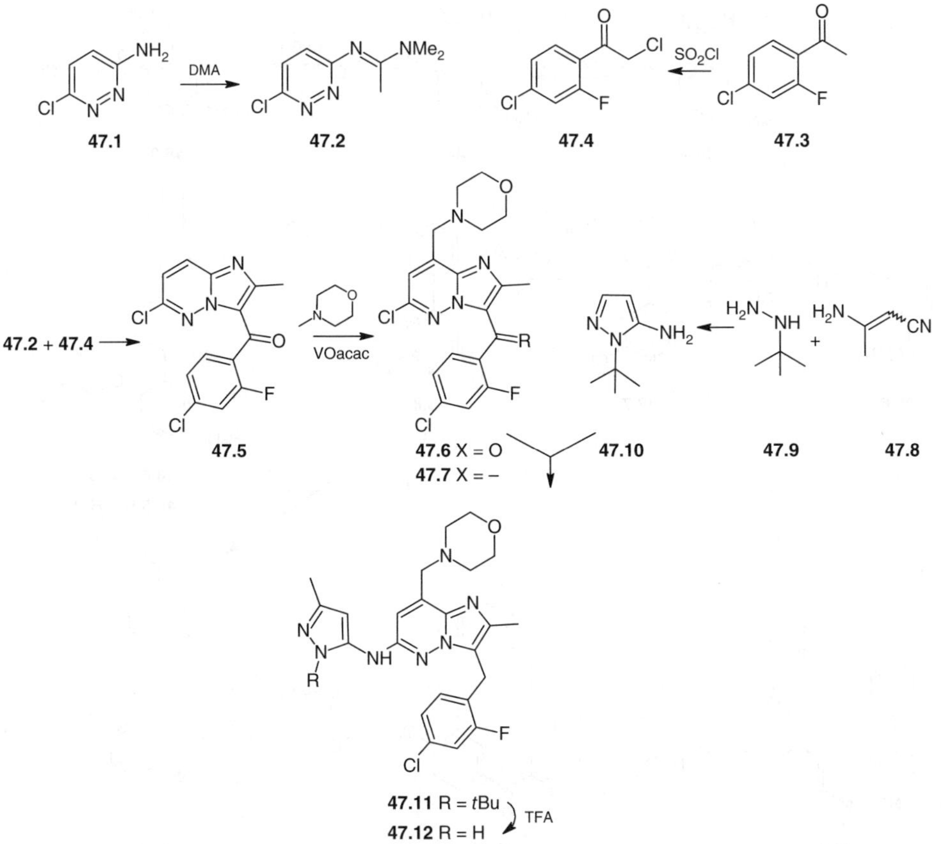

Scheme 8.47 *Gandotinib.*

will carry the future base. The treatment of that intermediate with sodium borohydride reduces the carbonyl group to a benzyl alcohol. The newly created hydroxyl group is then alkylated with allyl bromide. Synthesis of the other fragment begins with palladium-catalyzed cross coupling of aryl boric acid (**48.6**) with 2,4-dichloropyrimidine. The hydroxyl group in the resulting product is then alkylated with allyl bromide. The reduction of the nitro group by means of stannous chloride completes the synthesis of one half of the final product.

The displacement of the remaining chlorine in pyrimidine (**48.8**) by the primary amine in the other fragment (**48.5**) links the two moieties via an amine. The key reaction in this scheme comprises closing the 18-membered ring. Olefin metathesis is particularly suitable for this task since the reaction in this case is restricted to the two terminal olefins. Thus, the treatment of the linked product (**48.6**) with ruthenium-based Grubbs catalyst in the presence of acid causes the two terminal double-bonded groups to metastasize and eject ethylene; the macroscopic ring is closed in the process. The displacement of chlorine in the side chain by pyrrolidine completes the synthesis of **pacritinib** (**48.8**) [49, 50].

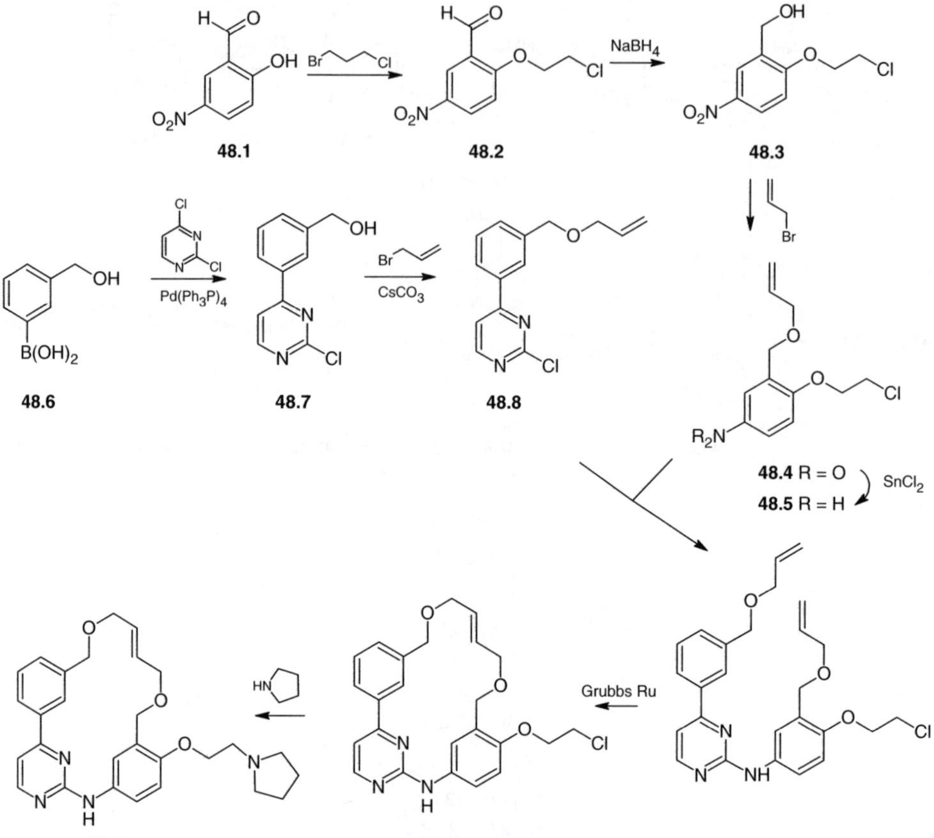

Scheme 8.48 *Pacritinib.*

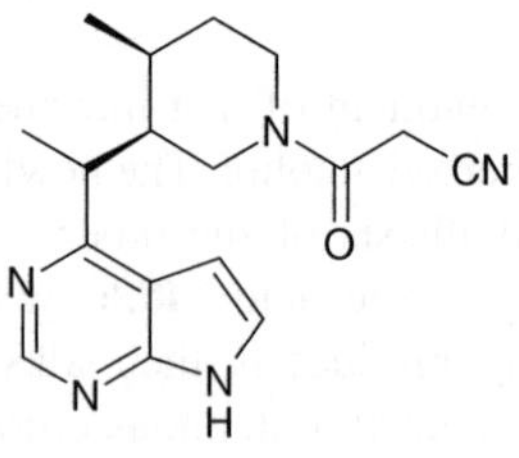

Scheme 8.49 *Tofacitinib.*

It is of interest to note in passing that the utility of Janus kinase inhibitors is not restricted to antineoplastic agents. The purine-based compound tofacitinib (**49**) (trade name Xeljanz®), for example, is approved for treating rheumatoid arthritis.

References

[1]   R.C. Schnuur, L.D. Arnold, U.S. Patent 5,747,498 (1998).
[2]   P. Knesl, D. Rösling, J. Jordis, *Molecules* **11**, 286 (2006).
[3]   A. Pandey, D.L. Volkots, J.M. Seroogy, J.W. Rose, J.-C. Yu, J.L. Lambing, A. Hutchaleelaha, S.J. Hollenbach, K. Abe, N.A. Giese, R.M. Scarborough, *J. Med. Chem.* **45**, 3772 (2002).
[4]   M.S. McClure, M.H. Osterhout, F. Roschangar, M.J Sacchetti, U.S. Patent 7,157,466 (2007)
[5]   E. Wallace, G. Topalov, J. Lyssikatos, A. Buckmelter, Q. Zhao, U.S. Patent 7,452,895, (2008).
[6]   H.-R. Tsou, E.G. Overbeek-Klumpers, W.A. Hallett, M.F. Reich, M. Floyd, B.D. Johnson, R. Michalak, R. Nilakantan, C. Discafani, J. Golas, S.K. Rabindran, R. Shen-Shi, Y.-F. Wang, J. Upeslacis, A. Wissner, *J. Med. Chem.* **48**, 1107 (2005).
[7]   S.A. Fakhoury, H.T. Lee, J.E. Reed, K.M. Schlosser, K.E. Sexton, H. Tecle, R.T. Winters, U.S. Patent 7,772,243, (2010).
[8]   M. Gao, C.M. Lola, M. Wang, K.D. Miller, G.W. Sledge, Q.-H. Zheng, *Bioorg. Med. Chem. Lett.* **21**, 3222 (2011).
[9]   Esai News Release 14-07, 2/3/2014.
[10] T. Naito, K. Yoshizawa, U.S. Patent 7,683,172 (2010).
[11] P.R. Patel, H. Sunc, S.Q. Li, M. Shenc, J. Khand, C.J. Thomas, M.I. Davis, *Bioorg. Med. Chem. Lett.* **23**, 4398 (2013).
[12] L.C. Bannen, S.-M. Chan, S.-M. Diva, T.P. Forsyth, R.G. Khoury, J.W. Leahy, M.B. Mac, L.W. Mann, J.M. Nuss, J.P. Parks, Y. Wang, W. Xu, U.S. Patent 7,579,473 (2009).
[13] L Sun, N. Tram, H. App, P. Hirth, G. MacMahon, C. Tan, *J. Med. Chem.* **41**, 2588 (1998).
[14] L. Sun, C. Liang, S. Shirazian, Y. Zhou, T. Miller, J. Cui, J. Fukuda, J.-Y. Chu, W.T. People, A. Nematalla, X. Wang, H. Chen, A. Sistla, T.C. Luu, F. Tang, J. Wei, Cho Tang, *J. Med. Chem.* **46**, 1116 (2003).
[15] G.J. Roth, A. Heckel, F. Colbatzky, S. Handschuh, J. Kley, T. Lehmann-Lintz, R. Lotz, I. Tontsch-Grunt, R. Walter, F. Hilberg, *J. Med. Chem.* **52**, 4466 (2009).
[16] J. Magano, J.R. Dunetz, Transition Metal Catalyzed Couplings in Process Chemistry, Wiley, Hoboken, NJ (2012), p.167.
[17] G.A. Crispino, M. Hamedi, Mourad, T.L. LaPorte, J.E. Thornton, J.A. Pesti, Z. Xu, P.C. Lobben, D.K. Leahy, J. Muslehiddinoglu, C. Lai, L.A. Spangler, R.P. Discordia, F.S., U.S. Patent, 7,671,199 (2010).
[18] C. Zhen-wei Cai, Y. Zhang, R.M. Borzilleri, L. Qian, S. Barbosa, D. Wei, X. Zheng, L. Wu, Junying Fan, Z. Shi, B. S. Wautlet, S. Mortillo, R. Jeyaseelan, Sr, D.W. Kukral, A. Kamath, P. Marathe, C. D'Arienzo, G. Derbin, J.C. Barrish, J.A. Robl, J.T. Hunt, L.J. Lombardo, J. Fargnoli, R.S. Bhide, *J. Med. Chem.* **51**, 1976 (2008).
[19] W.-S. Huang, C.A. Metcalf, R. Sundaramoorthi, Y. Wang, D. Zou, R.M. Thomas, X. Zhu, L. Cai, D. Wen, S. Liu, J. Romero, J. Qi, I. Chen, G. Banda, S.P. Lentini, S. Das, Q. Xu, J. Keats, F. Wang, S. Wardwell, Y. Ning, J.T. Snodgrass, M.I. Broudy, K. Russian, T. Zhou, L. Commodore, N.I. Narasimhan, Q.K. Mohemmad, J. Iuliucci, V.M. Rivera, D.C. Dalgarno, T.K. Sawyer, T. Clackson, W.C. Shakespear, *J. Med. Chem.* **53**, 4701 (2010).
[20] P. Betschmann, A.F. Burchat, D.J. Calderwood, M.L. Curtin; Davidsen, K. Steven, H.M. Davis, R.R. Frey, H.R. Heyman, G.C. Hirst, P. Hrnciar, M.R. Michaelides, M.A. Muckey, K.D. Mullen, D. Kelly, P. Rafferty, C.K. Wada, U.S. Patent 7,737,160 (2010).
[21] T. Matsushima, K. Takahashi, S. Funasaka, S. Shirotori, U.S. Patent 7,855,290 (2010).
[21*] U.R. Khire, D. Bankston, J. Barbosa, D.R. Brittelli, Y. Caringal, R. Carlson, J. Dumas, T. Gane, S.L. Heald, B. Hibner, J.S. Johnson, M.E. Katz, N. Kennure, J. Kingery-Wood, W. Lee, X.-G. Liu, T.B. Lowinger, I. McAlexander, M.-K. Monahan, R. Natero, J. Renick, B. Riedl, H. Rong, R.N. Sibley, R.A. Smitha, D. Wolanina, *Bioorg. Med. Chem.* **14**, 783 (2004).
[22] J. Das, P. Chen, D. Norris, R. Padmanabha, J. Lin, R.V. Moquin, Z. Shen, L.S. Cook, A.M. Doweyko, S. Pitt, S. Pang, D.R. Shen, Q. Fang, H.F. de Fex, K.W. McIntyre, D.J. Shuster, K.M. Gillooly, K. Behnia, G.L. Schieven, J. Wityak, J.C. Barrish, *J. Med. Chem.* **49**, 6465 (2006).
[23] D.H. Boschelli, F. Ye, Y.D. Wang, M. Dutia, S.L. Johnson, B. Wu, K. Miller, D.W. Powell, D. Yaczko, M. Young, M. Tischler, K. Arndt, C. Discafani, C. Etienne, J. Gibbons, J. Grod, J. Lucas, J.M. Weber, F. Boschelli, *J. Med. Chem.* **44**, 3965 (2001).

[24] F. Hennequin, J. Allen, J. Breed, J. Curwen, M. Fennell, T.P. Green, C. Lambert-van der Brempt, R. Morgentin, R.A. Norman, A. Olivier, L. Otterbein, P.A. Ple´, N. Warin, G. Costello, *J. Med. Chem.* **49**, 6465 (2006).

[25] A. Pandey, D.L. Volkots, J.M. Seroogy, J.W. Rose, J.-C. Yu, J.L. Lambing. A. Hutchaleelaha, S.J. Hollenbach, K. Abe, N. Giese, R.M. Scarborough, *J. Med. Chem.* **45**, 3772 (2002).

[26] P.A. Harris, A. Boloor, M. Cheung, R. Kumar, R.M. Crosby, R.G. Davis-Ward, A.H. Epperly, K.W. Hinkle, R.N. Hunter III, J.H. Johnson, V.B. Knick, C.P. Laudeman, D.K. Luttrell, R.A. Mook, R.T. Nolte, S.K. Rudolph, J.R. Szewczyk, A.T. Truesdale, J.M. Veal, L. Wang, J.A. Stafford, *J. Med. Chem.* **51**, 4632 (2008).

[27] W.-C. Shieh, J. McKenna, U.S. Patent 8,124,763 (2012).

[28] Xiaoyan Pan, Fang Wang, Yanmin Zhang, Hongping Gao, Zhigang Hu, Sicen Wang, Jie Zhang, *Bioorg. Med. Chem.* **21**, 2527 (2013).

[29] X. An, F. Lv, L. Yan, G. Wang, H. Li, WO Patent 2014022975 (2014).

[30] H. Liu, W. Zhu, Q. Guo, J. Wang, X. Wang, P. Gong, *Chin. J. Med. Chem.* **22**, 26 (2012).

[31] W. Ying, D. James, S. Zhang, T. Przewloka, J. Chae, D.U. Chimmanamada, C.W. Lee, E. Kostik, H.P. Ng, K. Foley, Z. Du, J. Barsoum, U.S. Patent 7,825,148 (2010).

[32] E.J. Pease, G.A. Breault, J. J.Morris, U.S. Patent 7,153.964 (2006); Anon (2013). http://www. yaopha.com/ (accessed October 21, 2014).

[33] A. Tasaka, T. Hitaka, E. Matsutani, U.S. Parent 6,716,863 (2004).

[34] J.J Cui, M. Tran-Dube, H. Shen, M. Nambu, P.-P. Kung, M. Pairish, L. Jia, J. Meng, L. Funk, I. Botrous, M. McTigue, N. Grodsky, K. Ryan, E. Padrique, G. Alton, S. Timofeevski, S. Yamazaki, Q. Li, H. Zou, J. Christensen, B. Mroczkowski, S. Bender, R.S. Kania, M.P. Edwards, *J. Med. Chem.* **51**, 6342 (2011).

[35] Y.-F. Liu, C.-L. Wang, Y.-J. Bai, N. Han, J.-P. Jiao, X. Qi, *Org. Proc. Res. Dev.* **12**, 490 (2008).

[36] S. Boyer, J. Dumas, B. Riedel, W. Scott, U.S. Patent Application 2005/0038080 (2005).

[37] L.C. Bannen, D.S.-M. Forsyth, P.T. Patrick, R.G. Khoury, J.W. Leahy, M.B. Mac, L.W. Mann, J.M. Nuss, J.J. Parks, Y. Wang, W. Xu, U.S. Patent 7,579,473 (2009).

[38] H. Kawasaki, H. Abe, K. Hayakawa, L.Tetsuya, S. Kikuchi, T. Yamaguchi, T. Nanayama, H. Kurachi, M. Tamaru, Y. Hori, M. Takahashi, T. Yoshida, T. Sakai, U.S. Patent 7,378,423, (2008).

[39] L. Honigberg, E. Verner, Z. Pan, U.S. Patent Appl.2011/0039868 (2005).

[40] L.D. Arnold, M.J. Mulvihill, U.S. Patent 7,534,797 (2009).

[41] N.J. Tom, M.J. Castaldi, D.B. Ripin, U.S. Patent 7,183,414 (2007).

[42] S. Cai, J. Chou, E. Harwood, Eric, T. Machajewski, D. Ryckman, X. Shang, S. Zhu, A.O. Okhamafe, M.S. Tesconi, U.S. Patent Application 2005/0137399 (2005).

[43] N. Matsunaga, S. Yoshida, A. Yoshino, T. Nakajima, U.S. Patent, 7,166,722 (2007).

[44] J. Smaill, G. Rewcastle, J.A. Loo, K.D. Greis, O.H. Chan, E.L. Reyner, E. Lipka, H.D.H. Showalter, P.W. Vincent, W.L. Elliott, W.A. Denny, *J. Med. Chem.* **43**, 1380 (2000).

[45] E.A. Arnott, J. Crosby, C. Evans, C. Matthew, J.G. Ford, M.F. Jones, K.W. Leslie, I.M. McFarlane, G.J. Sependa, U.S. Patent 7,851,623 (2010).

[46] C.J. Li, M.A. Ashwell, J. Hill, M.M. Mossa, M. Munshi, U.S. Patent 7,713,967 (2010).

[47] Q. Lin, D. Meloni, Y. Pan, M. Xia, J. Rodgers, S. Shepard, M. Li, L. Galya, B. Metcalf, T.Y. Yue, P. Liu, *J. Zhou. Org. Lett.* **11**, 1999 (2012).

[48] D. Mitchell, K.P. Cole, P.M. Pollock, D.M. Coppert, T.P. Burkholder, J.R. Clayton, *Org. Proc. Res. Dev.* **16**, 70 (2012).

[49] S. Blanchard, C. Lee, Cheng, H.K.M. Nagaraj, A. Poulsen, E.T. Sun, Y.L.E. Tan, A.D. William, U.S. Patent 8,153,632 (2012).

[50] Anon., *Drugs Fut.* **38**, 375 (2013).

9

Enzyme Inhibitors: Part II Additional Targets

The very large collection of tyrosine kinase inhibitors explored in the previous chapter makes it is easy to overlook compounds that inhibit other kinases.

9.1 Serine–Threonine Kinase Inhibitors

Some seven compounds that have generic names inhibit phosphorylation of the two additional hydroxyl-substituted amino acids: serine and threonine, rather than tyrosine. Judging by the dates of the publications this small collection of compounds that inhibits tyrosine kinase has, as a group, been developed more recently. The potential drugs that fall into this class present the same wide variation of chemical structures as the compounds that inhibit tyrosine kinases. These drug candidates are sorted on the basis of increasing structural complexity for lack of any other criterion.

The principal function of the enzyme poly(ADP ribose) polymerase (PARP) comprises repair of DNA damaged during replication. Tumor cells in various cancers are unusually dependent on PARP in large part because of their rapid proliferation. Inhibition of this factor leads to cell death. Iniparib (**1.1**) is said to be a prodrug for the nitroso derivative (**1.2**) that binds covalently to PARP-1. The compound has been tested in a sizeable number of clinical trials. In one Phase III trial, the drug failed to improve survival when tested in a "two-drug" protocol.

The compound is currently an article of commerce, available from chemical supply houses; those sources however probably list the compound as 3-nitro-4-iodobenzamide rather than its nonproprietary designation.

The kinase inhibitor rigosertib inhibits the same enzyme *in vitro*. The compound also shows activity against human tumors implanted in laboratory animals. The drug has shown

Antineoplastic Drugs: Organic Synthesis, First Edition. Daniel Lednicer.
© 2015 John Wiley & Sons, Ltd. Published 2015 by John Wiley & Sons, Ltd.

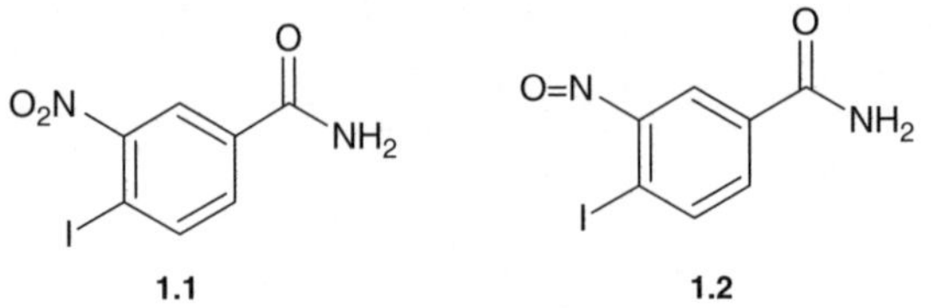

Scheme 9.1 *Iniparib.*

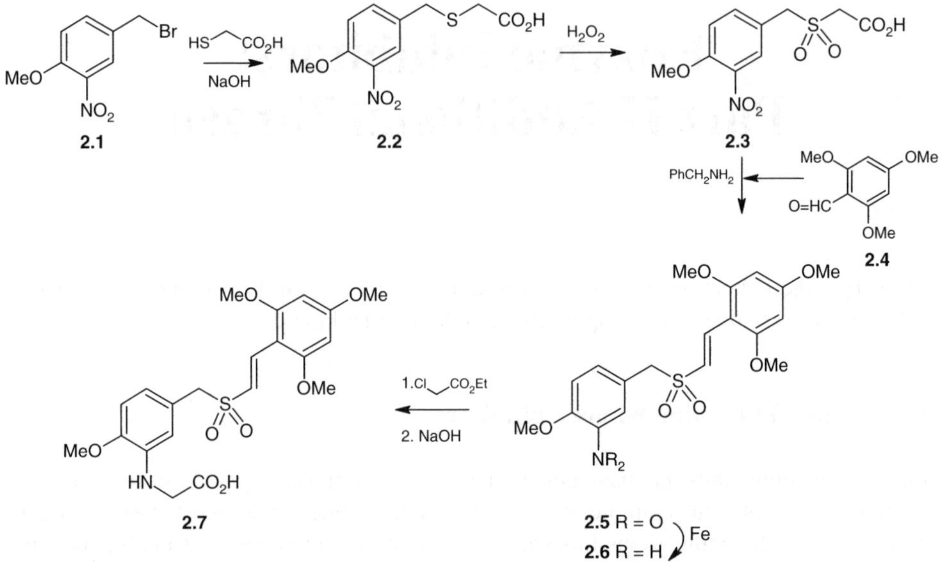

Scheme 9.2 *Rigosertib.*

activity in various solid tumors and hematological cancers. FDA has granted the drug orphan status for treating patients with myelodysplastic syndrome.

The synthesis of the compound begins by the displacement of halogen in benzyl bromide (**2.1**) by sulfur in thioglycolic acid. The sulfur in the product is then oxidized to a sulfone (**2.3**) by means of hydrogen peroxide. Aldol condensation of the now-activated **methylene** carbon in the product with the substituted benzaldehyde (**2.4**) leads initially to a hydroxyl acid. This intermediate dehydrates and then loses carbon dioxide to leave behind an E double bond linking the two fragments (**2.5**). The nitro group that has tagged along through the sequence is now reduced to an aniline by means of dissolving iron (**2.6**). The newly formed amine is next alkylated with ethyl 2-chloroacetate. The ester is then saponified, yielding the kinase inhibitor **rigosertib** (**2.7**) [1].

The kinase inhibitor pimasertib inhibits the PARP enzymes *in vitro*. The compound is also active against human tumors implanted on laboratory animals. Clinical trials of the compound are currently in early Phase II. The concise scheme for preparing this agent involves first linking isonicotinic acid (**3.1**) to the substituted aniline (**3.2**) by reacting the two components in the presence of the strong base LiHDMS. The reaction in effect

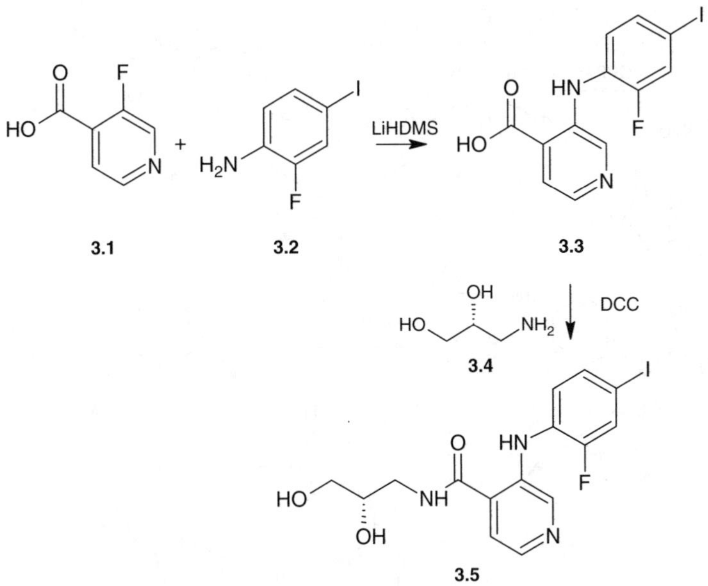

Scheme 9.3　*Pimasertib.*

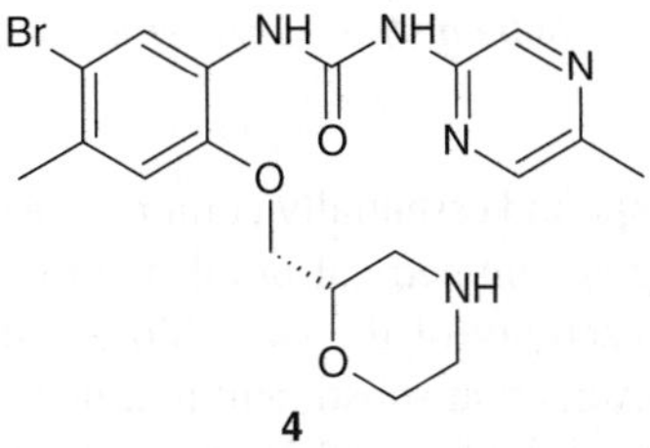

Scheme 9.4　*Rabusertib.*

comprises the displacement of fluorine by the nitrogen anion from the abstraction of an amine proton from (**3.2**). The carboxylic acid is next coupled with the defined enantiomer of 3,4-dihydroxypropylamine in the presence of dicylohexylcarbodiimide to afford the kinase inhibitor **pimasertib** (**3.5**) [2].

In vitro studies of the serine–threonine inhibitor **rabusertib**, also known as LY 2603618, show that the compound interferes with repair of DNA damaged by antineoplastic agents. The agent consequently potentiates the activity of a number of anticancer compounds known to damage DNA. Clinical trials are in early stages. Only the chemical structure is displayed in the absence of data on the preparation in the patent [3].

The potential antineoplastic agent tozasertib, also known as VX-680, binds to and inhibits the aurora kinase, one of the serine–threonine kinases. Inhibiting that enzyme halts cell division at the level of nuclear replication. The drug has been found to reduce tumor size in mice bearing acute myeloid leukemia (AML) tissue.

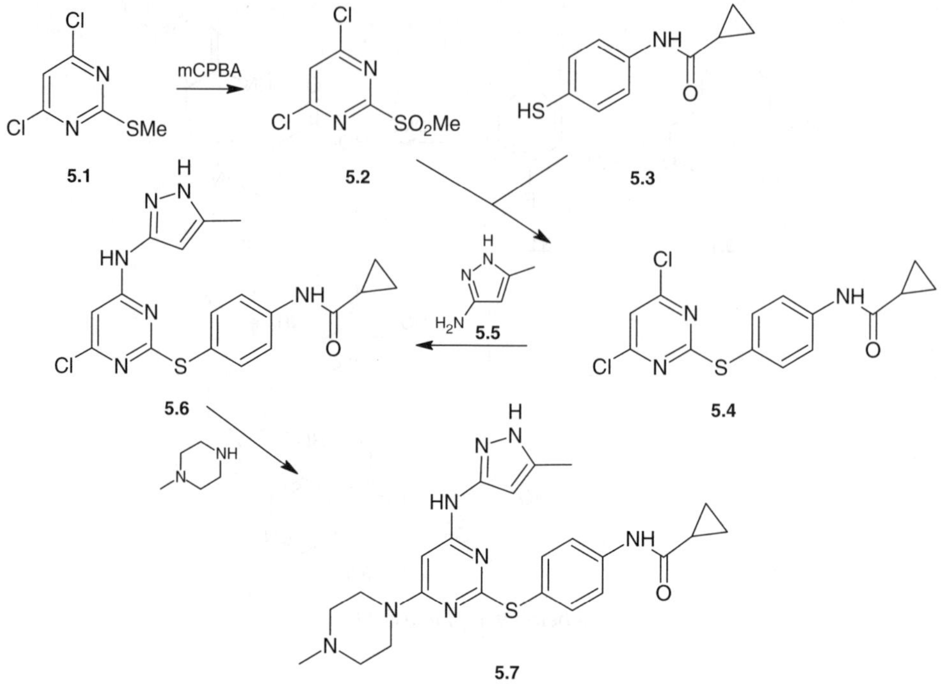

Scheme 9.5 *Tozasertib.*

The preparation of this compound essentially comprises a set of displacement reactions. Note that the pyrimidine ring is symmetrical in all intermediates and products, negating questions as to the relative reactivity of the two chlorine substituents. The oxidation of sulfur in pyrimidine (**5.1**) converts that substituent to a better leaving group. The reaction of that intermediate with benzothiophene (**5.3**) displaces the sulfoxide, linking the two moieties via a sulfide. The treatment of the product from that reaction with aminopyrazoline (**5.5**) displaces one of the pyrimidine chlorine atoms, attaching that heterocycle to pyrimidine (**5.6**). A second displacement reaction, this time using *N*-methylpiperazine, yields the Aurora kinin inhibitor **tozasertib** (**5.7**) [4].

Like the predecessor, volasertib binds to and inhibits serine–threonine kinins. Binding of that compound interferes with replication of the cell nucleus and as a result cellular proliferation. The compound shows good activity against a selection of human tumors implanted on laboratory animals. In 2013, in response to data from clinical trials, FDA granted volasertib the designation of "breakthrough therapy" for patients with AML. The following year, both FDA and the European Medicines Agency granted the agent orphan drug status for patients with AML.

One arm of the convergent synthesis starts with the displacement of the more reactive chlorine in pyrimidine (**6.1**) by one enantiomer of amine (**6.2**). The carboxylic acid of the product from that reaction is next treated with thionyl chloride to convert it to an acid chloride. The nitro group in the product is then reduced to the corresponding amine. Acylation

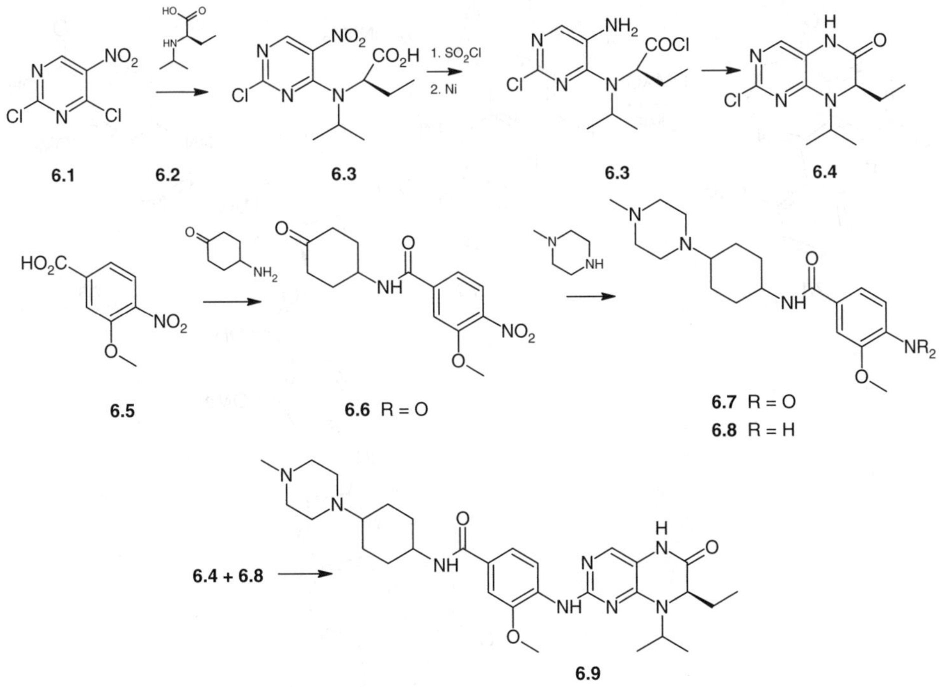

6.1   **6.2**   **6.3**   **6.3**   **6.4**

6.5   **6.6** R = O   **6.7** R = O
                           6.8 R = H

6.4 + 6.8 ⟶

6.9

Scheme 9.6   *Volasertib.*

of the newly formed amine by the adjacent acid chloride closes the ring, forming a benzopiperiinone (**6.4**).

The other arm of the synthesis begins with the acylation of 4-aminocyclohexanone by nitrobenzoic acid (**6.5**) in the presence of HBTU. Reductive amination of the carbonyl group by N-methylpiperazine incorporates that moiety, extending the side chain. The nitro group in the product in this fragment is then also reduced. The displacement of chlorine in the pyrimidine of moiety (**6.4**) by the newly formed amine in (**6.8**) links those two entities to afford the inhibitor **volasertib** (**6.9**) [5].

The compound alisertib is a highly specific inhibitor of the serine–threonine aurora A inhibitor. As a consequence, the agent disrupts the process whereby the cell nucleus replicates during cell proliferation. A Phase III trial of the drug in patients with relapsed or refractory T-cell lymphoma was initiated in early 2012.

The scheme for preparing the drug is quite straightforward despite the formidable appearance of its structure. The starting material (**7.1**) comprises one of the multitude of benzodiazepines prepared in programs aimed at preparing novel tranquilizers. Reaction with a combination of DMF and DMF acetal adds one nitrogen and one carbon atom next to the carbonyl group for the future fused pyrimidine ring. The condensation of the product (**7.2**) with 4-guanidyl benzoic acid (**7.3**) leads to the formation of a pyrimidine that links the two moieties. That product is the serine–threonine inhibitor **alisertib** (**7.4**) [6].

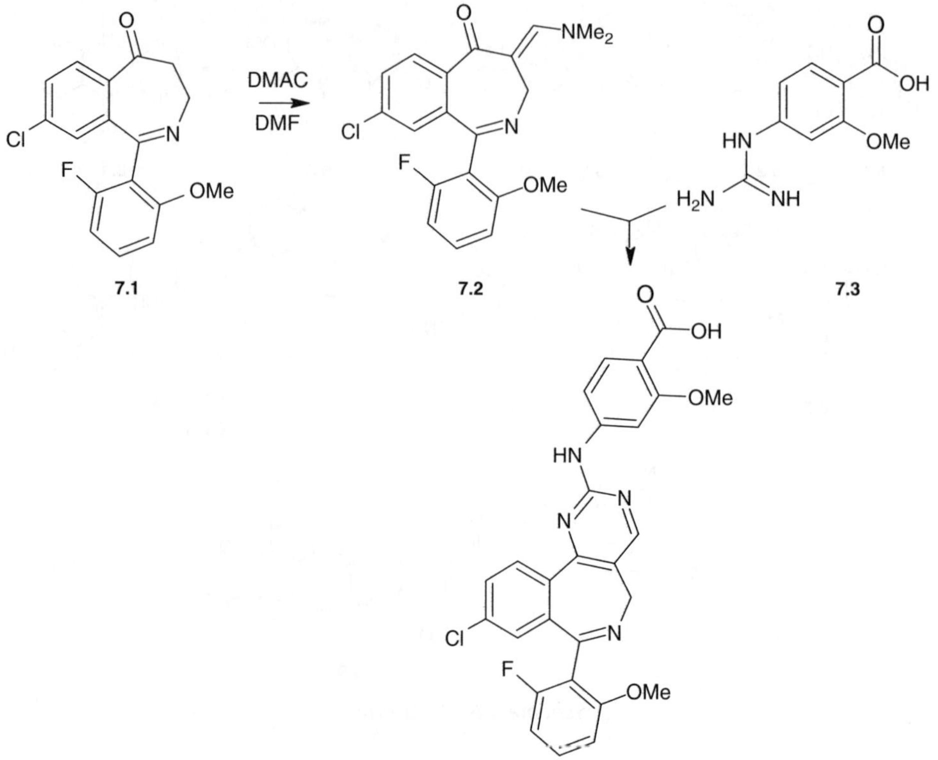

Scheme 9.7 *Alisertib.*

9.2 Additional Enzyme Inhibitors

9.2.1 Farnesyl Transferase Inhibitors

The protein Ras switches on other proteins involved in cell proliferation and growth. The attachment of the isoprenoid farnesyl to Ras oncoproteins by the enzyme farnesyl transferase leads to modified Ras that is then capable of attaching to cell membranes and thus transforming the cells. The overactive Ras can ultimately lead to cancer. Inhibiting farnesyl transferase would thus be expected to diminish, if not end, cancer cell proliferation.

The farnesyl transferase inhibitor tipifarnib was tested in the clinic against AML in the early years of the twenty-first century. The FDA granted the drug both orphan status and fast track status in 2004. Clinical studies were ongoing against the same cancer in 2011. Though the drug had not been approved by the agency as of early 2014, it was referred to by the name Zarnestra®.

The concise synthesis involves first acylation of *N*-methylaniline by cinnamoyl chloride (**8.1**). The treatment of the resulting amide with polyphosphoric acid leads to the attack of the protonated olefin onto the benzene ring and thus cyclization to a quinolone (**8.4**). The acylation of that product with 4-chlorobenzoyl chloride in the presence of aluminum

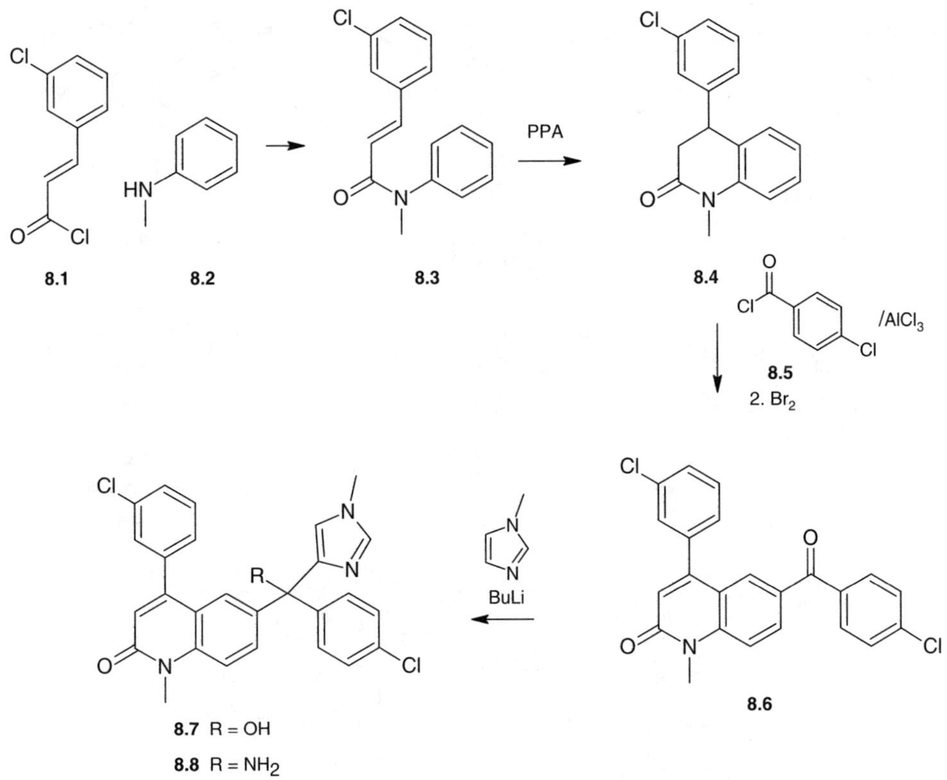

Scheme 9.8 *Tipifarnib.*

chloride attaches the benzene moiety. The treatment of the product with bromine then dehydrates the fused piperidine ring (**8.6**).

The anion from the reaction of *N*-methylimidazole with butyl lithium is then allowed to react with the carbonyl group to afford the tertiary carbinol (**8.7**). This product is then treated with thionyl chloride; this replaces the volatile hydroxyl group with chlorine. This is in turn replaced by an amine by treating the chloride with a solution of ammonia in an alcohol. This final product comprises **tipifarnib** (**8.8**) [7].

The farnesyl transferase inhibitor lonafarnib works well in assays against several human cancers as well as human lung cancer implanted in laboratory animals. The drug was used in a Phase I clinical trial in conjunction with gemcitabine as recently as 2011.

The dibenzocycloheptane starting material for preparing lonafarnib (**9.10**) is typical of the compounds used for the same purpose in research on classical antidepressants and muscle relaxants. The first step in the scheme for preparing lonafarnib comprises blocking reaction at a position favored for electrophilic substitution. Nitration of (**9.1**) with of a mixture of nitric and sulfuric acid indeed leads to nitration at that position. The nitro group is then reduced to an amine by means of iron filings. The reaction of that intermediate with bromine in acetic acid introduces bromine at the somewhat hindered position next to the cycloheptane ring (**9.4**). Heating that product with hydrochloric acid interestingly leads to

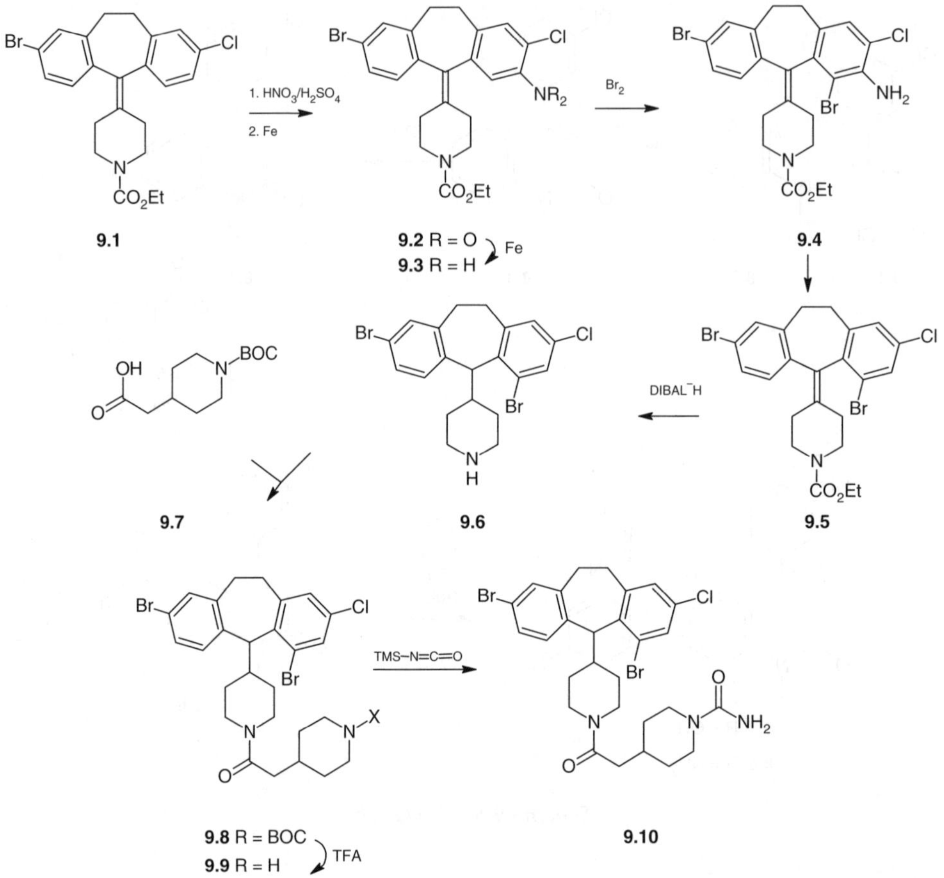

Scheme 9.9   Lonafarnib.

ejection of the blocking amine (**9.5**). Diisobutylaluminum hydride reduces the carbethoxy protecting group to an aminocarbinol. The hydride also, perhaps unexpectedly, reduces the double bond holding the piperidine ring. The next step comprises acylation of the newly freed amine by the acid (**9.7**) in the presence of ethyl-(*N*′,*N*′-dimethylamino)propylcarbodiimide and hydroxy benzotriazole. Trifluoroacetic acid then cleaves off the *tert*-butyloxycarbonyl protecting group. The reaction of the freed amine with trimethylsilylisocyanide then puts in place the terminal urea. **Lonafarnib (9.10)** is thus obtained [8].

9.2.2   Cyclin-Dependent Kinase Inhibitors

Cyclin-dependent kinases (CDKs) are a family of numerous proteins that regulate the cell cycle. Interference of CDK function will disrupt the cycle and cause the cell to die. The ideal inhibitor would comprise a compound that acted selectively on cancer cells. Short of that, an inhibitor would rely on the more rapid proliferation of cancer cells compared to normal ones.

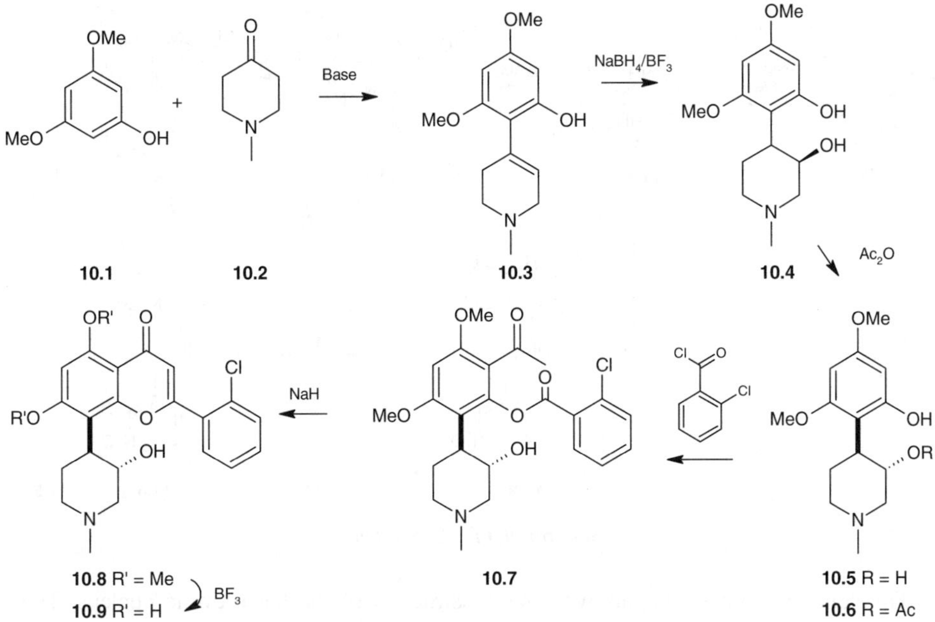

Scheme 9.10   *Alvocidib.*

The synthetic flavonoid alvocidib, formerly known as flavopiridol, has been found to inhibit CDKs. The compound was active against non-small cell lung cancer both *in vitro* and in trials using laboratory animals. Phase II trials against several cancers were conducted in the early 2000s.

One synthesis of this compound begins with an aldol-like condensation of phloroglucinol dimethyl ether (**10.1**) and *N*-methylpyrid-4-one to afford the adduct **10.3**. This transform can be rationalized by assuming that the anion from treatment of what is essentially the enol form of a β-dicarbonyl array adds to the ketone on the piperidine the initially formed hydroxyl will dehydrate to form the observed product (**10.3**). Hydroboration of that double bond with hydroborane generated *in situ* from sodium borohydride and boron trichloride followed by oxidation serves to introduce a hydroxyl group although with the wrong stereochemistry. The stereochemistry is the reversed by an unstated method (**10.4 → 10.5**). The hydroxyl group is next acylated with acetic anhydride (**10.6**) to protect that group. Reaction with 2-chlorobenzoyl chloride yields benzoate (**10.7**). Employing standard flavone chemistry, reaction with sodium hydride leads to another aldol condensation, this time between the anion from the acetyl methyl and the benzoate carbonyl group to yield the flavone ring (**10.8**); the ester covering the hydroxyl is lost in the process. Cleaving the methoxyl groups with boron trifluoride then yields the CDK inhibitor **alvocidib (10.9)** [9].

The CDK inhibitor dinaciclib (**11.9**) showed potent *in vitro* activity against several human cancers as well as activity against such cancer implanted in laboratory animals. The drug was being evaluated in a number of Phase II clinical trials against several malignancies including multiple myeloma.

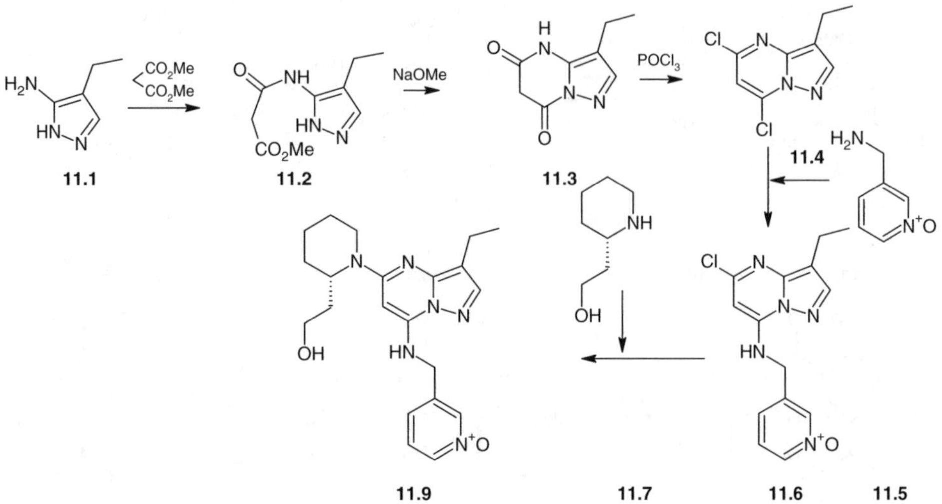

Scheme 9.11   *Dinaciclib.*

The concise synthesis begins with the construction of the heterocyclic nucleus. Thus, ester–amine interchange occurs between pyrazole (**11.1**) and methyl malonate leads to corresponding amide. The treatment of the product (**11.2**) with sodium methoxide then causes the resulting pyrazole anion to displace ethoxide from the remaining malonate ester, forming the pyrimidine ring. Reaction with the ubiquitous phosphorus oxychloride converts both carbonyl groups to enol chlorides, in the process freezing in the fully unsaturated tautomer (**11.4**). The displacement of the more labile halogen by the amine on 3-methylaminopridine oxide (**11.5**) leads to the intermediate (**11.6**). In a convergent step, commercial aminoalcohol (**11.7**) is resolved into its enantiomers by means of salts with d-10-camphorsulfonic acid. The resolved R isomer is next reacted with the intermediate (**11.6**); the amino group in the aminoalcohol displaces the remaining ring halogen. **Dinaciclib** (**11.9**) is thus obtained [10].

9.2.3   Proteasome Inhibitors

9.2.3.1   Peptidomimetic Boronic Acids

Proteasomes comprise protein complexes that reside both in the nucleus and cytoplasm of cells. The principal function of these complexes is to disassemble unneeded or damaged proteins such as misfolded proteins. Proteasomes also play a role in cell cycle regulation and cell death, or in other terms apoptosis. This process escapes regulation in tumor cells, leading to excess degradation of cell cycle regulating factors; this in turn causes tumor cell proliferation. Proteasome inhibitors disrupt these activities and induce apoptosis.

The proteasome inhibitor bortezomib is one of the rare therapeutic agents whose structure includes a boron atom. An inspection of the chemical structure of this drug reveals that it actually mimics the dipeptide leucine–phenylalanine in which the terminal carboxylic acid is replaced by a boronic acid. The drug was first approved by FDA as a parenteral agent for use in patients with multiple myeloma who had been pretreated with other antineoplastic agents. Additional indications for this drug, trade named Velcade®, have been added since. The enantioselective synthesis begins with the coupling of *tert*-butyloxycarbonyl-protected chiral

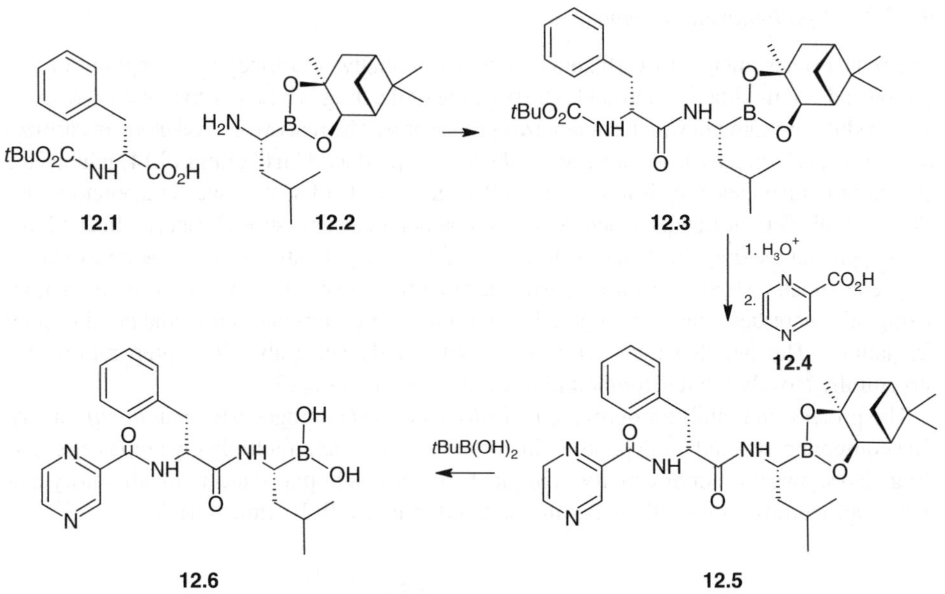

Scheme 9.12 *Bortezomib.*

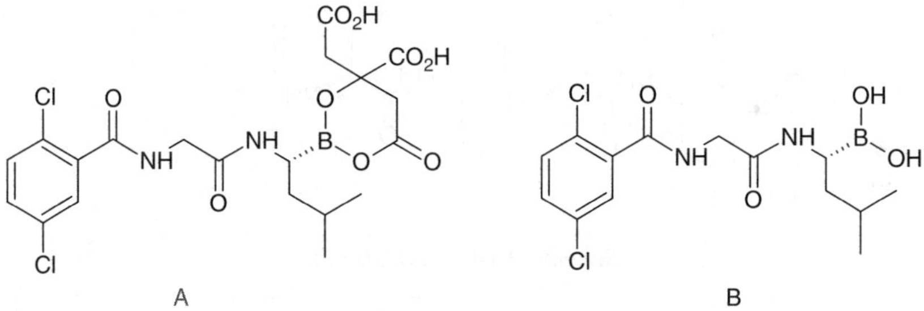

Scheme 9.13 *Ixazomib.*

phenylalanine (**21.1**) with the pinanediol cyclic ester of the boric acid analogue of leucine (**12.2**). The treatment of the resulting product with acid cleaves off the protecting group. The newly freed amine is then acylated with pyrazine carboxylic acid (**12.4**) to afford the corresponding amide (**12.5**). The pinanediol ester is then in turn cleaved by an exchange with *tert*-butylboronic acid. The free boronic acid comprises the proteasome inhibitor **bortezomib** (**12.6**) [11].

The proteasome inhibitor ixazomib (**13**) repeats the same theme: the structure of this compound in essence comprises a dipeptide, in this case alanine–leucine where the carboxylic acid is replaced by a boric acid. The latter is however tied up by a highly substituted hydroxy acid. (Note: some sources, notably PubChem, respond to a query by displaying version B.) This orally active compound exhibits better pharmacodynamic properties than its predecessor. It has shown good activity in both *in vitro* and *in vivo* preclinical antineoplastic activity. The drug entered a Phase III clinical trial in patients with multiple myeloma in 2013. Scheme 9.13 is constrained to the structure in the absence of sources for the detailed scheme for the synthesis [12].

9.2.3.2   Peptidomimetic Compounds

The proteasomes antagonist carfilzomib consist essentially of a tripeptide composed of two natural and one modified amino acid. The structure of the drug is based on that of the fermentation product epoxomicin which antagonizes proteasome. The reactive epoxyketone in carfilzomib traces back directly to its presence in the natural product. **Carfilzomib (9.14)** inhibits the growth of human cancer cells *in vitro* as well as in implants of such tissues in laboratory rats. Clinical trials demonstrated the activity of the compound against several cancers. In 2012, the FDA approved the drug, trade named Kyprolis®, for use in patients with multiple myeloma.

   The chemistry employed for preparing carfilzomib closely follows that used in synthesizing oligopeptides, that is, manipulation of protecting groups and sequential peptide bond formations. The detailed total synthesis is not readily accessible. The preparation of a structurally closely related compound illustrates the process [13].

   The proteasome inhibitor oprozomib **(15.6)** decreases angiogenesis in an *in vitro* assay. The compound also induces apoptosis in bortezomib-resistant multiple myeloma cells. The drug also shows antitumor activity in implants of various human cancers in laboratory animals. Oprozomib is currently being investigated in Phase Ib/II clinical trials.

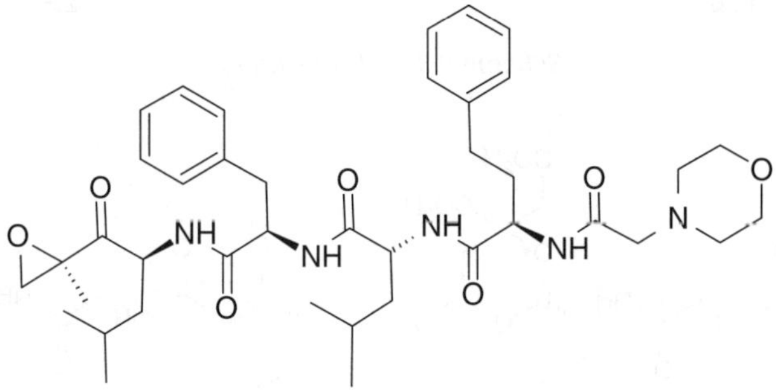

Scheme 9.14   Carfilzomib.

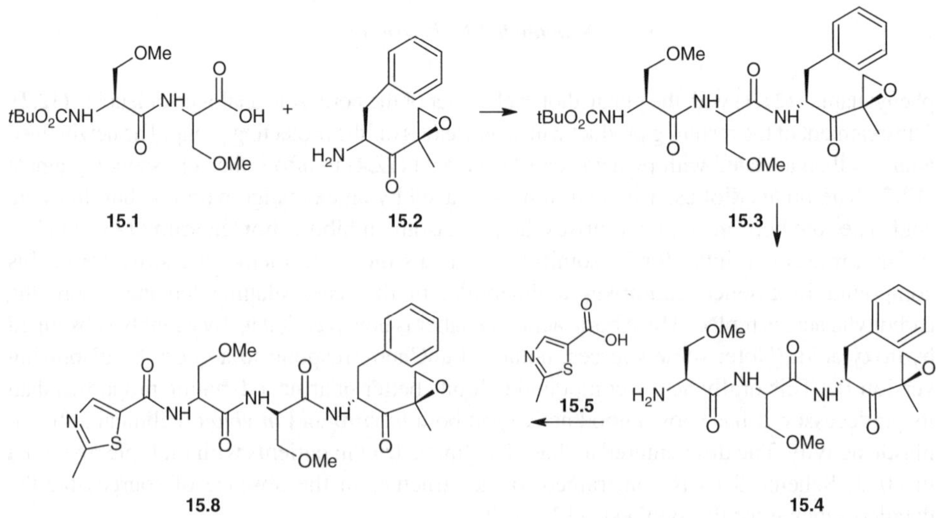

Scheme 9.15   Oprozomib.

The inhibitor oprozomib also consists of a peptide composed of both natural and modified amino acids; this molecule too incorporates an epoxyketone. The preparation of the compound thus consists of serial acylations of amines. The preparation starts by acylation of tyrosine by the BOC-protected tyrosine methyl ether in the presence of the peptide coupling reagent HBTU. The dipeptide (**15.1**) is then acylated with phenyl glycine. The reaction is then repeated using epoxy-amino acid (**15.4**), a reactive moiety that can form irreversible covalent bonds with amines in receptors. The reaction of the product (**15.6**) with the thiazole carboxylic acid in the presence of a coupling reagent affords the proteasome inhibitor **oprozomib** (**15.8**) [14].

9.2.3.3 A Natural Product

Compounds extracted from plants, fungi, microbes and the like have occupy a venerable place in medicinal chemistry. More than just a few natural products, as noted previously, went on to become antineoplastic drugs. It is thus of interest to note that a compound isolated quite recently from the marine bacterium *Salinispora tropica* is a potent proteasome inhibitor. The agent salinosporin, which is also known by the nonproprietary name marizomib, shows potent activity against cells from human cancers both *in vitro* and in implants in laboratory animals. The drug currently is being evaluated in the clinic in Phase II trials against malignancies such as multiple myeloma.

Salinosporamide, later identified by the USAN nonproprietary name marizomib, attracted the attention of the total synthesis community early in the twenty-first century in part because of the antineoplastic activity of the compound and, more likely, by the challenge posed by the number of functional groups crammed into a small space. The predictable groups each published total enantioselective total syntheses of salinosporamide. The synthesis displayed in Scheme 16 is arguably the shortest. This scheme starts by converting the substituted acetoacetate (**16.1**) to its acetal (**16.2**) by reaction with ethylene glycol. Saponification

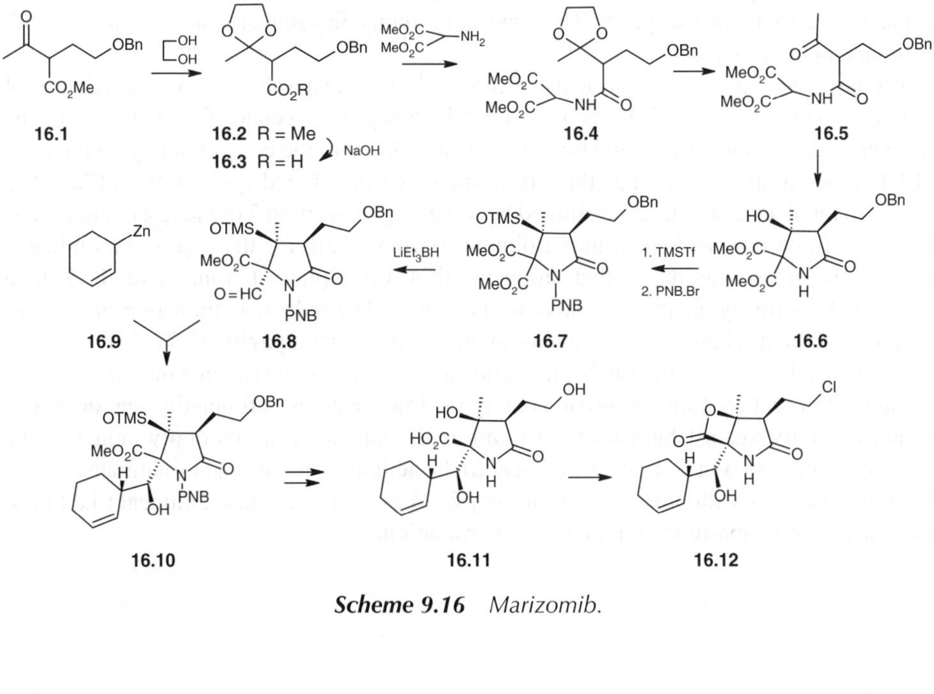

Scheme 9.16 *Marizomib.*

then affords to the corresponding carboxylic acid (**16.3**). The resulting intermediate is next treated with dimethyl-2-aminomalonate in the presence of a coupling reagent to afford amide (**16.5**). The key reaction in the preparation of the pyrrolidone comprises the enantioselective cyclization of the malonate. Allowing a solution of the amide to stand for several days accomplishes the desired internal aldol condensation, affording pyrrolidone (**16.6**) largely as a single diastereomer. The next several steps comprise protection of hydroxyl as a trimethylsylil ether (TMS) and pyrrolidine nitrogen as a 4-nitrobenzyl (PNB) derivative. Reaction with lithium triethylborohydride, the so-called superhydride, interestingly reduces just one of the malonate esters to an aldehyde, leaving the other untouched (**16.8**); this reduction proceeds stereoselectively. The requisite cyclohexene moiety is introduced by treating the aldehyde with cyclohex-3-ene zinc; note that this reaction too is stereoselective. The next several steps (not depicted) comprise the removal of protecting groups; special conditions were elaborated for each of those transforms. Hydroxy acid (**16.1**) is then converted to the required β-lactone by the peptide coupling agent BOP. The side chain hydroxyl is replaced by chlorine by the reaction of (**16.10**) with triphenylphosphonium chloride. The proteasome inhibitor **marizomib** (**16.11**) is thus obtained [15].

9.2.4 PARP Inhibitors

Messenger RNAs are the molecules that carry instructions from the DNA to the ribosomes, the machinery that uses that information to prepare proteins. These RNAs almost always include a polyadenine, a stretch of adenine not specified by DNA. When a nick or a break occurs in the DNA, PARP goes to that site and recruits DNA repair proteins that proceed to mend the fault. Chemotherapy drugs as well as radiation cause many nicks and breaks in quickly dividing cancer cells. Unrepaired damage makes it more likely that the cells will undergo apoptosis.

The PARP inhibitor rucaparib (**17.7**) is a potent inhibitor of human PARP-1 *in vitro* and binds to several regions of the complex. In animal models, the compound enhances the antitumor activity of several antineoplastic agents. Data also point to the radiation-enhancing activity of rucaparib. The drug is currently in clinical trials in patients with ovarian and breast cancer.

An early step in a synthesis of the PARP inhibitor rucaparib comprises alkylation of indole carbethoxy ester (**17.1**) with 1-nitro-4-hydroxyethyl acetate. The reduction of the nitro group by means of zinc and hydrochloric acid leads to the corresponding amino ester (**17.3**). Ester–amine interchange then forms the seven-membered cyclic amide (**17.3**). The next step involves attaching a substituted benzene ring to position 2 of the fused indole. The reaction of that intermediate with pyridinium tribromide affords the requisite brominated indole. The cross coupling of that product with 4-formylphenylboronic acid (**17.5**) then attaches the 4-formylphenyl derivative to the indole. The reductive amination of that last product with methylamine and cyanoborohydride affords **rucaparib** (**17.7**) [16].

The PARP inhibitor veliparib displays good activity against PARP enzymes *in vitro*. The compound potentiated the cytotoxic activity of drugs such as carboplatin. The drug also showed activity against human breast tumor cells implanted in laboratory animals. The drug has been tested in a variety of clinical trials; activity was sufficient to initiate a Phase III trial in patients with breast cancer in early 2014 as well as a Phase I clinical trial to test the efficacy of veliparib in conjunction with radiation.

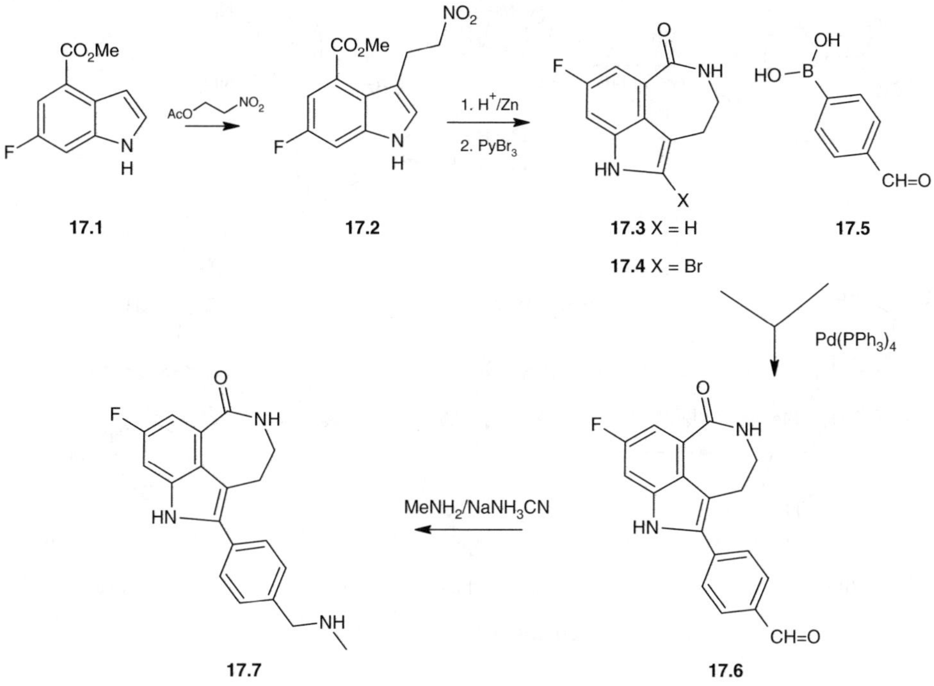

Scheme 9.17 *Rucaparib.*

The synthesis starts with the construction of the quaternary center required for inhibiting the PARP enzymes. The treatment of pyrrolidine carboxylic ester (**18.1**) in which the amine is present as its carbobenzyloxy derivative with sodium bis(trimethylsilyl)amide generates the corresponding carbanion. The treatment of that species with iodomethane leads to the methylated derivative (**18.2**). Saponification of the product affords the free carboxylic acid. The use of that intermediate to acylate the diamine (**18.4**) leads to attack of the less hindered of the two amines to afford the amide (**18.5**). The treatment of the resulting aminoamide with acid leads to attack of amide carbon on the unsubstituted amine. This leads to the formation of the benzimidazole ring. Catalytic hydrogenation then cleaves off the protecting group. The PARP inhibitor **veliparib** (**18.5**) is thus obtained [17].

The PARP inhibitor niraparib, also known as MK-4827, inhibits the proliferation of breast cancer cells *in vitro*. The compound also inhibits growth of human cancer cells implanted in laboratory animals. In common with the other PARP inhibitors, niraparib potentiates the activity of radiation therapy. The drug is currently being evaluated in several Phase III clinical trials.

There are two choices when faced with a product that incorporates an asymmetric center: the traditional approach comprises optical resolution at some point in the synthesis, and the more recent alternative that involves using an enzyme to catalyze the reaction that generates the chiral center. This last strategy is used in the synthesis of niraparib. Preparation of the key intermediate for an enantioselective synthesis comprises the addition of a methylene carbanion, generated from the reaction of DMSO and strong base to the carbonyl group in keto ester (**19.1**). The resulting epoxide is not isolated. Reaction with zinc bromide causes

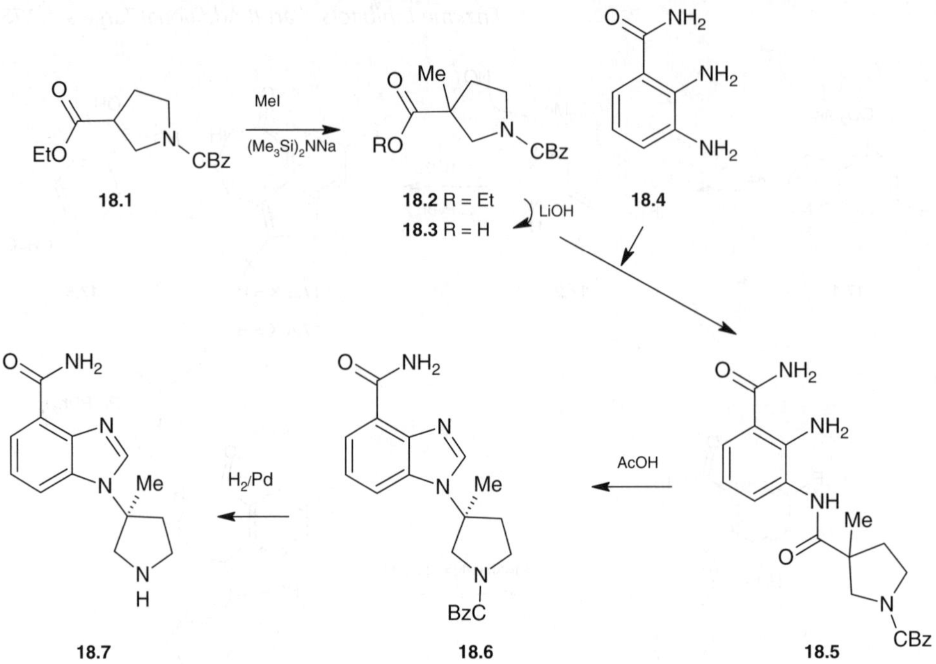

Scheme 9.18 Veliparib.

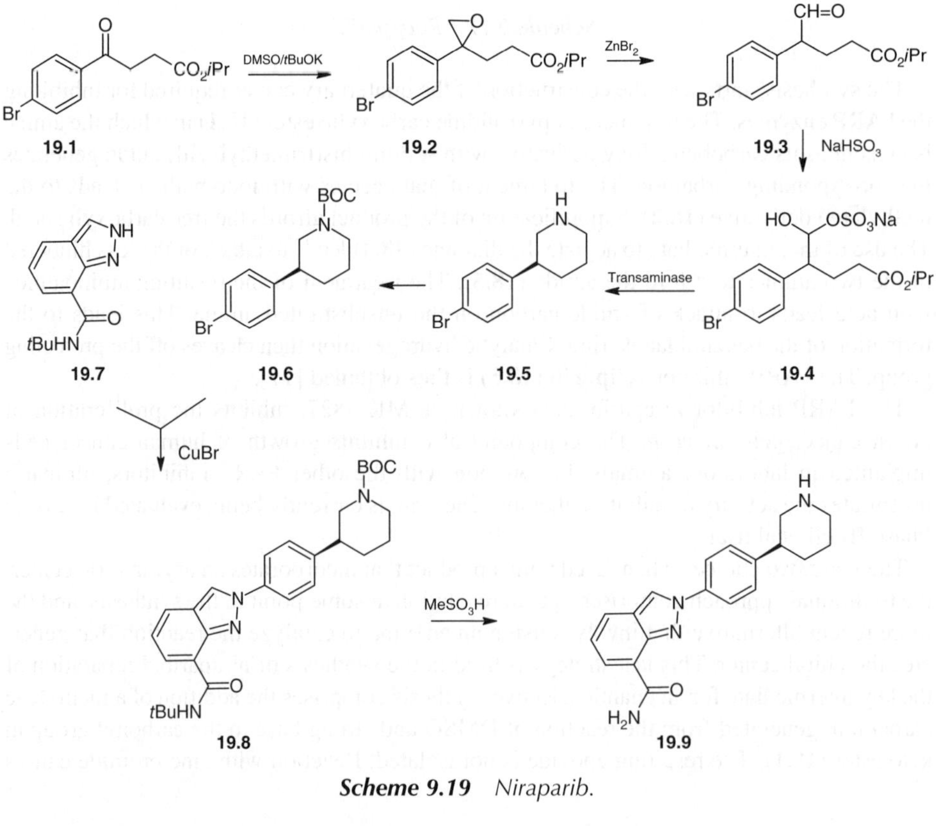

Scheme 9.19 Niraparib.

the oxirane to rearrange to an aldehyde (**19.2**). The treatment of that intermediate with sodium bisulfite then leads to the formation of the bisulfite adduct (**19.3**). The reductive amination/cyclization of the adduct with a selected transaminase enzyme first replaces the bisulfite function by an amine, which then causes that function to displace the ester cycliz- ing the side chain. This then provides the intermediate (**19.5**) as a single enantiomer. The final transform comprises reduction of the lactam carbonyl group. This series of reactions proceeds with high enantioselectivity. The amine is then treated with *tert*butyloxycarbonyl chloride to convert that function to its BOC derivative. This moiety is then coupled with indazone (**19.7**) in a reaction catalyzed by cuprous chloride. Treatment with methanesul- fonic acids cleaves off the *tert*butyl protecting groups. The PARP inhibitor **niraparib** (**19.9**) is thus obtained [18].

9.2.5   Various Other Enzyme Inhibitors

A collection of potential and FDA-approved enzyme inhibitors can't be classed in any of the previous considered categories. These are collected in this section, sorted by the date in which they were prepared as indicated by the CAS registry number—for the lack of any other criterion.

9.2.5.1   *RAS Inhibitor*

RAS proteins regulate cellular actions such as proliferation, migration, and eventually cell death. In the normal course of events, cells require that the terpene farnesol be bound to specific sites in order to function. Depriving cells of farnesol would be expected to inhibit proliferation. The modified analogue, farnesylthiosalicylic acid, will displace farnesol from those sites and downregulate cell proliferation.

This compound has been shown to block the ras activation in several liver cancer cell lines. The drug has been used in several clinical studies on patients with liver or pancreatic cancer. A synthetic route is not shown as Salirasib is an article of commerce and is listed in several chemical catalogues.

9.2.5.2   *Protein Kinase C Inhibitor*

Protein kinase C (PKC) has been implicated in cancer since activators of that enzyme can act as tumor promoters. The drug enzastaurin suppresses proliferation and in some cases induces apoptosis in human cancer cells *in vitro*. The compound acts as an angiogenesis inhibitor in laboratory animals. When the drug failed to show efficacy against lymphoma in a Phase III clinical study, the sponsor stopped developing the compound.

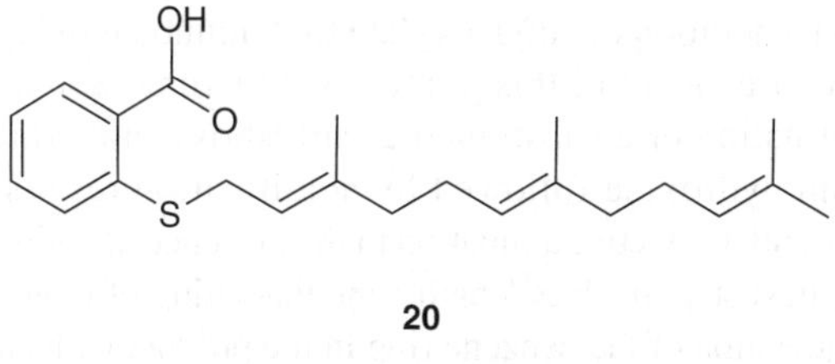

20

Scheme 9.20   *Salirasib.*

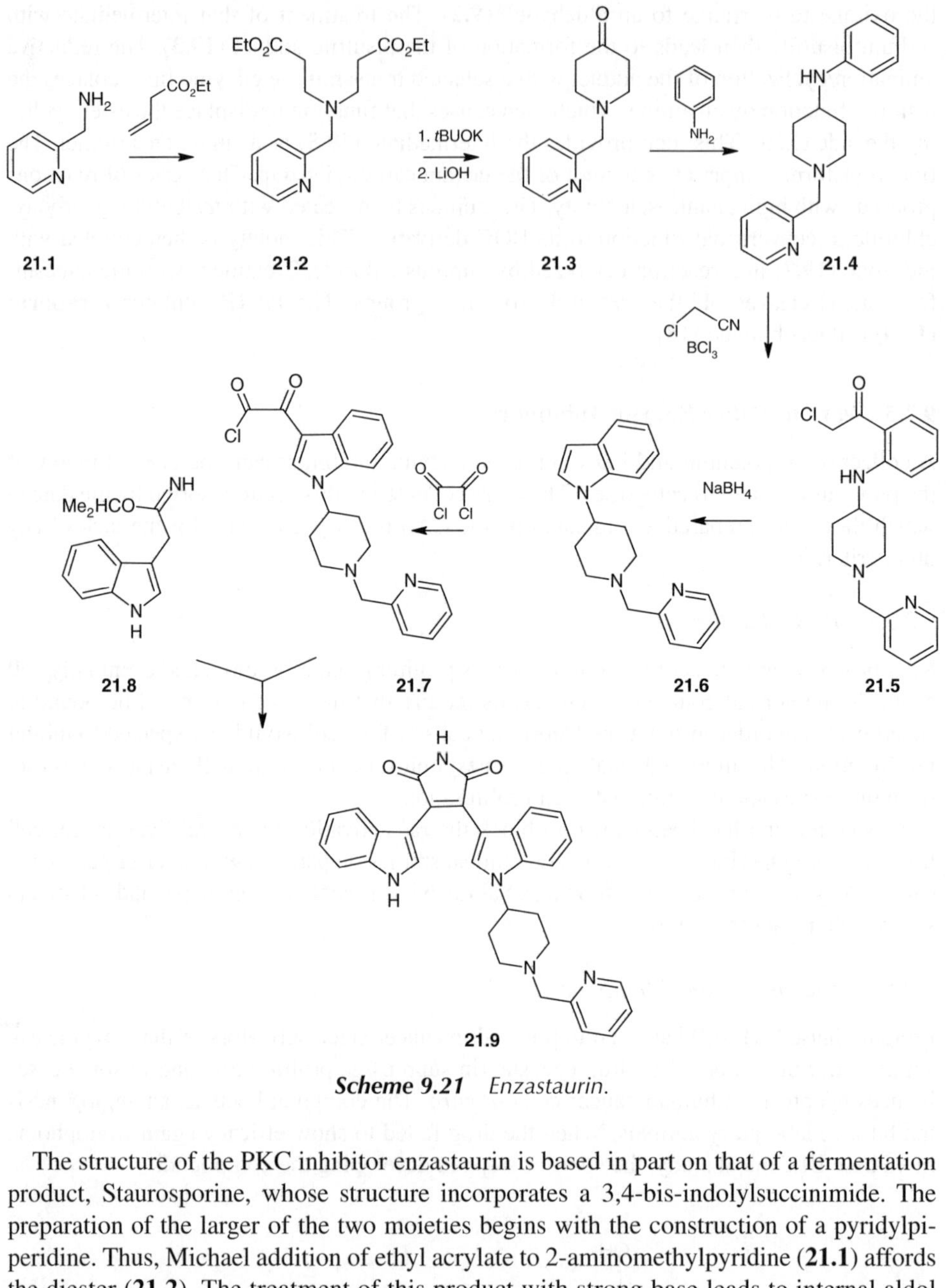

Scheme 9.21 *Enzastaurin.*

The structure of the PKC inhibitor enzastaurin is based in part on that of a fermentation product, Staurosporine, whose structure incorporates a 3,4-bis-indolylsuccinimide. The preparation of the larger of the two moieties begins with the construction of a pyridylpiperidine. Thus, Michael addition of ethyl acrylate to 2-aminomethylpyridine (**21.1**) affords the diester (**21.2**). The treatment of this product with strong base leads to internal aldol condensation and formation of a substituted 2-carbethoxy piperidine. Saponification of the ester with lithium hydroxide followed by acidification results in decarboxylation of the resulting β-keto acid. Reductive amination in the presence of aniline then adds a benzene ring to the string. The next step involves lengthening that string of moieties by an indole. The sequence starts with acylation of the benzene ring in the product with chloroacetonitrile with boron trichloride acting as the Lewis acid catalyst. The treatment of the product with sodium borohydride then serves to reduce the carbonyl group. On heating, the adjacent amine displaces the side chain chlorine to close the ring; the hydroxyl group is expelled at the same time to form an indole (**21.6**). The construction of the central five-membered ring involves first acylation of the indole with oxalyl chloride. That product is then condensed with the imidate from indole acetamide. The transform may be rationalized by assuming the initial formation of an amide from the attack of the acid chloride on the imide. The addition of the anion from the methylene group to the imidate and dehydration of the quaternary hydroxyl complete the formation of the succinimidate function. This sequence leads to the formation of the PKC inhibitor **enzastaurin** (**21.9**) [19].

9.2.5.3 *Mitosis Inhibitors*

Kinesins comprise a class of proteins that may be viewed as transport elements that move along microtubules to deliver specific cellular component. Mitotic spindles comprise the locus for attachment of the microtubules that will pull replicated genes apart during mitosis. The inhibition of kinesins brings mitosis, an essential process in cell proliferation, to a halt. Cancer cells will be preferentially affected due to their significantly higher growth rate. The kinesin protein inhibitor ispinesib shows potent reversible activity *in vitro* against a panel of human tumor cells and *in vivo* against such cells implanted in laboratory animals. A series of Phase I and II trials were carried out from about 2006 on. Results were at best on the modest side.

The preparation of racemic ispinesib (**22.11**) essentially consists of a series of sequential series of acylations of amines. Thus, the reaction of the benzyloxycarbonyl-protected diamino ester (**22.1**) with 4-toluylbenzoyl chloride (**22.2**) affords the corresponding amide. The ester function of the product (**22.3**) is then saponified. The resulting acid is then coupled with the substituted anthranilic ester (**22.5**), which then affords the chain-lengthened amide (**22.6**). The ester of that intermediate is again saponified and the resulting acid coupled with benzylamine. The functionality of the anthranilate ring is properly arranged for one of the standard quinazoline-forming reactions. The treatment of anthranilate (**22.9**) with base thus leads to the substituted quinazolone (**22.10**). The treatment with trifluoroacetic acid then cleaves off the protecting group on nitrogen to afford the spindle assembly inhibitor **ispinesib** (**22.11**) [20].

Like the very closely related mitotic spindle assembly kinesins, the kinesin-related motor proteins play an important part in preparing a cell nucleus for mitosis. Inhibiting that enzyme will disrupt mitosis and the resulting cell division. This in turn hinders cell proliferation. *In vitro* studies of have demonstrated that the inhibitor **litronesib** (**9.23**) prevents separation of the replicated nuclei during mitosis; the compound also blocks formation as well as the kinesin (Eg5) required for forming mitosis-related microtubules. The drug is currently under investigation in several clinical trials. The chemistry is of this drug is confined to the structure in the absence of sources for the synthesis.

9.2.5.4 *Mitogen-Activated Protein Kinase*

One of the series of proteins involved in cell response to various stimuli, mitogen-activated protein kinase (MEK) mediates, among other things, cell proliferation. Selumetinib is a potent inhibitor of that enzyme *in vitro*. The compound also inhibits growth of human tumors implanted in laboratory animals. The drug showed favorable responses in a series

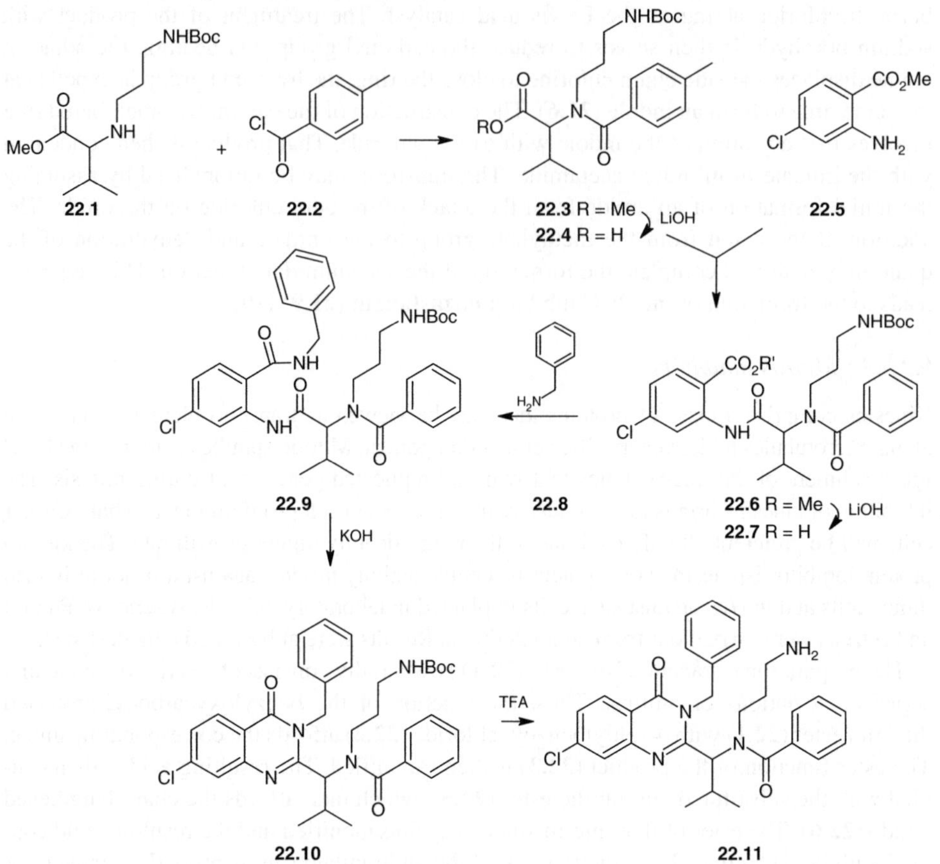

22.1 22.2 22.3 R = Me 22.4 R = H 22.5 22.6 R = Me 22.7 R = H 22.8 22.9 22.10 22.11

Scheme 9.22 *Ispinesib.*

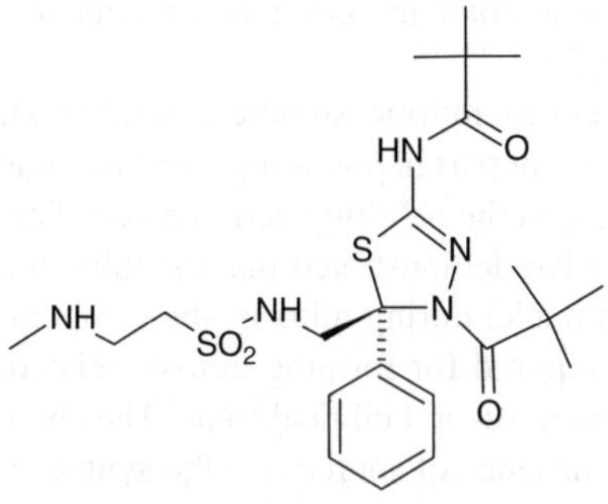

Scheme 9.23 *Litronesib.*

of Phase I and II clinical trials; a phase III trial against non-small cell lung cancer was initiated in late 2013.

The scheme for preparing selumetinib relies on the aromatic nucleophilic displacement of fluorine. Thus, the treatment of the trifluorinated benzoic acid 23.1 with ammonia leads

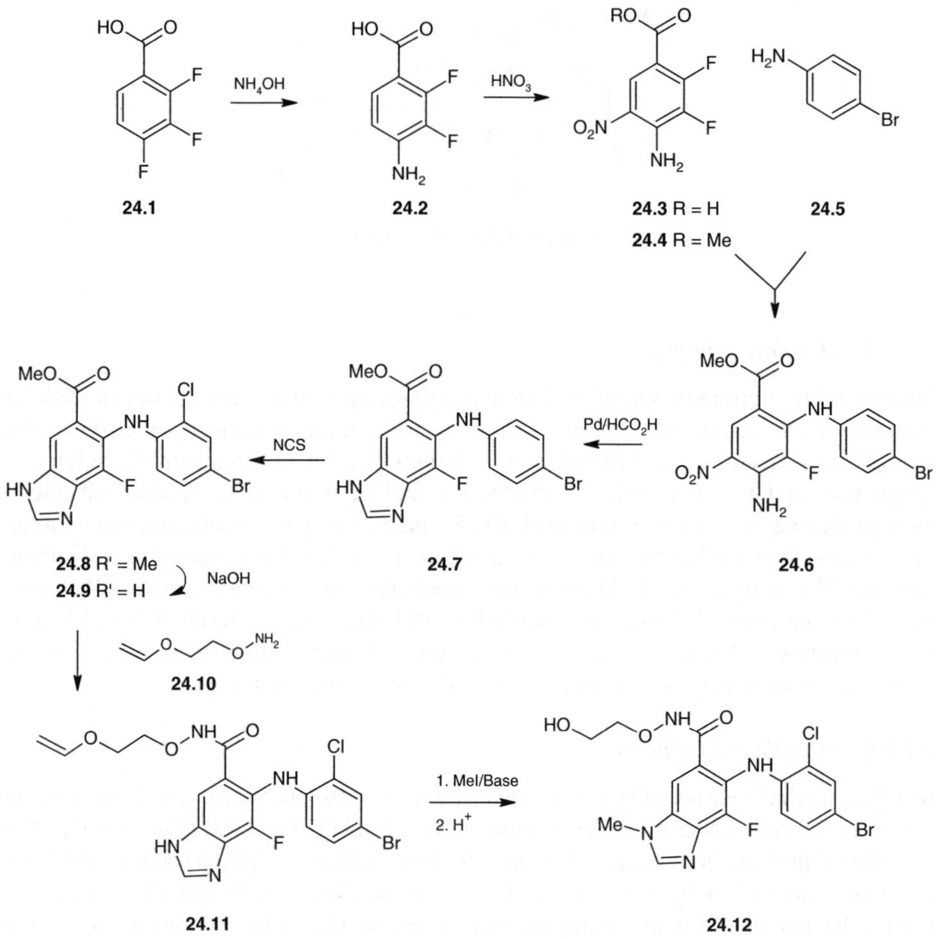

Scheme 9.24 *Selumetinib.*

to the displacement of fluorine at position 4 and formation of an aniline. The treatment of that product with fuming nitric acid installs a nitro group adjacent to the just-added amine. The carboxylic acid is next converted to the corresponding ester. The fluorine atom adjacent to the carboxylic acid is then displaced by 4-bromoaniline (**24.5**) to afford the biphenylamine (**24.6**). Somewhat unusual condition are used to generate the fused imidazole ring. The reaction of the ortho amino and nitro functions with palladium and formic acid leads first to the catalytic reduction of the nitro group using hydrogen from decomposition of formate. The carbon atom from the same reagent then serves as the carbon atom in the fused imidazole (**24.7**). The treatment of that product with *N*-chlorosuccinimide introduces the second halogen in the pendant ring (**24.8**). The ester is then saponified and then coupled with the O-substituted hydroxylamine **24.10** in the presence of the diimide EDCI. The penultimate reaction comprises alkylation of the free amine on the fused imidazole ring with methyl iodide and base. The treatment of that product with acid cleaves off the side chain protecting group, affording the MEK inhibitor **selumetinib** (**24.12**) [21].

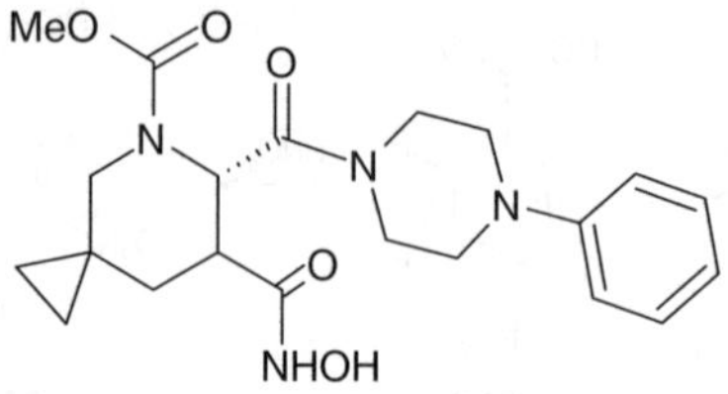

Scheme 9.25   *Aderbasib.*

9.2.5.5   Sheddase Inhibitor

Sheddases are a group of specialized proteases that cleave membranes near extracellular protrusions of membrane proteins. Tumor necrosis factor α-converting enzyme, for example, is a sheddase found in high concentrations near tumors and is implicated in carcinogenesis and tumor growth. An inhibitor would be expected to oppose that effect. Though the enzyme inhibitor **aderbasib** (**9.25**) shares the hydroxamate function with the histone deacetylase inhibitors, discussed in Chapter 7, it functions instead as a sheddase inhibitor. The orally active sheddase inhibitor aderbasib was tested in Phase I and II clinical trials. Development was however suspended in 2011 after negative results from a Phase II trial in patients with metastatic breast cancer. In the absence of published sources for the synthesis, the chemistry of this drug is confined to the chemical structure.

9.2.5.6   Smoothened Inhibitors

Rising to the challenge posed by theoretical nuclear physicists, pharmacologists have come up with nomenclature where the name of some enzyme is totally unrelated to its function. The smoothened protein, for example, is a receptor for proteins in the hedgehog pathway. The latter has been implicated in the development of some cancers. The smoothened inhibitor erismodegib exhibited good activity against prostate cancers in mice. Clinical trials of the drug in patients with locally advancer and metastatic basal cell carcinoma were very promising.

The relatively concise scheme for linking the four rings that comprise erismodegib starts with the displacement of chlorine in the substituted pyridine (**26.1**) by nitrogen in dimethyl morpholine (**26.2**). Catalytic hydrogenation then reduces the nitro group to an amine (**26.4**). That function is next acylated with bromotoluic acid (**26.5**). Suzuki coupling of that last intermediate with boronic acid (**26.7**) affords the smoothened inhibitor **erismodegib** (**26.8**) [22].

The first smoothened inhibitor to be approved for sale by FDA comprises a somewhat simpler string of rings. In early 2012, the agency approved the drug, trade named Erivedge®, to treat patients who have metastatic or locally advanced basal cell carcinoma. This compound, like erismodegib, inhibits the smoothened proteins on the hedgehog pathway.

This scheme employs an organozinc reagent for coupling a couple of rings. The first operation comprises forming the organozinc reagent from 2-bromopyridine by reacting that compound with zinc chloride and isopropylmagnesium chloride, and then treating the resulting intermediate with the substituted iodobenzene (**27.3**). The nitro group in the product is next reduced to an amine with iron in the presence of acid. Acylation of the newly formed amine with the substituted benzoic acid (**27.6**) affords the drug **vismodegib** (**27.7**) [23].

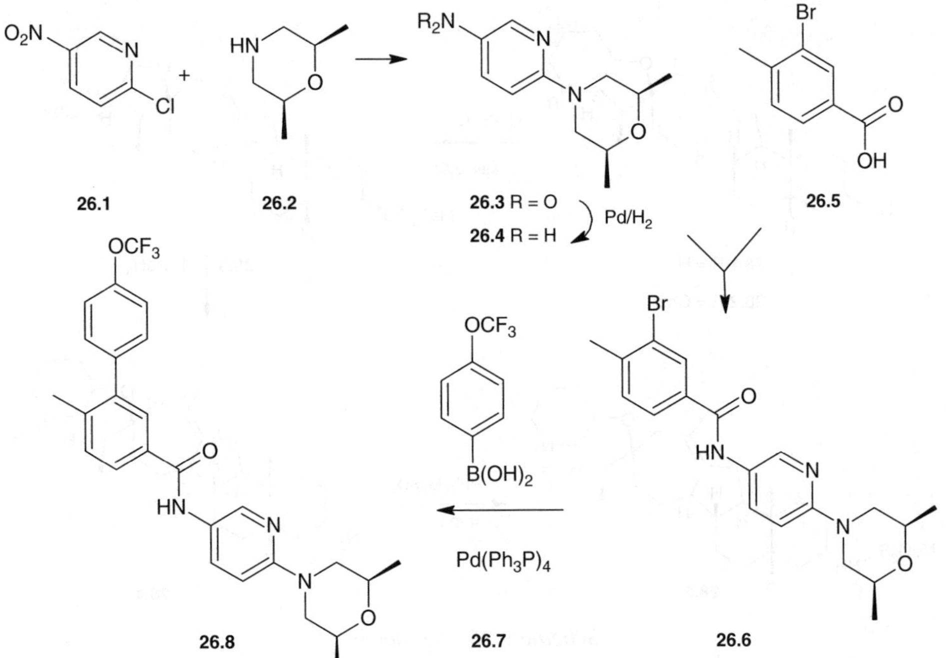

Scheme 9.26 *Erismodegib.*

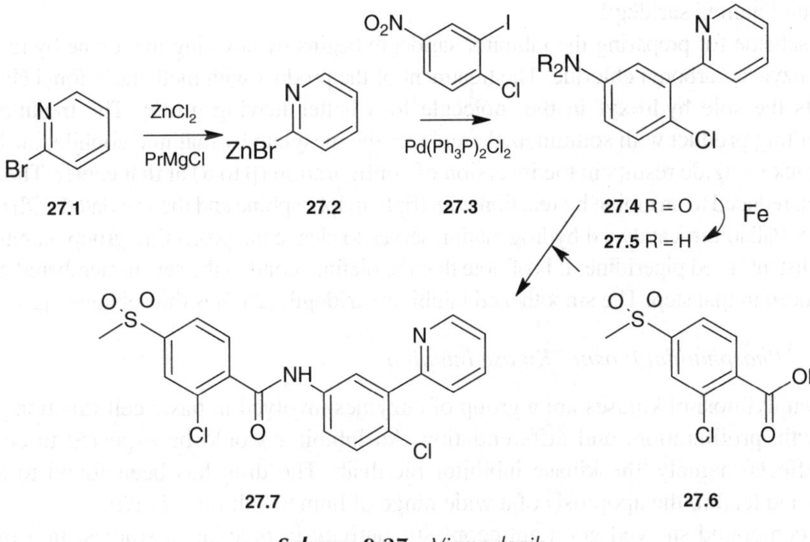

Scheme 9.27 *Vismodegib.*

Virtually, every compound listed in the preceding chapter and this one trace its origin to combinatorial chemistry. The smoothened inhibitor saridegib on the other hand represents the result of the once-traditional medicinal chemistry. The lead compound for this exercise is the naturally occurring triterpene cyclopamine (**28.1**). This compound is a teratogen that

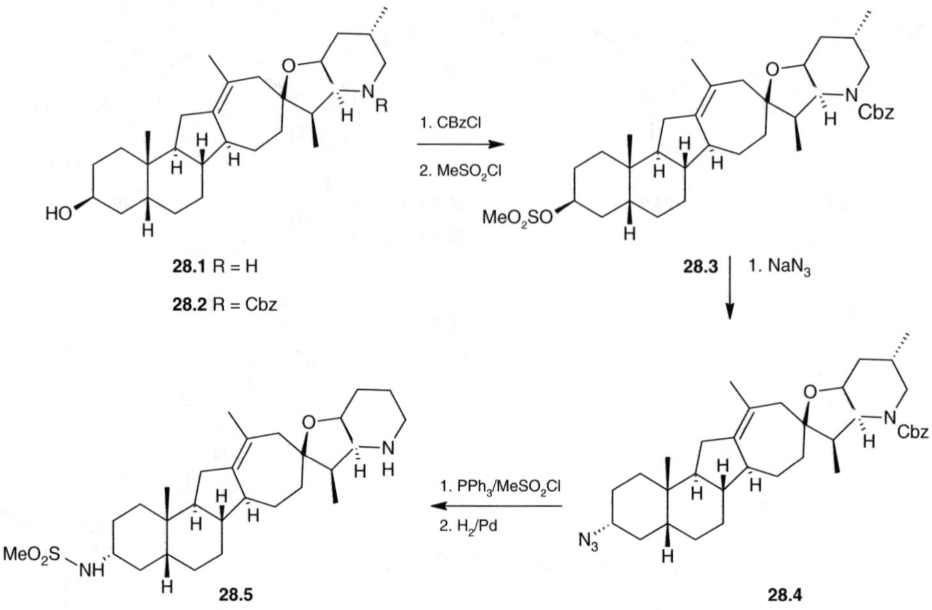

1. CBzCl
2. MeSO₂Cl

28.1 R = H
28.2 R = Cbz

28.3   1. NaN₃

1. PPh₃/MeSO₂Cl
2. H₂/Pd

28.5

28.4

Scheme 9.28   *Saridegib.*

causes lethal birth defects. A synthesis program aimed at exploring the effect of structural changes on activity and identifying a safer analogue led to the identification of the compound named saridegib.

The scheme for preparing the inhibitor saridegib begins by covering the amine by reaction with benzyloxycarbonyl chloride. The treatment of the product with methanesulfonyl chloride converts the sole hydroxyl in the molecule to a better leaving group. The treatment of the resulting product with sodium azide replaces the mesylate by that nucleophile; the backside attack by azide results in the inversion of configuration (β to α) at that center. The azide was then reduced to an amine by reaction with triphenylphosphine and then acylated with mesyl chloride. Palladium-catalyzed hydrogenation serves to cleave the protecting group on nitrogen on the distant fused piperidine. It is of note that the olefinic bond in the seven-membered ring is not reduced in that step. The smoothened inhibitor **saridegib** (**28.5**) is thus obtained [24].

9.2.5.7   *Phosphatidal Inositol Kinase Inhibitor*

Phosphatidylinositol kinases are a group of enzymes involved in basic cell functions such as growth, proliferation, and differentiation. An inhibitor would be expected to counter those effects, namely, the kinase inhibitor pictilisib. The drug has been found to arrest growth and lead to the apoptosis of a wide range of human cell lines *in vitro*.

The compound showed good antineoplastic activity in preclinical studies. In a mouse glioblastoma, models of pictilisib significantly decreased the growth of tumors by 90+%. Clinical trials are underway against a selection of solid and blood cancers as of early 2014.

One arm of the convergent synthesis involves attachment of a basic side chain to the thiazole moiety. This involves first introduction of a formyl group; the treatment of the thiazolopyrimidine (**29.1**) with butyl lithium generates an anion on the thiazole ring;

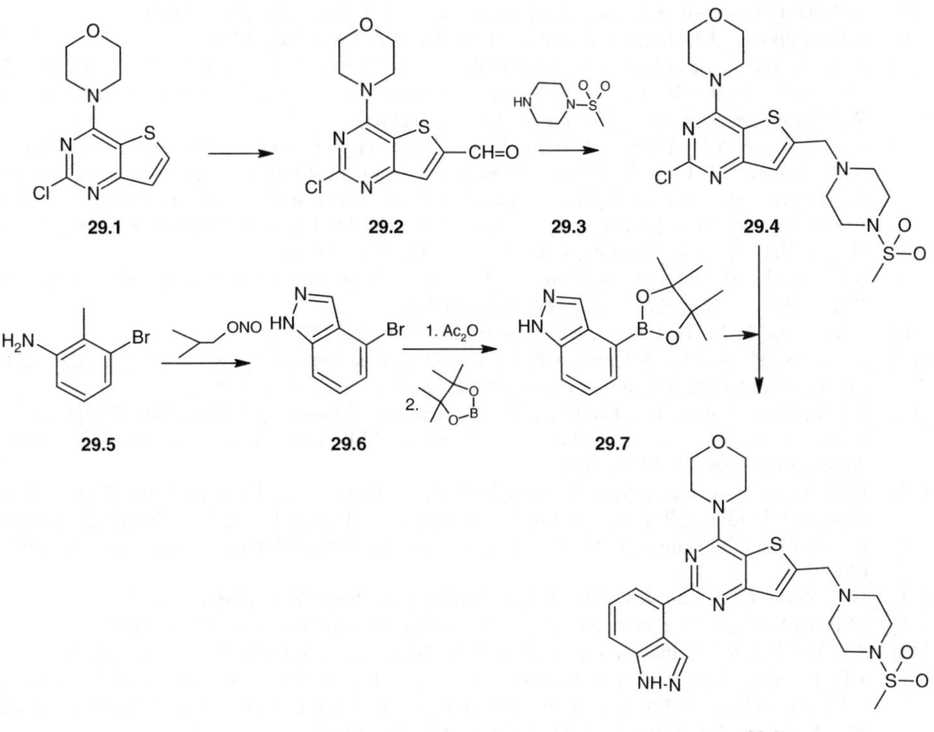

Scheme 9.29 *Pictilisib.*

the addition of DMF transfers the formyl group to the thiazole. Reductive alkylation of that group with *N*-methylsulfonylpiperidine completes that subroutine. The reaction of the substituted aniline (**29.5**) with amyl nitrite leads to the formation of the fused pyrazole (**29.6**). The first-formed diazonium salt apparently attacks the adjacent methyl group to form the ring, The secondary amine on the newly formed ring is next protected as an amide by means of acetic anhydride. The treatment of that intermediate (**29.6**) with pinacolborane replaces the bromine substituent to afford the requisite Suzuki reagent. Combining the two reagents (**29.7** and **29.4**) couples the two moieties to form the phosphatidylinositol kinase inhibitor **pictilisib** (**29.8**) [[25]].

References

[1] E.P. Reddy, M.V. Reddy, S.C. Bell, U.S. Patent 7,598,232 (2009).
[2] U. Abel, H. Deppe, A. Feurer, U. Gradler, K. Otte, R. Sekul, M. Thiemann, A. Goutopoulos, M. Schwarz, X. Jiang, U.S. Patent 7,956,191 (2011).
[3] W.A. Wisdom, A.A. Colvin, S. Koppenol, U.S. Patent 8,455,471 (2013).
[4] D. Bebbington, H. Binch, J.–D. Charrier, S. Everitt, D. Fraysse, J. Golec, D. Kay, R. Knegtel, C. Mak, F. Mazzei, A. Miller, M. Mortimore, M. O'Donnell, S. Patel, W.T. People, F. Pierard, J. Pinder, J. Pollard, S. Ramaya, D. Robinson, A. Rutherford, J. Studley, J. Westcot, *Bioorg. Med. Chem. Lett.* **19**, 3586 (2009).

[5] G. Linz, G.F. Kraemer, L. Gutschera, G. Asche, U.S. Patent 7,759,485 (2020).

[6] C.F. Claiborne, T.B. Sells, S.G. Stroud, U.S. Patent 8,026,246 (2011).

[7] P. Angibaud, X. Bourdrez, A. Devine, D.W. End, E. Freyne, Y. Ligny, P. Muller, P. Muller, G. Mannens, I. Pilatte, V. Poncelet, S. Skrzat, G. Smets, J. Van Dun, P. Van Remoortere, M. Venet, W. Wouters, *Bioorg. Med. Chem. Lett.* **13**, 1543 (2003).

[8] F.G. Njoroge, A.G. Taveras, J. Kelly, S. Remiszewski, A.K. Mallams, R. Wolin, A. Afonso, A.B. Cooper, D.F. Rane, Y.-T. Liu, J. Wong, B. Vibulbhan, P. Pinto, J. Deskus, C.S. Alvarez, J. del Rosario, M. Connolly, J. Wang, J. Desai, R.R. Rossman, W.R. Bishop, R. Patton, L. Wang, P. Kirschmeier, M.S. Bryant, A.A. Nomeir, C.-C. Lin, M. Liu, A.T. McPhail, R.J. Doll, V.M. Girijavallabhan, A.K. Ganguly, *J. Med. Chem.* **41**, 4890 (1998).

[9] K.K. Murthi, M. Dubay, C. McClure, L. Brizuela, M.D. Boisclair, P.J. Worland, M.M. Mansuri, K. Pal, *Bioorg. Med. Chem. Lett.* **10**, 1037 (2000).

[10] F.X.Cheng, M.M. Tamarez, J. Xie, U.S. Patent 7,786,306 (2010).

[11] J. Adams, M. Behnke, S. Chen, A.A. Cruickshank, L.R. Dick, L. Grenier, J.M. Klunder, Y.-T. Ma, L. Plamondon, R.L. Stein, *Bioorg. Med. Chem. Lett.* **8**, 333 (1998).

[12] E.L. Ferdous, J. Abu, M.J. Kaufman, S.A. Komar, D.L. Mazaik, Q.J. McCubbin, P. Nguyen, V. Palaniappan, R.D. Skwierczynski, N.T. Truong, C.M. Varga, P.N. Zawaneh, U.S. Patent Application 2009/0325903 (2009).

[13] H.–J. Zhou, M.A. Aujay, M.K. Bennett, M. Dajee, S.D. Demo, Y. Fang, M.N. Ho, J. Jiang, C.J. Kirk, G.J. Laidig, E.R. Lewis, Y. Lu, T. Muchamuel, F. Parlati, E. ing, K.D. Shenk, J. Shields, P.J. Shwonek, T. Stanton, C.M. Sun, C. Sylvain, T.M. Woo, J. Yang, *J. Med. Chem.* **52**, 3028 (2009).

[14] H.-J. Zhou, C.M. Sun, K.D. Shenk, G.J. Laidig, U.S. Patent 7,687,456 (2010).

[15] M.P. Mulholland, G. Pattenden, I.A.S. Walter, *Org. Biomol. Chem.* **6**, 2728 (2008).

[16] S.E. Webber, S.S. Canan-Koch, J. Tikhe, L.H. Thoresen, U.S. Patent 6,495,541 (2002).

[17] T.D. Penning, G.D. Zhu, V.B. Gandhi, J. Gong, X. Liu, Y. Shi, V. Klinghofer, E.F. Johnson, C.K. Donawho, D.J. Frost, V. Bontcheva-Diaz, J.J. Bouska, D.J. Osterling, A.M. Olson, K.C. Marsh, Y. Luo, V.L. Giranda, *J. Med. Chem.* **52**, 514 (2009).

[18] C.K. Chung, P.G. Bulger, B. Kosjek, K.M. Belyk, N. Rivera, M.E. Scott, G.R. Humphrey, J. Limanto, D.C. Bachert, K.M. Emerson, *Org. Process. Res. Dev.* **18**, 215 (2014).

[19] M.M. Faul, J.R. Gillig, M.R. Jirousek, L.M. Ballas, T. Schotten, A. Kahl, M. Mohr, *Bioorg. Med. Chem. Lett.* **13**, 1857 (2003).

[20] *After* L. Sobrera, J. Boles, N. Serradel, M. Bayes, *Drugs Future*, **31**, 778 (2006).

[21] E.M. Wallace, J.P. Lyssikatos, A.L. Marlow, T.B. Hurley, T. Brian, U.S. Patent 7,576,114 (2009).

[22] C. Dierks, M. Warmuth, X. Wu, U.S. Patent Application 2010/0,197,659 (2010).

[23] J.L. Gunzner, D. Sutherlin, M.S. Stanley, L. Bao, G.M. Castanedo, R.L. Lalonde, S. Wang, M.E. Reynolds, S.J. Savage, K. Malesky, M.S. Dina, U.S. Patent 7,888,364 (2011).

[24] B. Austad, M.L. Behnke, A.C. Castro, A.B. Charette, M.J. Grogan, S. Janardanannair, A. Lescarbeau, S. Peluso, M. Tremblay, U.S. Patent Application 2009/0,012,109 (2009).

[25] A.J. Folkes, K. Ahmadi, W.K. Alderton, S. Alix, S.J. Baker, G. Box, I.S. Chuckowree, P.A. Clarke, P. Depledge, S.A. Eccles, L.S. Friedman, A. Hayes, T.C. Hancox, A. Kugendradas, L. Lensun, P. Moore, A.G. Olivero, J. Pang, S. Patel, W.T. People, G.H. Pergl-Wilson, F.I. Raynaud, A. Robson, N. Saghir, L. Salphati, S. Sohal, M.H. Ultsch, M. Valenti, H.J.A. Wallweber, N.C. Wan, C. Wiesmann, P. Workman, A. Zhyvoloup, M.J. Zvelebil, S.J. Shuttleworth, *J. Med. Chem.* **51**, 5523 (2008).

10

Miscellaneous Antineoplastic Agents

The preceding chapters are testimony to the effort that is currently devoted to discover new antineoplastic drugs—new not only as to chemical structure but also the mechanism whereby they attack the malignant cells. It is probably inevitable that not all the existent and potential antineoplastic compounds fall neatly into one of the categories to be found in Chapters 1 through 9. The compounds that did not fit are found in this chapter. Both the chemical structures and the mechanism whereby these outliers bring a particular cancer to a halt have little in common. The compounds marked miscellaneous are consequently presented simply in order of their increasing structural complexity.

10.1   Acyclic

The antineoplastic compound darinaparsin in some ways recaptures the distant past. Some may recall that the first truly effective small-molecule antibacterial drug Salvarsan (arsphenamine) included covalently bound arsenic in its structure. More currently, it has been established that the arsenical antineoplastic candidate darinaparsin owes activity to the slow release of cytotoxic arsenic species; these in turn generate active oxygen that induce tumor cell apoptosis. The drug has been administered to patients with various malignant tumors. In 2010, the FDA designated the drug, trade named Zinapar®, an orphan drug for treating patients with T-cell lymphoma.

The drug can be obtained in straightforward fashion by the reaction of glutathione (**1.1**) with dimethylarsenic chloride. **Darinaparsin (1.2)** is thus obtained [1].

Antineoplastic Drugs: Organic Synthesis, First Edition. Daniel Lednicer.
© 2015 John Wiley & Sons, Ltd. Published 2015 by John Wiley & Sons, Ltd.

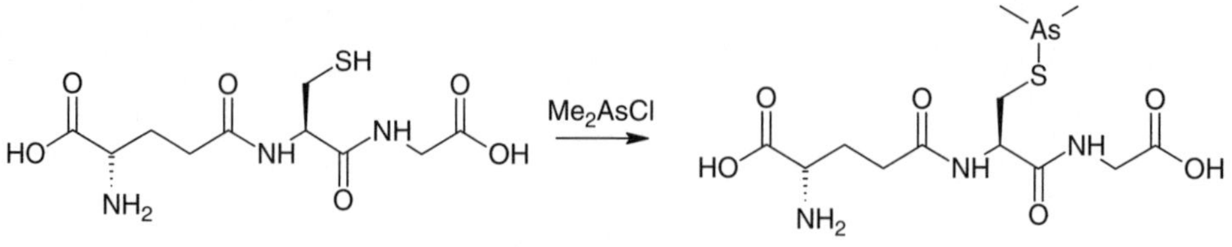

Scheme 10.1   *Darinaparsin.*

10.2   Monocyclic

The reactive acivicin isoxazole elaborated by *Streptomyces sviceus* was marked active, in the 1970s, by an *in vitro* screen designed to identify cytotoxic products. It was subsequently found that this compound interfered with glutamate metabolism. The structure of this compound suggests that it can be viewed as a cyclized version of glutamic acid where the second carboxylic acid is replaced by the reactive imino chloride. The antineoplastic activity was confirmed by *in vitro* assays against a series of human malignancies. The drug has been administered to patients with various cancers in Phase II clinical trials.

The key reaction in one synthesis of acivicin comprises a 1,3-dipolar cycloaddition; the scheme further relies on the sugar D-ribose as a template for passing on chirality. The reaction of hydroxylamine (**2.1**) that replaces the anomeric hydroxyl group in that sugar with paraformaldehyde affords the corresponding transient nitrone (**2.3**). The treatment of that intermediate with vinyl pyrazolone (**2.4**) leads to the formation of the requisite isoxazole ring via an electrocyclic addition reaction. Formic acid then breaks open what is essentially an acid-labile carbinolamine. This then leads to loss of the sugar moiety (**2.6**) that has imposed chirality. *N*-Chlorosuccinimide introduces chlorine at the vinylic position. Boron trichloride is next used to open the pyrazolone revealing the α-amino acid. **Acivicin (2.8)** is thus obtained [2]. It is interesting to note in passing that the isoxazole ring is not affected to this reagent.

Though boron-substituted drugs are relatively rare, two proteasome inhibitors in which that element occurs are discussed in the preceding chapter. Boron is attached covalently in both cases to carbon in an α-amino acid. The structure of the peptidomimetic compound talabostat differs in that boron is placed on a terminal pyrrolidyl amide where it seems to serve as a surrogate acid. The drug inhibits dipeptidyl peptidases, and in addition, it stimulates the production of factors that increase the immune response. Talabostat showed promising activity in patients afflicted with various cancers. In 2006, the drug was accorded fast track designation by the FDA for treating non-small cell lung cancer. In 2007, however, the agency closed down clinical studies of the compound.

The occurrence of two chiral centers in the molecule virtually mandated an enantioselective synthesis in order to avoid the complication that arises from forming a pair of diastereomers. The synthesis thus starts with the conversion of pyrolidine to its N-protected *tert*butoxycarbonyl derivative. Treatment with strong base, followed by triethyl borate, attaches that group to the heterocycle; acid hydrolysis then affords the corresponding boric acid (**3.3**). Both enantiomers are formed in equal amounts. The reaction of the mixture with (+) pinanediol will afford a 1:1 mixture of what are now differing diastereomers, separable by recrystallization. Nitrogen in a single isomer is then deprotected

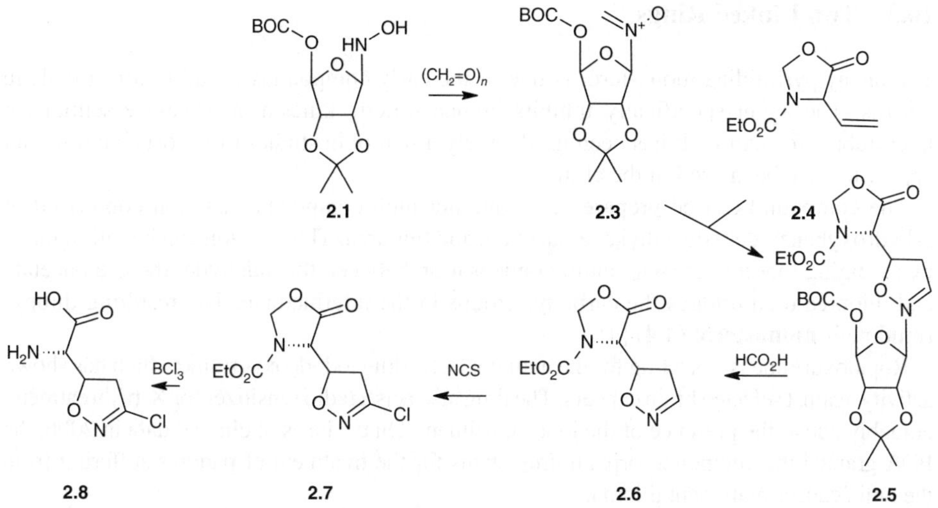

Scheme 10.2 Acivicin.

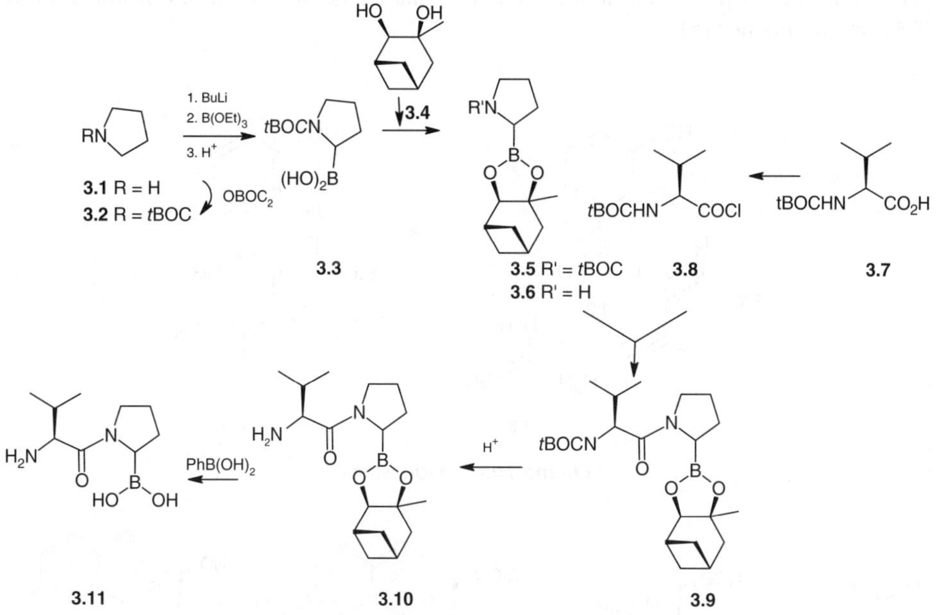

Scheme 10.3 Talabostat.

with acid to afford the free base. In a separate arm of this converging scheme, the naturally occurring α-amino acid leucine (**3.7**) is converted to the BOC-protected acid chloride. The acylation of the boron-containing moiety with the leucine derivative leads to amide (**3.9**). Treatment with acid cleaves the *t*-BOC protecting group. The terpenoid diol is next removed by exchange with excess phenylboronic acid to afford **talabostat** (**3.11**) [3].

10.3 Two Linked Rings

The phenylpyrimidine monasterol is one of the early compounds found to arrest cells in mitosis. The agent specifically inhibits human mitotic kinesin, a protein essential for microtubule formation. It has been used widely as a tool in studies of mitosis but does not appear to have been used in the clinic.

The compound can be prepared in a one-pot multicomponent reaction comprised of 3-hydroxybenzaldehyde, ethyl acetoacetate, and thiourea. The reaction can be rationalized by assuming that it starts with aldol condensation between the aldehyde: the acetoacetate and thiourea then bridges the carbonyl groups in the intermediate. The resulting thiopyrimidine is **monasterol (4.4)** [4].

Ropidoxuridine is a prodrug for the antineoplastic drug iododeoxyuridine which has shown activity against selected brain cancers. The drug also acts a radio sensitizer for X-ray treatment, arguably due to the presence of the iodo substituent. On the basis of clinical data in 2006, the FDA granted the compound orphan drug status for the treatment of patients suffering from the brain cancer malignant glioma.

This compound is prepared by a standard glycosidation sequence. Thus, 5-iodouridine is converted to the silyl derivative by reaction with trimethylsilyl chloride. This is then allowed to react with the protected 3-deoxyribosyl chloride (**5.3**). The nitrogen atom on the pyrimidone then displaces the anomeric chlorine linking the two moieties. **Ropidoxuridine (5.5)** is thus obtained [5].

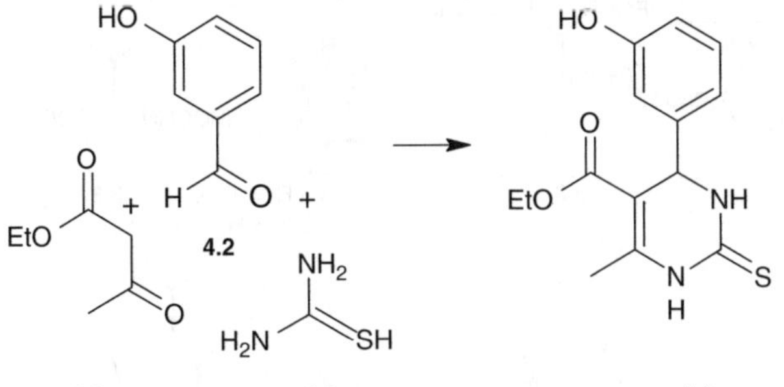

4.1 **4.2** **4.3** **4.4**

Scheme 10.4 *Monasterol.*

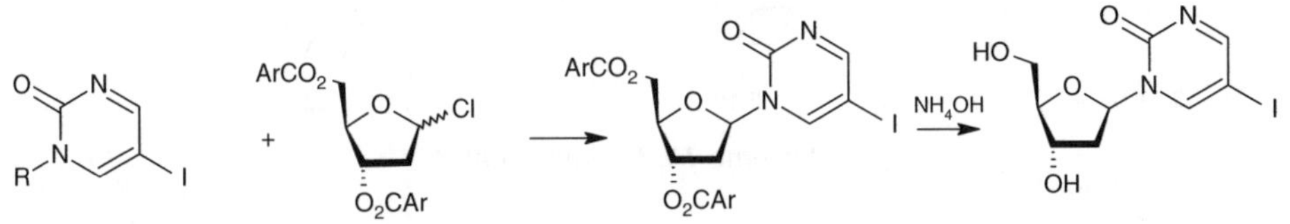

5.1 R = H
5.2 R = SiMe3 (Me3SiCl)
5.3 **5.4** **5.5**

Ar = C_6H_4pCl

Scheme 10.5 *Ropidoxuridine.*

10.4   Rings on a Chain

10.4.1   Two Rings

Virtually every antineoplastic agent dealt with so far, with the exception of some of the hormone antagonists in Chapter 3, contains a nitrogen atom in its structure. The modified plant-derived compound terameprocol presents a notable exception. This compound inhibits a group of proteins involved in the regulation of the cell cycle. The inhibition of those proteins in effect shuts down proliferation. This agent has shown good activity against human cancers in cell culture as well as animal models. The drug has shown activity against selected cancers in clinical trials.

Nordihydroguaiaretic acid, the starting material for preparing terameprocol, is a natural product found in the creosote bush. Note that this compound is the *meso* isomer, that is to say that the two halves are mirror images precluding the existence of enantiomers. A published synthesis provided a mixture of the isomers [6]. Exhaustive methylation of that commercially available phenol typically with dimethyl sulfate in the presence of base affords **terameprocol (6.2)** [6].

The compound elesclomol induces a disturbance of the balance between the production of free radicals, such as peroxides, and antioxidant defenses in cells. The higher levels of free radicals in tumor cells compared to normal cells exhaust the defenses, inducing selective apoptosis in tumor cells. Elesclomol interestingly showed activity in human cancer tissues implanted in laboratory animals only when administered in conjunction with proven antineoplastic drugs. The FDA granted the compound orphan drug status in 2008 for treating patients with metastatic melanoma.

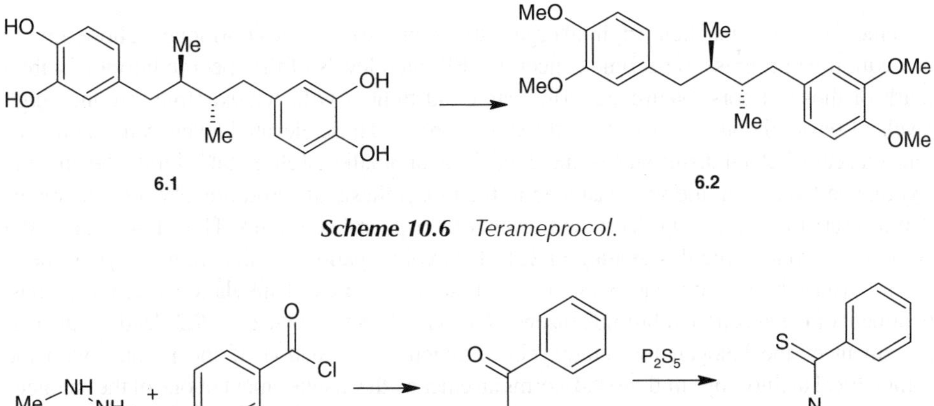

Scheme 10.6   *Terameprocol.*

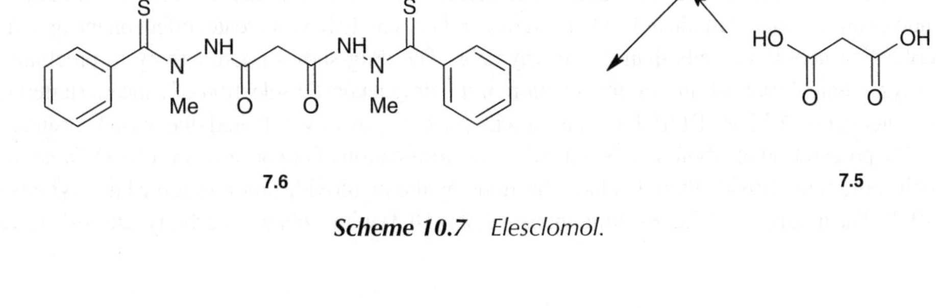

Scheme 10.7   *Elesclomol.*

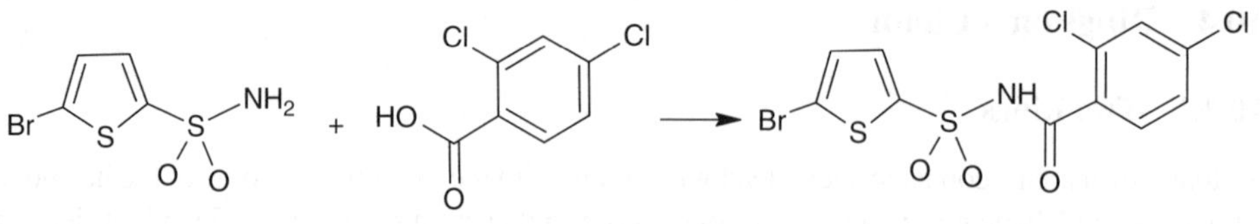

Scheme 10.8   *Tasisulam.*

The concise scheme for preparing elesclomol (**7.6**) starts with the acylation of methylhydrazine with benzoyl chloride. Heating the product (**7.3**) with phosphorus sulfide then replaces the oxygen in the amide carbonyl group by sulfur. The reaction of the resulting thioamide with malonic acid in the presence of the diimide EDC results in the acylation of the remaining open position of the hydrazine by each of the malonic acid carboxyl groups. The resulting symmetrical product comprises **elesclomol** (**7.6**) [7].

The structurally relatively simple compound tasisulam (**8.3**) induces apoptosis in cancer cells by more than one mechanism. The principal means of halting proliferation involves mitotic inhibition. Phase II clinical trials against a selection of cancers seemed to produce encouraging data. A Phase III trial in patients with advanced metastatic melanoma was halted because of safety concerns. **Tasisulam** (**8.**3) is prepared by the acylation of thiophene sulfonamide (**8.1**) with 2,4-dichlorobenzoic acid [8].

10.4.2   Four and More Rings

Carcinoid tumors have their origin in organs that synthesize various hormones. Those tumors can with time progress to become cancerous. Elevated levels of the specific hormones are a mark of those tumors. Neuroendocrine carcinoid tumors will, for example, produce high levels of serotonin, also known as hydroxytryptophan. Those elevated levels can lead to various musculoskeletal disturbances and behavioral anomalies. Telotristat inhibits the enzyme tryptophan hydrolase, the very last step in the biosynthesis of serotonin. Preclinical studies demonstrated that the drug decreased levels of peripheral serotonin. The FDA granted the compound orphan drug designation in 2012 for treating patients with carcinoid syndrome.

The synthesis of telotristat essentially consists of a series of alkylation reactions. Thus, treatment of the substituted bromobenzene (**9.1**) with 3-methyl pyrazole (**9.2**) leads to the displacement of the halogen by heterocyclic nitrogen. The reaction of the product with the symmetrical dichloropyrimidine (**9.4**) forms an ether by the displacement of one of the halogens by the alcohol oxygen. The fourth ring is then tacked on by the Suzuki coupling reaction. The treatment of the last intermediate with the boronic acid from 4-bromophenylalanine (**9.6**) affords **telotristat** (**9.7**) [9].

The receptor antagonist zibotentan binds selectively to the endothelin A receptor inhibiting endothelin-dependent mediated cell proliferation. This should have a greater effect on the quickly replicating neoplastic cells than on healthy cells. The drug shows good activity in preclinical assays. Clinical trials of an oral formulation of the drug against a selection of cancers started in the midaughts. A Phase III trial in patients with prostate cancer was halted due to poor results.

The preparation of zibotentan begins with the bromination of aminopyrazine (**10.1**). Reaction with sodium methoxide then displaces bromine by the methoxide anion to afford methyl ether (**10.3**). The treatment of the resulting intermediate (**10.3**) with *iso*butoxycarbonyl chloride leads

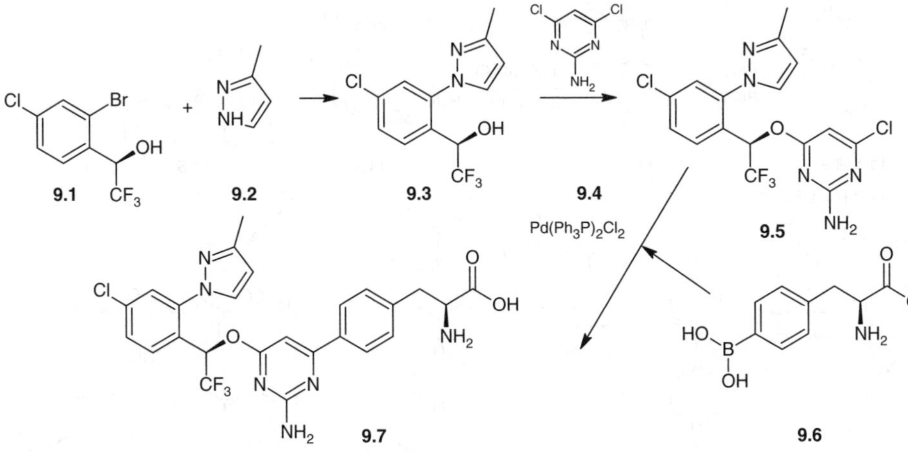

Scheme 10.9 *Telotristat.*

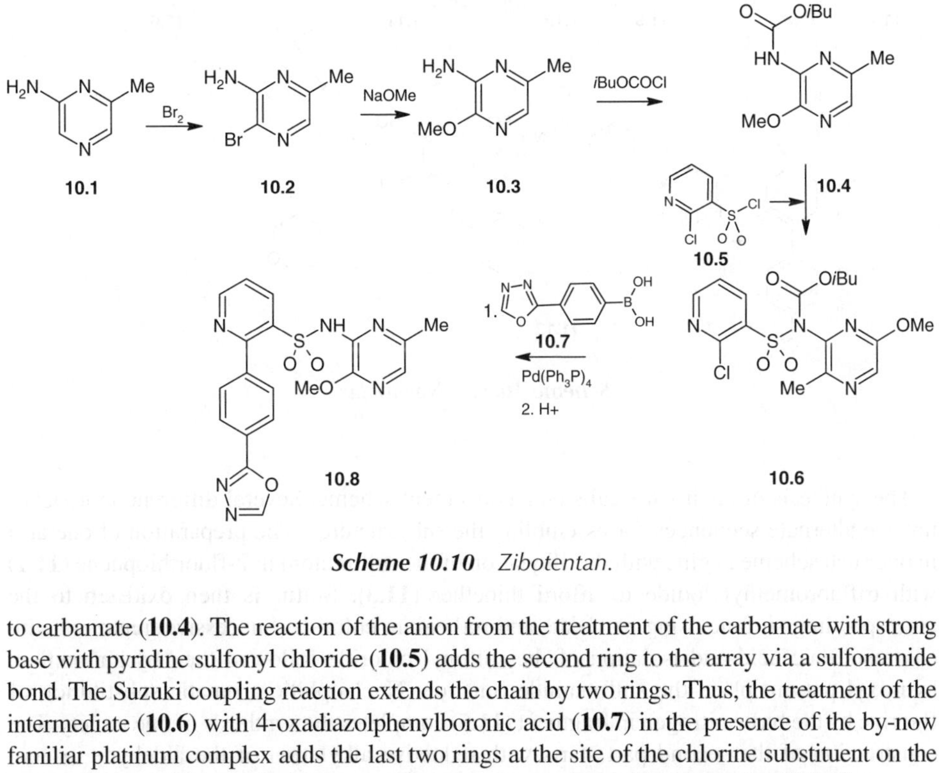

Scheme 10.10 *Zibotentan.*

to carbamate (**10.4**). The reaction of the anion from the treatment of the carbamate with strong base with pyridine sulfonyl chloride (**10.5**) adds the second ring to the array via a sulfonamide bond. The Suzuki coupling reaction extends the chain by two rings. Thus, the treatment of the intermediate (**10.6**) with 4-oxadiazolphenylboronic acid (**10.7**) in the presence of the by-now familiar platinum complex adds the last two rings at the site of the chlorine substituent on the pyridine ring. Mild hydrolysis cleaves the carbamate to afford **zibotentan** (**10.8**) [10].

Navitoclax may well be the longest chain of linked carbo- and heterocyclic rings to be tested for its antineoplastic activity. The compound acts by inhibiting a complex of the so-called "prosurvival" proteins present in large quantities in tumor cells. The compound was active against a selection of human cancer tissues *in vitro* and those implanted in laboratory animals. Phase II clinical trials against a selection of cancers are in progress.

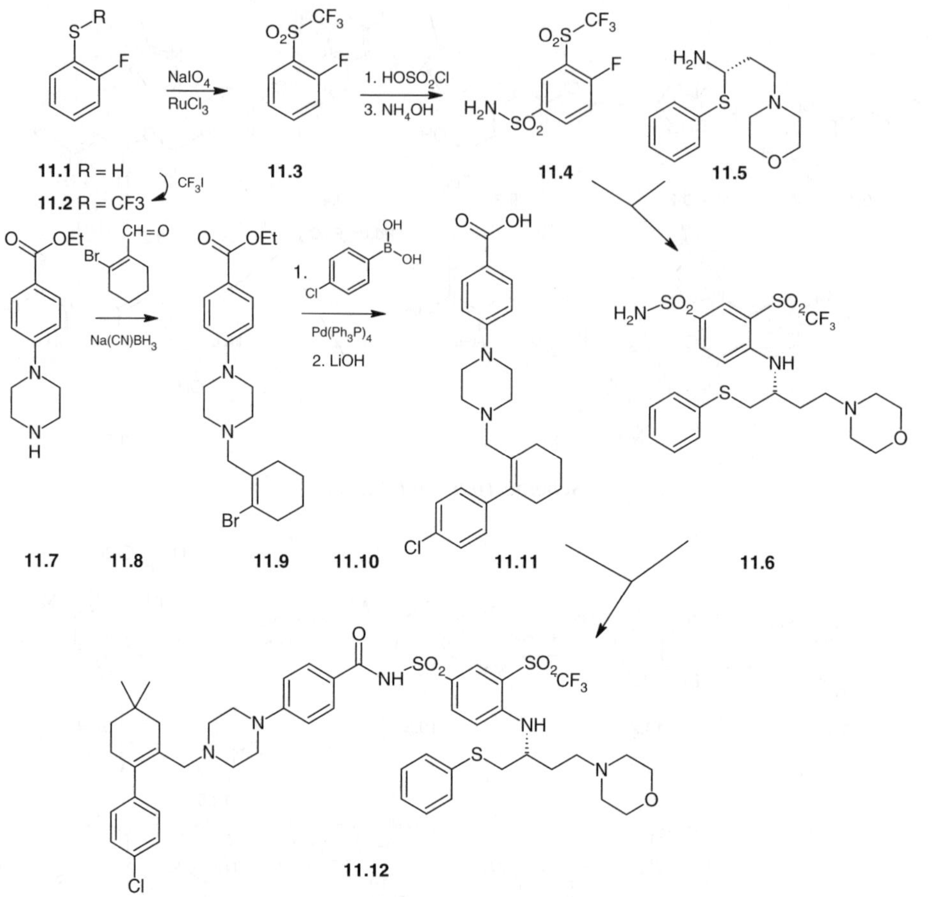

Scheme 10.11   *Navitoclax.*

The synthesis of navitoclax calls on a convergent scheme. Several different approaches involve alternate sequences for assembling the substructures. The preparation of one arm in one such scheme begins with the alkylation of the sulfur atom in 2-fluorthiophene (**11.1**) with trifluoromethyl iodide to afford thioether (**11.3**). Sulfur is then oxidized to the corresponding sulfone using periodate and ruthenium chloride. Successive treatment with chlorosulfonic acid and isolation of the sulfonyl chloride followed then by ammonium hydroxyl puts in place the sulfonamide function. Nucleophilic aromatic substitution of fluorine by the basic nitrogen in fragment (**11.5**) completes the synthesis of one moiety. The construction of the second part starts with reductive alkylation of the aldehyde in enol bromide (**11.9**) with the open amine in arylpiperazine (**11.7**) by means of sodium cyanobo-rohydride. The Suzuki cross coupling of the enol bromide in (**11.9**) with 4-chlorophenyl boronic acid in the presence of the palladium complex completes the preparation of the other moiety. The ester group at one end of that compound is then saponified. Amide formation mediated by a diimide then completes the synthesis of the inhibitor **navitoclax** (**11.12**) [11].

10.5   Fused Rings

10.5.1   Indoles

The indolone-based antineoplastic agent lenalidomide (**12.6**), a structurally very close relative of the ill-reputed sedative thalidomide, is a compound used in the clinic as an antineoplastic agent. The impetus for its preparation came from the observation that thalidomide exhibited antineoplastic activity. Lenalidomide has been shown to act largely as an immunomodulator; the compound also exhibits some antiangiogenic activity. As early as 2006, the FDA approved lenalidomide capsules, under the trade name Revlimid®, for the treatment of patients with multiple myeloma when administered in conjunction with the corticosteroid dexamethasone. In 2013, this indication was expanded to the use of the drug alone in a selected group of patients.

One scheme for preparing the compound involves first bromination of the methyl group in nitrotoluic ester (**12.1**). The displacement of the newly introduced bromine by the amine in glutaramide (**12.3**) installs the atoms for the next ring. Thus, the treatment of that intermediate (**12.4**) with triethylamine closes the succinamide ring. Catalytic hydrogenation then reduces the nitro group to afford **lenalidomide** (**12.6**) [12].

Indole-based obatoclax is an antagonist to a group of proteins that control both cell proliferation and death. The compound exhibits antineoplastic activity in preclinical studies. In 2006, the drug entered Phase I clinical trials. In due course, this was followed by Phase II trials in patients with leukemia, lymphoma, or myelofibrosis. Studies were discontinued in 2012 due to poor results.

The synthetic scheme for the preparation of obatoclax is yet another case in which the Suzuki cross coupling plays a key role. The reaction of pyrrolone (**13.1**) with phosphorus oxybromide and diethylaminoformamide both converts the carbonyl group to its enol

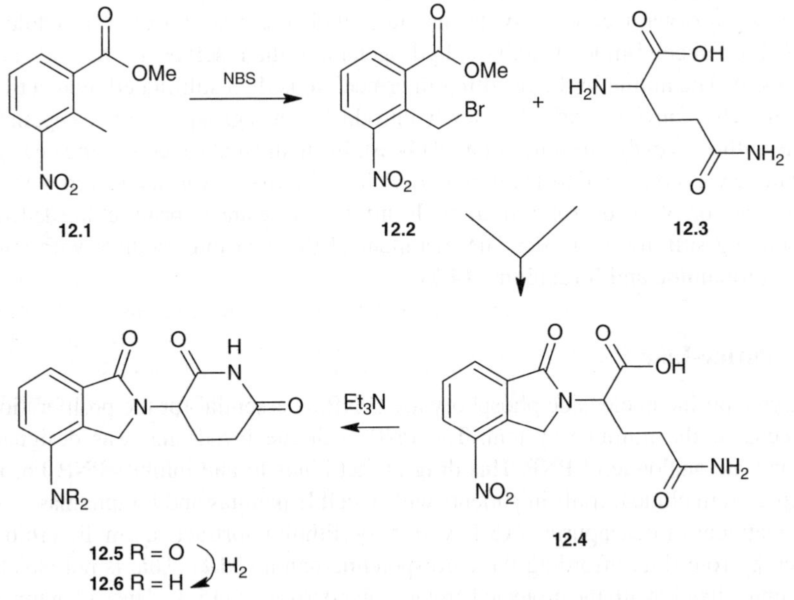

Scheme 10.12   *Lenalidomide.*

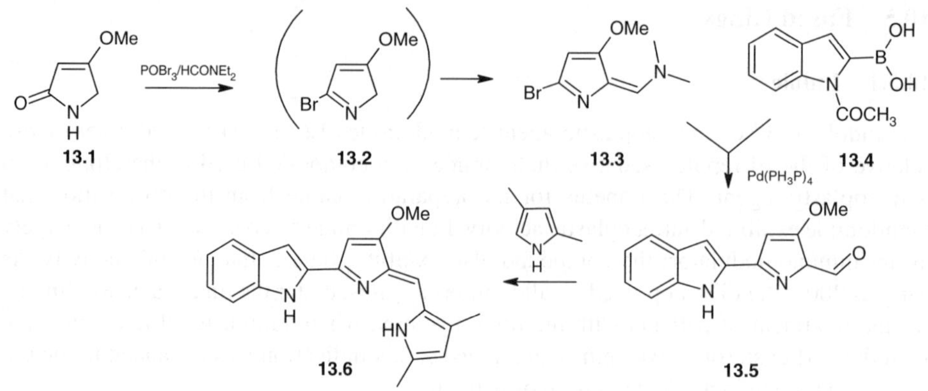

Scheme 10.13   Obatoclax.

bromide (**13.2**) and adds a formyl group at the position adjacent to ring nitrogen (**13.3**) though probably in the reverse order. The treatment of that intermediate with indole-2-boronic acid in the presence of the palladium complex affords the cross-coupling product 13.5. Hydrolysis then reveals the formyl group on the pyrrole ring; the acetyl group on indole nitrogen is lost in the process. Aldol condensation between the nucleophilic position on 1,3-dimethylpyrole and the aldehyde in **13.5** adds the last ring. **Obatoclax (13.6)** is thus obtained [13].

Several discrete mechanisms can account for the antineoplastic activity of indole-based indisulam. One of these interestingly comprises the inhibition of carbonic anhydrate, a well-known activity of sulfonamide diuretic drugs. The compound shows activity against human cancers both in cell culture and when implanted in laboratory animals. Indisulam entered clinical trials in the midaughts.

The straightforward concise synthesis starts with the reaction of nitroindole (**14.1**) with *N*-chlorosuccinimide. Catalytic hydrogenation then serves to reduce the nitro group (**14.3**). The amine in the resulting intermediate is then sulfonated by reaction with 4-nitrophenylsulfonyl chloride. The newly introduced nitro group is reduced again though this time with zinc in the presence of acid (**14.5**). In an unusual sequence, the newly introduced amine was converted to a sulfonamide. Thus, the amine was diazotized with nitrous acid; the reaction of the diazonium salt with sulfur dioxide and cuprous chloride led to the corresponding sulfonyl chloride. The treatment of this last intermediate with ammonia affords sulfonamide and **indisulam (14.7)** [14].

10.5.2   Purine-Like

The enzyme purine nucleoside phosphorylase (PNP) is essential for the proliferation of B and T cells of the immune system. The PNP analogue forodesine was designed as a transition state analogue of PNP. This drug in fact binds to and inhibits PNP. Forodesine has been used in clinical trials in patients with T-cell leukemias and lymphomas.

The treatment of deazapurine (**15.1**) with butyl lithium abstracts a vinylic proton from the fused pyrrole ring, affording the corresponding anion (**15.2**). This is not isolated but reacted immediately with the protected amino sugar derivative (**15.3**). The carbanion adds to

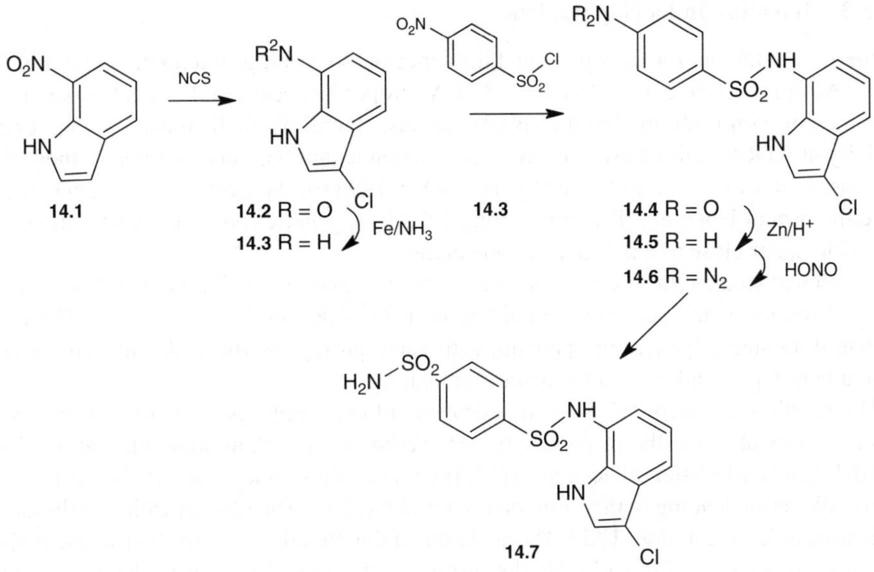

Scheme 10.14 *Indisulam.*

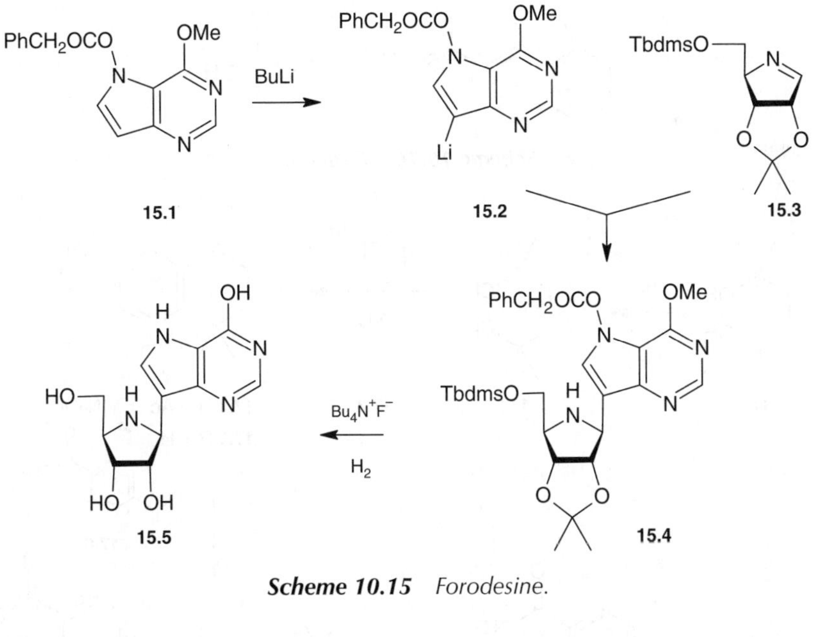

Scheme 10.15 *Forodesine.*

the imine function of the sugar linking the two moieties by means of a carbon to carbon bond (the analogous bond in natural saccharides usually comprises a carbon to nitrogen bond). Treatment with tetrabutyl ammonium fluoride cleaves off the silicon protecting group. Catalytic hydrogenation then frees the hydroxyl at C5 in the amino sugar, affording **forodesine** (**15.5**) [15].

10.5.3 Tetralins and a Naphthalene

Retinoids are known for their profound influence on both cell differentiation and proliferation. All-trans retinoic acid (tretinoin, ATRA) proper has been used to treat leukemias in conjunction with traditional antineoplastic agents. Rapid metabolism and to some extent toxicity of ATRA limit its use. The synthetic analogue tamibarotene overcomes those limitations. ATRA itself was approved by the FDA in 1995 for the treatment of patients with a specific form of leukemia. The drug is long off patent; generic forms (USAN tretinoin) are available under close to a half dozen trade names.

Preclinical studies showed that the tetralin-based analogue tamibarotene (**17.8**) binds to retinoid receptors and acts as a retinoid agonist. In 2009, the FDA granted tamibarotene orphan drug status for treating patients with acute promyelocytic leukemia. The drug is available in Japan under the trade name Amnolake®.

The reaction of glycol (**17.1**) with hydrogen chloride replaces the hydroxyl groups by chlorines probably via the respective tertiary carbocations. Aluminum chloride-mediated Friedel–Crafts alkylation of acetanilid (**17.3**) with the dichloride results in double Friedel–Crafts alkylation leading to the formation of tetralin (**17.4**). The saponification of the amide then affords the free amine (**17.5**). The acylation of that function with the half-acid chloride/half ester of terephthalic acid adds the last ring. The saponification of the ester affords **tamibarotene (17.8)** [16].

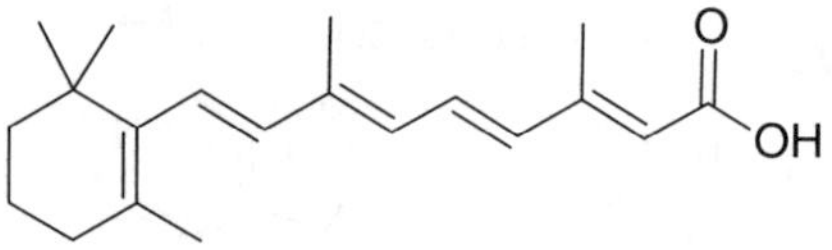

Scheme 10.16   Tretinoin.

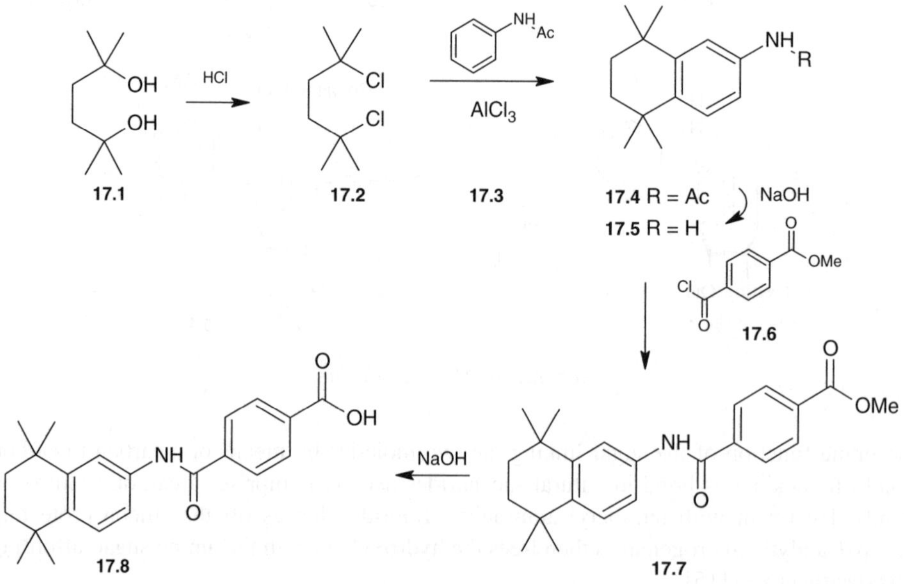

Scheme 10.17   Tamibarotene.

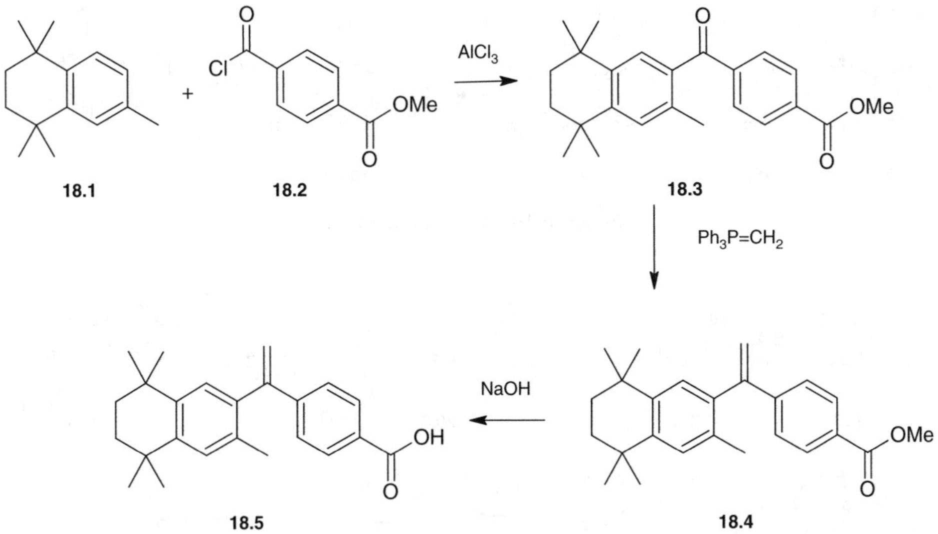

Scheme 10.18 *Bexarotene.*

The structurally related tetralin bexarotene (**18.5**) binds to the equivalent set of retinoid receptors though to a slightly different set than tamibarotene. The FDA approved the drug for treating the skin lesions due to T-cell lymphomas. The drug is available under the trade name Targretin®.

The very concise synthesis involves first the acylation of the tetramethylated tetralin (**18.1**) with the same acid chloride (**18.2**) as that used above adds the required extra ring. Condensation with the ylide from methyltriphenylphosphonium halide and base replaces the carbonyl oxygen by carbon. The saponification of the ester then affords **bexarotene (18.5)** [17].

10.5.4   Etc.

It is a given that solid tumors are poorly supplied with blood vessels. Almost by definition, the interior of such a tumor will lack oxygen. There have been a number of attempts to take advantage of this hypoxic environment by designing compounds that are chemically non-reactive until they are reduced to cytotoxic species. One such candidate, tirapazamine, has been shown to generate reactive nitroxyl species when reduced. More detailed examination has shown that only the radicals formed within the cell nucleus contribute to the drug's cytotoxicity. The drug has been evaluated in Phase II and II clinical trials both alone and in conjunction with well-known antineoplastic drugs.

The addition of cyanamid to ortho nitroaniline leads to the fused heterocyclic product (**19.4**). The sequence most likely first involves the addition of aniline nitrogen to form the transient guanidine such as (**19.3**). One of the very basic nitrogen atoms then adds to the nitro group, closing the fused ring. Treatment with hydrogen peroxide oxidizes the other nitrogen adjacent to benzene to yield **tirapazamine (19.5)** [18].

The use of metabolic inhibitors as antineoplastic agents, especially agents that disrupt folate synthesis, is discussed in some detail in Chapter 2. The antineoplastic agent metesind (**20.9**)

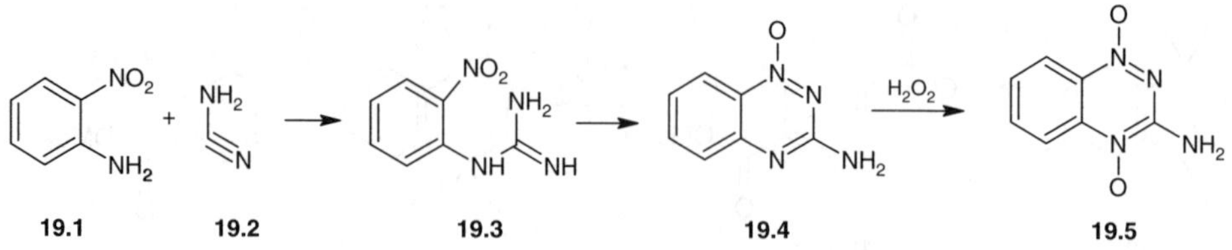

Scheme 10.19 *Tirapazamine.*

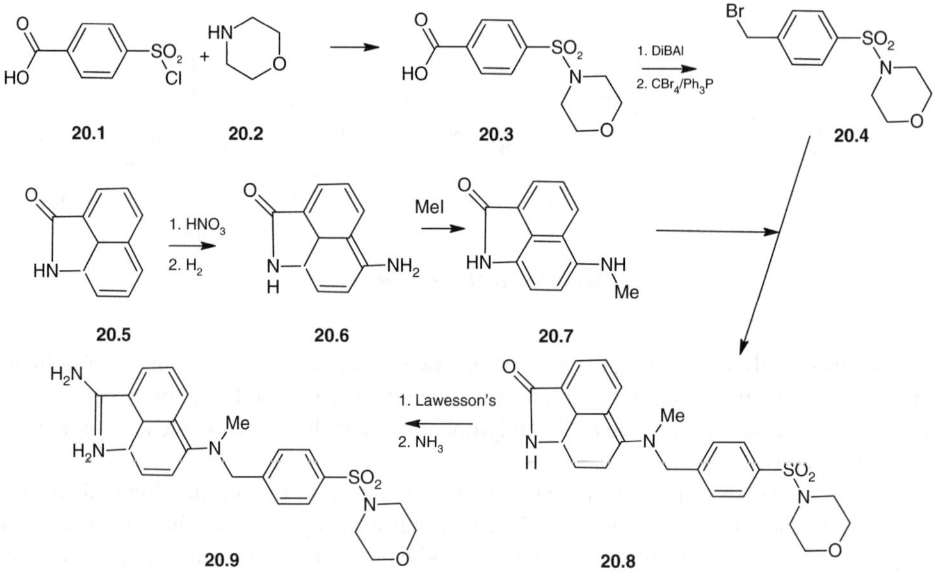

Scheme 10.20 *Metesind.*

represents a current approach to the same problem. The structures of prior folate antagonists generally retained features of the natural folates. Metesind on the other hand was designed as a topological analogue to fit the active site of the enzyme as predicted by X-ray crystallography. Though the agent showed activity in preclinical assays, it may not have entered clinical trials.

The convergent scheme begins with the reaction of arylsulfonyl chloride (**20.1**) with morpholine. The carboxylic acid in the intermediate (**20.3**) is next esterified, and the resulting ester reduced to a carbinol by means of diisobutylaluminum hydride; the Mitsunobu reaction with carbon tetrabromide in the presence of triphenylphosphine affords benzyl bromide (**20.4**). The nitration of the naphthalene derivative **20.5** proceeds as expected at the position para to nitrogen. The nitro group is then reduced by catalytic hydrogenation, and the resulting amine (**20.6**) alkylated with methyl iodide (**20.7**). The alkylation of this last intermediate with the bromobenzene in (**20.4**) connects the two moieties (**20.8**). The reaction of the sole carbonyl group in the molecule with Lawesson's reagent converts this to the corresponding thiocarbonyl group. The alkylation with methyl iodide in the presence of base forms an enol thioether. The reaction of this last derivative with ammonia replaces the ether with an enol amine. **Metesind (20.9)** is thus obtained [19].

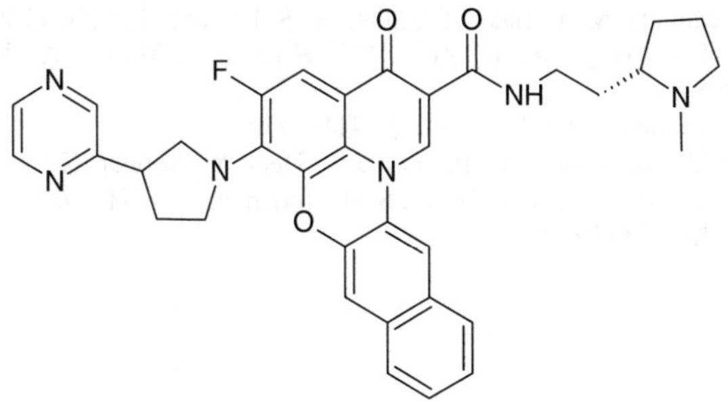

Scheme 10.21   *Quarfloxacin.*

The numerous quinolone antibacterial agents all kill bacteria by interfering with the action of bacterial gyrase, an obligate enzyme for replication with a function analogous to eukaryotic topoisomerase. An informed viewer confronted with the structure of **Quarfloxacin** might well conclude that this molecule was a gyrase inhibitor that had finally been adapted to eukaryotes. Biochemical studies confound that conclusion. Quarfloxacin instead disrupts a critical interaction required for biogenesis of ribosomal RNA which in turn is required for protein synthesis. That form of RNA is particularly very active in cancer cells contributing to their proliferation. The drug was being tested in Phase II clinical trials in 2011.

References

[1]   C. Campas, R. Castaner, *Drugs Future.* **34**, 97 (2009).
[2]   S. Mzenga, C.M. Yang, R.A. Whitney, *J. Am. Chem. Soc.* **109**, 276 (1987).
[3]   S.J. Coutts, T.A. Kelly, R.J. Snow, C.A. Kennedy, R.W. Barton, J. Adams, D.A. Krolikowski, S.J. Campbell, J.F. Ksiazek, W.W. Bachovchin, *J. Med. Chem.* **39**, 2087(1996).
[4]   R. Fazaeli, S. Tangestaninejad, H. Aliyan, M. Moghadam, *Appl. Catal.* **309**, 44 (2006).
[5]   S.M. Efange, E.M. Alessi, H.C. Shih, Y.C. Cheng, *J. Med. Chem.* **28**, 904 (1985).
[6]   J.-K. Son, S.H. Lee, L. Nagarapu, Y. Jahng, *Bull. Korean Chem. Soc.* **26**, 1117 (2005).
[7]   S. Chen, L. Sun, K. Koya, N. Tatsuta, Z. Xia, T. Korbut, Z. Du, J. Wu, G. Liang, J. Jiang, M. Ono, D. Zhou, A. Sonderfan, *Bioorg. Med. Chem. Lett.* **23**, 5070 (2013).
[8]   C.S. Grossman, P.A. Hipskind, H.S. Lin, K.L. Lobb, G.B. Lopez de Uralde, J.E. Lopez, M.M. Mader, M.E. Richett, C. Shih, A. De Dios, U.S. Patent 7,084,170 (2006).
[9]   A. Devasagayaraj, Jin, Z.-C. H., A. Tunoori, Y. Wang, C. Zhang, U.S. Patent, 7,553,840 (2009).
[10] R.H. Bradbury, R.J. Butlin, R. James, U.S. Patent 6,060,475 (2000).
[11] C.-M. Park, M. Bruncko, J. Adickes, J. Bauch, H. Ding, A. Kunzer, K.C. Marsh, P. Nimmer, A.R. Shoemaker, X. Song, S.K. Tahir, C. Tse, X Wang, M.D. Wendt, X. Yang, H. Zhang, S.W. Fesik, S.H. Rosenberg, S.W. Elmore, *J. Med. Chem.* **51**, 6902 (2008).
[12] G.W. Muller, R. Chen, S.-Y. Huang, L.G. Corral, L.M. Wong, R.T. Patterson, Y. Chen, G. Kaplan, D.I. Stirling, *Bioorg. Med. Med. Chem. Lett.* **9**, 1625 (1999).
[13] K. Daïri, Y. Yao, M. Faley, S. Tripathy, E. Rioux, X. Billot, D. Rabouin, G. Gonzalez, J.-F. Lavallée, G. Attardo, *Org. Proc. Res. Dev.* **11**, 1051 (2007).
[14] T. Owa, H. Yoshino, T. Okauchi, K. Yoshimatsu, Y. Ozawa, N. Hata, S. T. Nagasu, N. Koyanagi, K. Kitoh, *J. Med. Chem.* **42**, 3789 (1999).
[15] G.B. Evans, R.H. Fourneaux, V.L. Schramm, V. Singh, P.C. Tyler, *J. Med. Chem.* **47**, 3275 (2004).

[16]   Y. Hamada, I. Yamada, M. Uenaka, T. Sakata, U.S. Patent 5,214,202 (1993).

[17]   M.F. Boehm, R.A. Heyman, L. Zhi, C.K. Hwang, S. White, A. Nadzan, U.S. Patent 5,780,676 (1998).

[18]   J.C. Mason, C. Tennnant, *J. Chem. Soc. B.* **911**(1970).

[19]   M.D. Varney, G.P. Marzoni, C.L. Palmer, J.G. Deal, S. Webber, K.M. Welsh, R.J. Bacquet, C.A. Bartlett, C.A. Morse, C.L.J. Booth, S.M. Herrmann, E.F. Howland, R.W. Ward, J. White, *J. Med. Chem.* **35**, 663 (1992).

Appendix A

Combinatorial Chemistry

A high proportion of the compounds described in this compendium are the fruits of combinatorial chemistry. The impressive proportion of tyrosine kinase inhibitors, found in Chapter 8, that have been approved by the FDA demonstrate the power of this approach to drug discovery.

The standard model for drug discovery up until the 1980s began with the synthesis of organic compounds by a chemist. The structures of those compounds varied from those based on a scientist's informed hunch at one extreme all the way to compounds designed to get around the patents that covered marketed drugs. The product was usually purified to meet criteria accepted by journals that published chemical articles. Most pharmacology departments in pharmaceutical concerns had on tap a series of assays, many of which involved the use of test animals. Accordingly, a very large portion of the assays designed to uncover the biological activity of test candidates involved animal models that hopefully represented the target disease. The sizeable samples of test substances required for many of those tests dictated the scale of syntheses for preparing test compounds. These assays fell into one of two categories: a relatively high turnover screen that tested almost all compounds that had been synthesized by the chemists as well as some from outside sources and more detailed assays for targeted agents. The broad screens were intended to organize serendipity; they were designed to uncover lead compounds with hitherto unexplored structures. Those labs also maintained more discriminating tests for studying in greater depth compounds selected by the initial screen. The now common *in vitro* screens were at the time usually relegated to antibiotic, antiviral, and antineoplastic programs. That model has led to the discovery of many very important drugs. The poorly populated or even empty pipelines that became evident at the turn of the twenty-first century, however, led to the search for a new and different approach to drug discovery. The expense of acquiring and maintaining test animal colonies provided an additional motive for seeking a new model.

The era that began in the mid-1980s saw the identification and isolation of an increasing number of enzymes involved in biological processes in both normal and diseased states. That in turn permitted the development of an array of high-capacity biochemical *in vitro* assays that measured such enzymes. This provided the methodology for measuring the effect of a test substance on those enzymes, be it inhibition, potentiation, or no effect. Another benefit of *in vitro* assays was the markedly decreased amount of compounds required for tests. Contributing to the consideration of a different way of producing libraries

Antineoplastic Drugs: Organic Synthesis, First Edition. Daniel Lednicer.
© 2015 John Wiley & Sons, Ltd. Published 2015 by John Wiley & Sons, Ltd.

of novel compounds was the growing expense of producing pure samples of novel compounds by synthesis.

One alternative to the then-standard approach for drug discovery comprised replacing the initial *in vivo* screen by an *in vitro* assay. The high throughput capacity of such screens by far exceeded the number of novel compounds turned out by the organic chemists. An unusually prolific chemist, for example, produced about 100 novel compounds per year, a mere drop in the bucket for a screen that ran of 96-well plates.

The time-proven Merrifield solid-phase method for synthesizing proteins, developed in the early 1960s, provided the tool for the new methodology. Investigators accordingly adapted solid-phase-based chemistry for creating large collections of test candidates. There would be no effort to separate the compounds that comprised the mixtures prior to testing the effect of the resulting collection against a given enzyme. The initial step for creating such a collection of compounds, dubbed library, by combinatorial chemistry comprises attaching a starting compound to a polymer bead by a bond that can be cleaved when the process is complete. The compound that is attached to the bead must incorporate a function that can be used to link the next unit, which should itself have a reactive function. Various strategies have been used for proceeding further. In an approach dubbed "split and mix," the reagent-linked beads from the first reaction are first collected and then divided into several lots. Each of these lots is then treated with a different second reagent. The beads are then recombined and again split into new lots, and each of those is treated with a different third reagent. An alternate scheme denoted as "parallel synthesis" differs from the former in that the lots are not recombined; each is carried through the sequence as an individual unit. The numbers of cycles in either approach are predicated on the need to limit the molecular weight of final products between 750 and 1000.

The other segment of the new approach comprises the high-capacity *in vitro* screen. In the most general terms, the interaction of a test compound with an enzyme, or the like, needs to be reduced to an event that can measured by an instrument. This requirement has led to the development of a sizeable number of assays where the response to a test agent is linked to a color change. The dyes selected for that purpose most often change color when exposed to either reductive or alternatively oxidative milieus Purpose-designed automated colorimeters that can read and record changes on 96-well plates are often used to measure assay results. Collections are commonly screened after they have been detached from the beads. In some cases, the results can be obtained from agents still attached to the beads.

A positive response to the assay of one of the libraries turns on a research program to identify the specific compound in the mixture that interacted with the enzyme. This is not quite akin to finding the needle in a haystack since the investigator has a record of the compounds that went into the production of that library. This aspect of the program, which is akin to natural product isolation, obviously leans heavily on separation science; the *in vitro* screen provides the means for steering this aspect of the program. Commonly used instrumental methods such as IR, NMR, and mass spectrometry provide the tools for determining the structure of the compound that gave the positive assay result. That active compound is then prepared by synthetic chemistry to confirm its structure and activity. A second round of combinatorial chemistry may be invoked in those cases where the active compound lacks some property such as oral activity.

Index of Heterocycle Syntheses

Preparation of heterocyclic moieties comprises an important tool in an organic chemists repetoire. This page lists those transforms in the preceding discussions. Names are the rough and ready designations used in everyday practice rather than those dictated by CAS or IUPAC.

Antineoplastic Drugs: Organic Synthesis, First Edition. Daniel Lednicer.
© 2015 John Wiley & Sons, Ltd. Published 2015 by John Wiley & Sons, Ltd.

Subject Index

Antineoplastic Drugs: Organic Synthesis, First Edition. Daniel Lednicer.
© 2015 John Wiley & Sons, Ltd. Published 2015 by John Wiley & Sons, Ltd.